Squeezed and Nonclassical Light

NATO ASI Series

Advanced Science Institutes Series

A series presenting the results of activities sponsored by the NATO Science Committee, which aims at the dissemination of advanced scientific and technological knowledge, with a view to strengthening links between scientific communities.

The series is published by an international board of publishers in conjunction with the NATO Scientific Affairs Division

A	**Life Sciences**	Plenum Publishing Corporation
B	**Physics**	New York and London
C	**Mathematical and Physical Sciences**	Kluwer Academic Publishers Dordrecht, Boston, and London
D	**Behavioral and Social Sciences**	
E	**Applied Sciences**	
F	**Computer and Systems Sciences**	Springer-Verlag
G	**Ecological Sciences**	Berlin, Heidelberg, New York, London,
H	**Cell Biology**	Paris, and Tokyo

Recent Volumes in this Series

Volume 184—Narrow-Band Phenomena—Influence of Electrons with Both Band and Localized Character
edited by J. C. Fuggle, G. A. Sawatzky, and J. W. Allen

Volume 185—Nonperturbative Quantum Field Theory
edited by G. 't Hooft, A. Jaffe, G. Mack, P. K. Mitter, and R. Stora

Volume 186—Simple Molecular Systems at Very High Density
edited by A. Polian, P. Loubeyre, and N. Boccara

Volume 187—X-Ray Spectroscopy in Atomic and Solid State Physics
edited by J. Gomes Ferreira and M. Teresa Ramos

Volume 188—Reflection High-Energy Electron Diffraction and Reflection Electron Imaging of Surfaces
edited by P. K. Larsen and P. J. Dobson

Volume 189—Band Structure Engineering in Semiconductor Microstructures
edited by R. A. Abram and M. Jaros

Volume 190—Squeezed and Nonclassical Light
edited by P. Tombesi and E. R. Pike

Volume 191—Surface and Interface Characterization by Electron Optical Methods
edited by A. Howie and U. Valdrè

Series B: Physics

Squeezed and Nonclassical Light

Edited by

P. Tombesi

La Sapienza University
Rome, Italy

and

E. R. Pike

Royal Signals and Radar Establishment
Great Malvern, United Kingdom
and
King's College
London, United Kingdom

Springer Science+Business Media, LLC

Proceedings of a NATO Advanced Research Workshop
on Squeezed and Non-Classical Light,
held January 25-29, 1988,
in Cortina d'Ampezzo, Italy

Library of Congress Cataloging in Publication Data

NATO Advanced Research Workshop on Squeezed and Non-classical Light (1988: Cortina d'Ampezzo, Italy)
Squeezed and nonclassical light / edited by P. Tombesi and E. R. Pike.
p. cm.—(NATO ASI series. Series B, Physics; v. 190)
"Proceedings of a NATO Advanced Research Workshop on Squeezed and Non-classical Light, held January 25-29, 1988, in Cortina d'Ampezzo, Italy"—T.p. verso.
"Published in cooperation with NATO Scientific Affairs Division."
Includes bibliographies and index.
ISBN 978-1-4757-6576-2 ISBN 978-1-4757-6574-8 (eBook)
DOI 10.1007/978-1-4757-6574-8
1. Quantum optics—Congresses. 2. Light sources—Congresses. 3. Electric discharge lighting—Congresses. I. Tombesi, P. (Paolo) II. Pike, E. R. (Edward Roy), 1929- . III. North Atlantic Treaty Organization. Scientific Affairs Division. IV. Title. V. Title: Squeezed and non-classical light. VI. Series.
QC446.15.N38 1988 88-29431
535—dc19 CIP

Originally published by Plenum Press, New York in 1989

ORGANIZING COMMITTEE

P. Tombesi	University of Rome "La Sapienza"
E.R. Pike	King's College London
H. Yuen	Northwestern University
D. Walls	University of Auckland
R. Slusher	AT&T Bell Laboratories

INVITED SPEAKERS

G.S.Agarwal	University of Hyderabad, India
G.Bjork	The Royal Institute of Technology, Sweden
R.K.Bullough	The University of Manchester, U.K.
C.M.Caves	University of Southern California, U.S.A.
F.De Martini	University of Rome "La Sapienza", Italy
R.Glauber	Harvard University, U.S.A.
R.Graham	Universität - Gesamthochschule Essen, W. Germany
P.Grangier	Institut d'Optique, France
J. Kimble	The University of Texas at Austin, U.S.A.
P.Kumar	Northwestern University, U.S.A.
G.Leuchs	Max-Planch-Institut fur Quantenoptik, W. Germany
M.Lewenstein	Polish Academy of Sciences, Poland
L.Lugiato	Politecnico di Torino, Italy
P.Meystre	University of Arizona, U.S.A.
G.J.Milburn	Australian National University, Australia
M.Ozawa	Nagoya University, Japan
M.Rasetti	Politecnico di Torino, Italy
W.Schleich	Max-Planck Institut fur Quantenoptik, W. Germany
B.Yurke	AT&T Bell Laboratories, U.S.A.

PREFACE

The recent generation in the laboratory of phase squeezed and intensity squeezed light beams has brought to fruition the theoretical predictions of such non-classical phenomena which have been made and developed in recent years by a number of workers in the field of quantum optics. A vigorous development is now underway of both theory and experiment and the first measurements have been confirmed and extended already in some half dozen laboratories.

Although the fields of application of these novel light sources are as yet somewhat hazy in our minds and some inspired thinking is required along these lines, the pace and excitement of the research is very clear. It is to he hoped that the new possibilities of making measurements below the quantum shot noise limit which is made possible by these squeezed states of light will lead to further fundamental advances in the near future.

In this NATO ARW a number of the leaders in the field met in the extremely pleasant surroundings of Cortina d'Ampezzo and their contributions are recorded in this volume. The meeting was held at the Istituto d'Arte which was enjoying its 100th anniversary celebrations. This ARW was preceeded by an ONR Special Seminar on "Photons and Quantum Fluctuations", the proceedings of which will be published by Adam Hilger Ltd. The timeliness of the meeting was acknowledged by the support of the NATO Scientific Affairs Division which we would like to acknowledge on behalf of all the participants.

We would also like to acknowledge additional financial assistance provided by Ministero Pubblica Istruzione, Consiglio Nazionale delle Ricerche (CNR), Gruppo Nazionale di Struttura della Materia (CNR) and local sponsorship from Olivetti Spa, Consorzio per lo Sviluppo e Turismo di Cortina d'Ampezzo, Municipio di Cortina d'Ampezzo, Istituto Statale d'Arte and Hotel Europa. Particular thanks are due to Dr. Dick Slusher, Professor Danny Walls and Professor Horace Yuen for assistance in organising the meeting, and also special thanks are due to Professor M. Spampani, Mr. G.Milani, Ing. E.Cardazzi, Professor G. Demenego, Mr. E.Demenego and, in particular, Professor G. Olivieri.

We have had full collaboration from Plenum Press in the publication of this volume, and the success of the meeting and the not inconsiderable work before and afterwards owes much to our uncomplaining secretaries Angela Di Silvestro, Marcella Mastrofini and Beverley James.

Roy Pike

Paolo Tombesi

June 1988

CONTENTS

QUANTUM NOISE REDUCTION IN OPTICAL SYSTEMS

D.F. Walls and P.D. Drummond,* A.S. Lane, M.A. Marte, M.D. Reid,** and H. Ritsch***

* *Physics Department, University of Auckland, Auckland, New Zealand*
** *Physics Department, University of Waikato, Hamilton, New Zealand*
*** *Institute for Theoretical Physics, University of Innsbruck Innsbruck, Austria*

1. INTRODUCTION

There is considerable activity in the generation of light fields with less quantum fluctuations than a coherent field, for example squeezed or sub-Poissonian light.(1-3) With such light, it is possible to beat the usual shot noise limit which leads to an enhanced sensitivity in many optical measurements. Such sensitivity is required in measurements such as the detection of gravitational radiation by optical interferometry where the signal is comparable to the quantum noise.

Squeezed states of light have now been produced by four wave mixing in atomic vapours(4,5) and optical fibres(6) and in optical parametric oscillation.(7) Sub-Poissonian light has been produced in resonance fluorescence(8) and parametric down conversion.(9) In this paper we consider two systems: the parametric oscillator, and lasers driven with squeezed pumps.

In a parametric down conversion process a pump photon produces a pair of correlated signal and idler photons.(10,11) In the parametric oscillator there is a strong correlation between the intensities and phases of the signal and idler beam. Reynaud *et al.*(12,13) have shown that the spectrum of fluctuations in the difference of the signal and idler intensities is reduced to zero in an ideal situation with equal cavity decay rates and no losses. In this paper we wish to consider the non-ideal situation where the cavity decay rates are unequal and losses are present in the cavity. Reduced intensity fluctuations below the vacuum level have recently been seen experimentally by Heidmann *et al.*(29)

We also consider a new feature of the nondegenerate parametric oscillator. This is the correlation of the phases in the above threshold signal and idler beams. These are correlated in a nonclassical way, just as are the intensities. The correlations are unusual in that neither phase is stable, owing to an overall phase diffusion in the output. Thus the signal and idler phases have a correlated phase diffusion, in which the phase sum is stable but the phase difference is random. Our analysis also predicts E.P.R.-type behaviour in quantum measurements on the output fields.

All the systems used to produce squeezed states of light have been passive systems. In an active system, such as a laser, the noise from the pumping mechanism together with the associated spontaneous emission from the atoms is the dominant noise source and hence the emitted light has Poissonian photon statistics. Yamamoto *et al.*[14] analysed a system where the pump fluctuations are reduced and where the output from the laser has sub-Poissonian photon statistics. Machida *et al.*[15] have recently demonstrated the operation of a semiconductor laser pumped with an electron beam with reduced fluctuations and shown that the output beam has a 7.3% reduction below the standard quantum limit in number fluctuations.

A related system is found in studies of the micromaser.[16,17] In the micromaser, single excited atoms are fed into the laser cavity. If there is a constant velocity of the atoms and hence a constant interaction time the output of the micromaser is found to have sub-Poissonian statistics. When a velocity distribution and hence a spread of interaction times equivalent to a stochasticity of the pump is included, the usual Poisson statistics of a coherent laser are recovered.

In this paper we shall investigate a new class of laser which is pumped with squeezed light. This differs from the system considered by Yamamoto *et al.*[14] where the pump is in a near photon number eigenstate. We shall also compare the squeezed pump laser with the laser studied by Gea-Banacloche[18] where a squeezed field replaces the vacuum entering the cavity.

2. THE NON DEGENERATE PARAMETRIC OSCILLATOR

The non degenerate parametric oscillator consists of three interacting field modes, a pump signal and idler within an optical cavity which is coherently driven at the pump frequency. The Hamiltonian describing this interaction is[19]

$$\hat{H} = \sum_{j=0}^{3} \hat{H}_j$$

$$\hat{H}_0 = \sum_{i=0}^{2} \hbar\omega_i a_i^+ a_i$$

$$\hat{H}_1 = i\hbar \left(Ee^{-i\omega_0 t} a_0^+ - a_0 E^* e^{i\omega_0 t}\right) \tag{2.1}$$

$$\hat{H}_2 = i\hbar\chi \left(a_0 a_1^+ a_2^+ - a_0^+ a_1 a_2\right)$$

$$\hat{H}_3 = \Sigma a_i \hat{\Gamma}_i^+ + h.c.$$

where a_0, a_1 and a_2 are the boson annihilation operators for the pump, signal and idler modes respectively. We assume resonance: $\omega_0 = \omega_1 + \omega_2$. The pump mode is driven by a coherent driving field with amplitude E and central frequency ω_0, χ is the strength of the parametric interaction. $\hat{\Gamma}_0$, $\hat{\Gamma}_1$ and $\hat{\Gamma}_2$ are the heat bath operators for the cavity damping of the pump, signal and idler modes respectively. A set of stochastic differential equations for the stochastic variables in the generalized P-representation[20] may be derived. These have the form

$$\dot{\alpha}_0 = -\kappa_0\alpha_0 + E - \chi\alpha_1\alpha_2$$

$$\dot{\alpha}_1 = -\kappa_1\alpha_1 + \chi\alpha_0\alpha_2^+ + (\alpha_0\chi)^{1/2}\zeta_1(t) \tag{2.2}$$

$$\dot{\alpha}_2 = -\kappa_2\alpha_2 + \chi\alpha_0\alpha_1^+ + (\alpha_0\chi)^{1/2}\zeta_2(t)$$

where the noise terms have the following correlations

$$<\zeta_1(t)\zeta_2(t')> = \delta(t - t')$$
$$<\zeta_1^+(t)\zeta_2^+(t')> = \delta(t - t') \qquad (2.3)$$

and κ_0, κ_1, κ_2 are the cavity decay rates of the pump, signal and idler modes respectively.

The steady state solutions to these equations in terms of the intensity and phase variables $<\alpha_j>_{ss} = \sqrt{I_j^0}e^{-i\phi_j^0}$ are given below threshold by

$$|E|<E_T = \frac{\kappa_0\sqrt{\kappa_1\kappa_2}}{\chi}$$

$$I_1^0 = I_2^0 = 0 \qquad (2.4)$$

$$I_0^0 = \frac{|E|^2}{\kappa_0^2}, \quad \phi_o^0 = \phi_p$$

where $E = |E|e^{-i\phi_p}$ and ϕ_p is the phase of the coherent pump. Above threshold we have $|E| > E_T$ and

$$I_0^0 = \frac{\kappa_1\kappa_2}{\chi^2}, \quad \phi_0^0 = \phi_p$$

$$(2.5)$$

$$I_{1,2}^0 = \frac{C}{\kappa_{1,2}}, \quad \phi_1^0 + \phi_2^0 = \phi_p$$

where

$$C = \frac{\kappa_0\kappa_1\kappa_2}{\chi^2}(P - 1), \quad P = |E|/E_T$$

The interesting point in the above threshold solution is that only the sum of the phases $\phi_1 + \phi_2$ is defined. No steady state exists for the phase difference which may freely wander with noise fluctuations.[21]

In order to calculate the noise properties of the parametric oscillator above threshold, we adopt a linearized analysis about the steady state solutions. We introduce generalised phase and intensity variables, defined as follows:

$$I_j = \alpha_j^+\alpha_j; \quad \phi_j = (1/2i)\,ln(\alpha_j^+/\alpha_j) \qquad (2.6)$$

and write the linearised equations in terms of the following variables

$$\Delta I_0 = I_0 - I_0^0$$

$$\Delta I_\pm = [\kappa_1(I_1 - I_1^0) \pm \kappa_2(I_2 - I_2^0)]/\kappa_+$$

$$\Delta\phi_0 = \phi_0 - \phi_p$$

$$\Delta\phi_+ = \phi_2 + \phi_1 - \phi_p$$

$$\phi_- = \sqrt{\frac{\kappa_2}{\kappa_1}}\,\phi_1 - \sqrt{\frac{\kappa_1}{\kappa_2}}\,\phi_2$$

The equations read

$$\dot{\Delta I_0} = -\kappa_0 \Delta I_0 - \kappa_+ \Delta I_+$$

$$\dot{\Delta I_+} = -2\kappa_- \Delta I_- + 2\kappa_0 (P - 1)\Delta I_0 + \Gamma_+(t)$$

$$\dot{\Delta I_-} = -2\kappa_+ \Delta I_- + 2\delta\kappa_0 (P - 1)\Delta I_0 + \Gamma_-(t)$$

$$\dot{\Delta\phi_0} = -\kappa_0 (\Delta\phi_0 + (P - 1)\Delta\phi_+)$$

$$\dot{\Delta\phi_+} = -2\kappa_+ (\Delta\phi_+ - \Delta\phi_0) + \Gamma_+^{\phi}(t)$$

$$\dot{\Delta\phi_-} = \Gamma_-^{\phi}(t) \qquad (2.7)$$

where $\kappa_\pm = (\kappa_1 \pm \kappa_2)/2$, $\delta = \kappa_-/\kappa_+$

and the noise terms have the following correlations

$$\langle\Gamma_+(t)\Gamma_+(t')\rangle = -\langle\Gamma_-(t)\Gamma_-(t')\rangle$$

$$= (4\kappa_1\kappa_2 C/\kappa_+^2)\delta(t - t')$$

$$\langle\Gamma_+(t)\Gamma_-(t')\rangle = 0$$

$$\langle\Gamma_-^{\phi}(t)\Gamma_-^{\phi}(t')\rangle = -\langle\Gamma_+^{\phi}(t)\Gamma_+^{\phi}(t')\rangle$$

$$= (\kappa_1\kappa_2/C)\delta(t - t')$$

$$\langle\Gamma_+^{\phi}(t)\Gamma_-^{\phi}(t')\rangle = \kappa_-\frac{\sqrt{\kappa_1\kappa_2}}{C}\delta(t - t')$$

A stability analysis shows that the intensity system is stable. We also find that the $\Delta\phi_0, \Delta\phi_+$ system is stable. The ϕ_- variable, however has a zero eigenvalue and moves about randomly under the influence of the stochastic force.

3. SPECTRUM OF INTENSITY FLUCTUATIONS

We wish to calculate the spectrum of fluctuations in the intensity difference between the signal and idler modes. We assume a single ported cavity where the output fields from the cavity at the signal and idler frequencies, $b_1(t)$ and $b_2(t)$ are related to the cavity modes by the boundary conditions[22]

$$b_j(t) = \sqrt{2\kappa_j}\,a_j(t) + b_j^{in}(t) \qquad (3.1)$$

The b_n^{in} specify the fields input to the cavity boundary. The spectrum of fluctuations in the intensity difference is defined by

$$S_D(\omega) = \int_{-\infty}^{\infty} d\tau \; e^{-i\omega\tau} < \hat{I}_-(\tau), \hat{I}_-(0)> \tag{3.2}$$

where $\hat{I}_- = b_1^+ b_1 - b_2^+ b_2$, and we define $<A,B> \equiv <AB> - <A><B>$. Noting that $\hat{I}_-$ here is an operator for the external photon intensity, we define a normalized spectrum

$$\bar{S}_D(\omega) = \frac{S_D(\omega)}{4\kappa_i I_i^0} \qquad (i = 1 \text{ or } 2) \tag{3.3}$$

so that the vacuum or shot noise level is $\bar{S}_D(\omega) = 1$ and perfect noise suppression is $\bar{S}_D(\omega) = 0$.

The results for the spectrum will be published in detail in ref.(23). The form of the spectrum for different parameter values is shown in Figs.(1-3). In the case of the equal signal and idler decay rates the spectrum is a simple inverted Lorentzian with half width $2\kappa_+$ and maximum noise suppression at zero frequency,[12,13]

$$\bar{S}_D(\omega) = 1 - \frac{4}{4 + (\omega/\kappa_+)^2} \tag{3.4}$$

In this case the fluctuations in the signal-idler intensity difference are reduced below the shot noise level and are independent of the pump decay rate κ_0 and the pump power E. It may be seen from Eq.(2.7) that when $\kappa_1 = \kappa_2$, the signal idler intensity difference variable decouples from the equations for the signal-idler intensity sum and pump fluctuations ΔI_0 and the single solution Eq.(3.4) follows directly. Thus, although the individual signal and idler intensities show significant fluctuations sensitive to pump parameters, the intensities are correlated so that the signal-idler intensity difference fluctuations are reduced according to Eq.(3.4).

When the signal and idler decay rates are different, the spectrum of fluctuations changes. The perfect noise suppression in the external signal and idler difference intensity is obtained at zero frequency, regardless of the relative size of decay rates, since the zero frequency result is for detection times long compared to both signal and idler cavity escape times. However, with non-equal signal and idler decay rates it is clear, from Eq.(2.7) that the fluctuations in the difference intensity couple to the fluctuations in the pump and signal-idler intensity sum. The bandwidth of the $\bar{S}_D(\omega)$ noise reduction is now sensitive to the ratios of decay rates of the cavity modes and to the pump power.

We shall consider first the limit where the pump cavity decay rate κ_0 is much larger than the signal and idler decay rates κ_1 and κ_2. Then adiabatically eliminating the pump we obtain

$$\Delta\dot{I}_+ = -2\kappa_-\Delta I_- - 2\kappa_+(P-1)\,\Delta I_+ + \Gamma_+(t)$$

$$\Delta\dot{I}_- = -2\kappa_+\Delta I_- - 2\kappa_-(P-1)\,\Delta I_+ + \Gamma_-(t) \tag{3.5}$$

The eigenvalues of these equations are

$$\lambda_\pm = -P_+\left(E \pm \sqrt{(P-2)^2 + 4(P-1)\delta^2}\right) \tag{3.6}$$

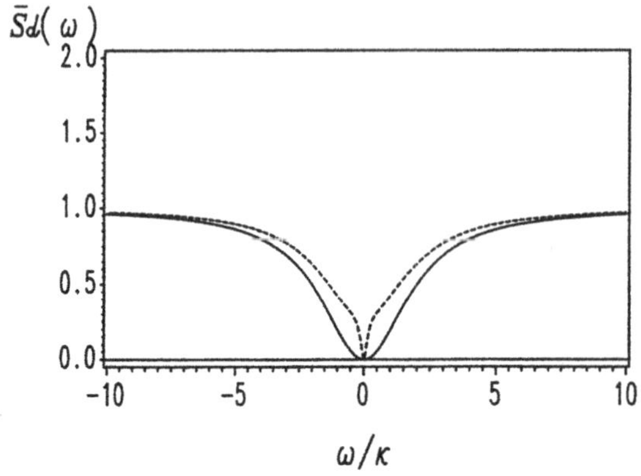

Figure 1a. Plot of $\bar{S}_D(\omega)$ above threshold for high pump cavity damping:

$P = 1.1, \frac{\kappa_0}{\kappa_+} = 10,$

$\delta = 0$ ——— ; $\delta = 0.5$— —

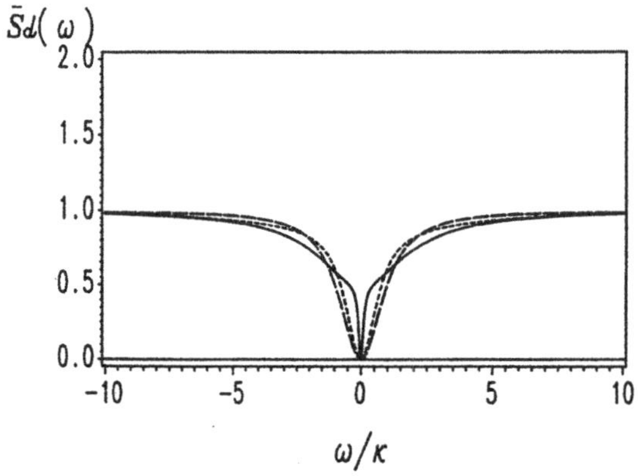

Figure 1b. Plot of $\bar{S}_D(\omega)$ above threshold for high pump cavity damping:

$\delta = 0.6, \frac{\kappa_0}{\kappa_+} = 10$

$P = 1.1$——— ; $P = 2$- - - ; $P = 5$— —

The spectrum is the sum of two Lorentzians both centred at $\omega = 0$ and with half widths λ_+ and λ_- respectively. For small δ^2 such that $4(P-1)\delta^2 \ll (P-2)^2$ the eigenvalues become $\lambda_+ \to -2\kappa_+(P-1)$ and $\lambda_- \to -2\kappa_+$ for $P < 2$. Thus near to threshold $P \gtrsim 1$ we see (Fig.1a) a narrow Lorentzian component with half width $2\kappa_+(P-1)$ and a broader component with width $2\kappa_+$. On moving further above threshold the narrow line broadens significantly (Fig.1b). The effect of larger values of δ^2 is still to reduce the bandwidth of effective noise suppression.

In the opposite limit where the pump decay rate is much smaller than the sum of the signal and idler decay rates ($\kappa_0 \ll 2\kappa_+$) we may adiabatically eliminate the intensity difference variable ΔI_-. This gives the coupled system

$$\Delta\dot{I}_0 = -\kappa_0 \Delta I_0 - \Delta I_s$$

$$\Delta\dot{I}_+ = \left[-2\delta^2 \kappa_0 (P-1) + 2\kappa_0(P-1)\right] \Delta I_0 + \Gamma_+(t) - \delta\Gamma_-(t) \qquad (3.7)$$

In this case the spectrum in the limit $(\kappa_0/\kappa_+) \to 0$ has a flat floor level δ^2 with Lorentzian components characterized by the eigenvalues $\lambda_\pm$, where

$$\lambda_\pm = -\frac{\kappa_0}{2} \pm \frac{\kappa_0}{2}\sqrt{1 + 8(P-1)[\delta^2 - (\kappa_+/\kappa_0)]} \qquad (3.8)$$

With (κ_+/κ_0) large the eigenvalues $\lambda_\pm$ are complex for sufficiently large P and we see noisy sidepeaks at approximately

$$\frac{\omega}{\kappa_0} \sim \pm\sqrt{2(\kappa_+/\kappa_0)(P-1)} \qquad (3.9)$$

The peak separation increases with pump power. In Fig.(2a) we see that the central frequencies between the noisy peaks have reduced fluctuations with perfect reduction at zero frequency. In essence then, for reasonable pump powers, the bandwidth of noise reduction is reduced below $2\kappa_+$ especially at lower intensities near threshold.

An interesting parameter regime of possible experimental interest is where the signal cavity decay rate is much greater than the signal or idler. Figure (2b) shows such correlation spectra, demonstrating the small (κ_+/κ_0) (good pump) features discussed above in the limit of larger κ_-. Comparison of Figs.(2a) and (2b) shows that the effect of increasing κ_- is to increase the base noise level prominent at outer frequencies over the bandwidth $2\kappa_+$. The noisy sidepeaks are positioned at approximately the frequencies given by

$$\frac{\omega}{\kappa_0} \sim \pm\sqrt{1 + 8(P-1)[\delta^2 - (\kappa_+/\kappa_0)]} \qquad (3.10)$$

and there is little noise reduction for higher frequencies.

4. SPECTRUM OF QUADRATURE FLUCTUATIONS

The preceding section describes observations of intensities only. In order to observe squeezing, it is necessary to use a local oscillator - normally this is synchronised with the pump oscillator. In degenerate parametric oscillator squeezing it is usual to generate the pump input via frequency doubling from the laser that also produces the local oscillator. In an above threshold situation which we are treating here, it is crucial to consider the effects of laser bandwidth in this type of experiment. The case of a Lorentzian laser line is the simplest case, since it corresponds to a phase-diffusion model of the input laser amplitude. We assume that the input

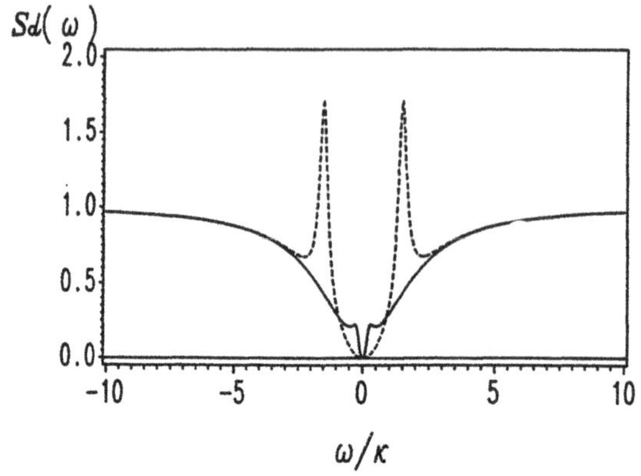

Figure 2a. Plot of $\bar{S}_D(\omega)$ above threshold for low pump cavity damping: $\frac{\kappa_0}{\kappa_+} = 3$, $\frac{\kappa_1}{\kappa_2} = 2$

P = 1.1—— ; P = 5 — —

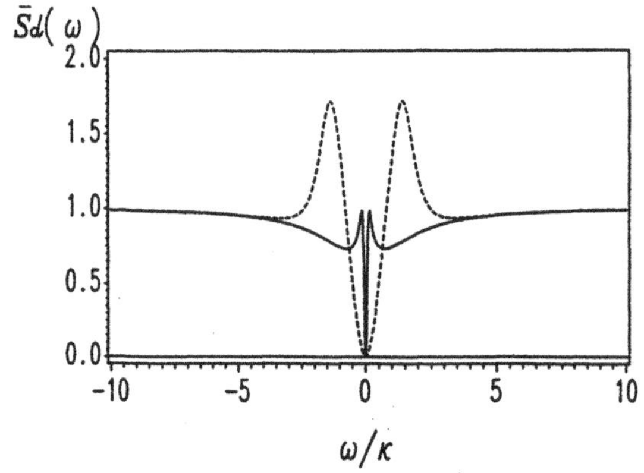

Figure 2b. Plot of $\bar{S}_D(\omega)$ above threshold for low pump cavity damping: $\frac{\kappa_0}{\kappa_+} = 0.182$, $\frac{\kappa_2}{\kappa_1} = 10$

P = 1.1—— ; P = 10- - -

is well-stabilized in intensity. In this Lorentzian limit, the driving field E is replaced by a stochastic input E(t), where

$$E(t) = |E|e^{-i(\omega_0 t+\phi_p(t))}$$

$$<\dot{\phi}_p(t)\dot{\phi}_p(t')> = \gamma_p\delta(t - t') \qquad (4.1)$$

This corresponds to an input laser with FWHM of $(\gamma_p/2\pi)$ Hz. The resulting stochastic equations for the intracavity field can then be analysed as before, except in a rotating frame which rotates with the input phase $\phi_p(t)$.

This choice of rotating frame has the useful feature that when the input laser is synchronised to the local oscillator, there is a cancellation of phase between the intracavity fields and the local oscillator fields. However, there is an extra term in the stochastic equations due to the rotating-frame having a time-dependence. This extra term is a type of memory effect, since the intracavity field stores both the current phase of the pump as well as previous phases. The extra term in the stochastic equation for variable (α_i) is $(\omega_i/\omega_0)\alpha_i\dot{\phi}_p(t)$.

To detect the squeezing, a local oscillator is used, typically with frequency $(\omega_0/2)$ and phase $\left[\theta + \phi_p(t)/2\right]$. This can be most readily analysed in the case of a degenerate parametric amplifier, where the signal and idler modes are identical. In this simple case, the squeezing spectrum is the spectrum of intensity fluctuations in the detected photo-current of local oscillator plus squeezed field. This spectrum is defined as

$$V(\theta,\omega) = 1 + 2\kappa_1 \int e^{i\omega t} \left\{\cos^2\theta<\Delta I_1(t)\Delta I_1(0)>/I_1^0 + 4I_1^0\sin^2\theta<\Delta\phi_1(t)\Delta\phi_1(0)>\right\} dt \qquad (4.2)$$

The result is a sum of the two terms, one corresponding to phase and the other corresponding to intensity fluctuations:

$$V(\theta,\omega) = 1 + 4S_I(\omega)\cos^2\theta + 4S_\phi(\omega)\sin^2\theta \qquad (4.3)$$

where

$$S_I(\omega) = \frac{r^2 + \bar{\omega}^2}{(2r(P-1) - \bar{\omega}^2)^2 + r^2\bar{\omega}^2}$$

$$S_\phi(\omega) = \frac{-(r^2 + \bar{\omega}^2) + (I_1^0\gamma_p/2\kappa_1)[(2+r)^2 + \bar{\omega}^2]}{(2rP - \bar{\omega}^2)^2 + \bar{\omega}^2(2+r)^2}$$

$$\bar{\omega} = \omega/\kappa_1$$

$$P = |E/E_T$$

$$r = \kappa_0/\kappa_1$$

We note the following results from this spectrum. Firstly, the term in $S_I(\omega)$ which corresponds to intensity fluctuations, always has a noise level above the shot-noise background. This background is the noise-level for a coherent signal, which gives $V(\theta,\omega) = 1$. Any squeezing in this degenerate case is due to the phase-noise term only. In the limit of $\gamma_p \to 0$ (zero bandwidth), the phase-noise term tends to reduce fluctuations below the shot-noise level, giving rise to squeezing.

In order to have squeezing, it is clearly essential to have a well-stabilized input laser. In fact, the requirement is that

$$\gamma_p I_1^0 < 2\kappa_1 \ [r^2 + \bar{\omega}^2]/[(2 + r)^2 + \bar{\omega}^2] \tag{4.4}$$

This means that $\gamma_p << (2\kappa_1/I_1^0)$ is required to have a sufficiently well-stabilized line-width over a wide range of values of frequency ($\bar{\omega}$) and relaxation times (r). That is, the laser must have a much narrower line-width than the cavity line-width (κ_1). In fact the laser must be stabilized to at least $[I_1^0]$ times better than the cavity-width, where I_1^0 is the intracavity photon number in the signal field.

We now wish to turn to the full non-degenerate calculation. For simplicity, it is clearly preferable to omit the complications of laser bandwidth effects. In view of Eq.(4.4), it is sufficient to require that $\gamma_p << (2\kappa_1/I_1^0)$ for this to be a good approximation. The external spectra in this case are now defined as

$$\begin{aligned} S_{ij}(\omega) &= \int_{-\infty}^{\infty} e^{i\omega\tau} \langle b_i^{+}(t+\tau)\, b_j(t)\rangle d\tau \\ C_{ij}(\omega) &= \int_{-\infty}^{\infty} e^{i\omega\tau} \langle b_i^{+}(t+\tau)\, b_j(t)\rangle d\tau \end{aligned} \tag{4.5}$$

where, as before, b_j indicates the external field of the cavity incident on a detector. These are readily calculated from the stochastic equations already derived. The elements of form $S_{12}(\omega)$, $C_{11}(\omega)$ are zero. For the case of equal decay rates, $\kappa_1 = \kappa_2 = \kappa$, the nonzero elements of the signal and idler spectra are

$$\begin{aligned} S_{11}(\omega) &= S_{22}(\omega) = 2I_1^0 L_0 - \tfrac{1}{8}L_1 - \tfrac{1}{8}C_{\phi^+} + \tfrac{1}{8}C_{I^+} \\ C_{12}(\omega) &= C_{21}^{+}(\omega) = 2I_1^0 L_0 + \tfrac{1}{8}L_1 + \tfrac{1}{8}C_{\phi^+} + \tfrac{1}{8}C_{I^+} \end{aligned} \tag{4.6}$$

The solutions are linear combinations of the Lorentzian components associated with the ϕ_-, ΔI_-, $\Delta\phi_+$ and ΔI_+ variables respectively, and are functions of the scaled parameters $\bar{\omega} = \omega/\kappa$, $r = \kappa_0/\kappa$ and $P = |E/E_T$, where E_T is the threshold pump amplitude. The term

$$L_0 = \frac{2(1/8I_1^0)}{\left[(1/8I_1^0)^2 + \bar{\omega}^2\right]} \tag{4.7}$$

is the large but narrow Lorentzian due to the phase diffusion of ϕ_-. The remaining terms describe the small fluctuations due to the stable subset, which appear as small but broad components in the spectrum. The second term

$$L_1 = \frac{2(1/8I_1^0 + 2)}{\left[(1/8I^0 + 2)^2 + \bar{\omega}^2\right]} \tag{4.8}$$

describes fluctuations in the signal-idler intensity difference. The third term describes fluctuations in the signal-idler phase sum and simplifies for large I^0 to

$$C_{\phi+} = \frac{4(r^2 + \bar{\omega}^2)}{\left[(2rP - \bar{\omega}^2)^2 + \bar{\omega}^2(2 + r)^2\right]} \tag{4.9}$$

The fourth term describing the fluctuations in the signal-idler intensity sum is positive and will be given elsewhere. The fluctuations in ΔI_- and $\Delta\phi_+$ are negative in the P-representation, implying noise levels below the vacuum or shot noise limit.

Consider the quadrature phase amplitudes defined as:

$$X_i^\theta(t) = e^{-i\theta} b_i + e^{i\theta} b_i^\dagger$$

Here we abbreviate $X_i^0(t) = X_i$ and $X_i^{\pi/2}(t) = Y_i$. The average value $\langle X_i^\theta \rangle$ of any of the quadrature amplitudes is zero, because of the phase diffusion. The large phase fluctuations occur on a long timescale so that one may envisage measuring an instantaneous amplitude $X_1^\theta(t)$. The intensity undergoes small stable fluctuations on a much shorter timescale.

Although we cannot say *a priori* what the projection X_1 will be at a particular time, we note the signal and idler phases ϕ_1 and ϕ_2 are correlated, since $\phi_1 + \phi_2$ has minimal fluctuations. The much smaller intensity fluctuations are <u>also</u> correlated, since $I_1 - I_2$ has minimal fluctuations. Thus we expect to infer quadrature phase information of the signal by measuring the quadrature phase of the idler.

The quantity $2V(\theta,\phi)$ where

$$V(\theta,\phi) = \tfrac{1}{2}\langle\left(X_1^\theta(t) - X_2^\phi(t)\right)^2\rangle = 1 + 2\left(\langle a_1^\dagger a_1\rangle - \cos(\theta + \phi)\langle a_1 a_2\rangle\right) \tag{4.10}$$

is a direct measure of the error in inferring the signal amplitude $X_1^\theta(t)$, given an experimental determination $X_2^\phi(t)$ of idler amplitude. $V(\theta,\phi)$ is minimum for $\theta = -\phi$: thus a measurement of X_2 implies X_1, <u>and</u> Y_2 implies $-Y_1$. When $V(\theta,\phi) = 0$ there is a perfect correlation between $X_1^\theta(t)$ and $X_2^\phi(t)$.

We point out that the ability to infer *at a distance* either of two non-commuting signal observables, with a precision below the vacuum noise level, is a direct example of the Einstein-Podolsky-Rosen paradox. The minimum uncertainty relation for the signal conjugate variables is $\Delta X_1 \Delta Y_1 = 1$. Thus observation of $V(0,0) < 0.5$ <u>and</u> $V(\pi/2,-\pi/2) < .5$ constitutes an EPR experiment.

One may more easily measure experimentally the spectrum of fluctuations in the difference $X_1^\theta - X_2^\phi$ [11], defined as

$$V(\theta,\phi,\omega) = \tfrac{1}{2}\int_{-\infty}^{\infty} e^{i\omega\tau}\langle X_1^\theta(t+\tau) - X_2^\phi(t+\tau),\, X_1^\theta(t) - X_2^\phi(t)\rangle d\tau$$

$$= 1 + 2\left(S_{11}(\omega) - \cos(\theta + \phi)C_{12}(\omega)\right) \tag{4.11}$$

The expression with the optimal choice $\theta + \phi = 0$ is

$$V(\theta,-\theta,\omega) = 1 - \tfrac{1}{2}L_1 - \tfrac{1}{2}C_{\phi+}$$

determined only by the quantum fluctuations ΔI_- and $\Delta\phi_+$. Maximum suppression of noise in ΔI_- (perfect signal-idler intensity correlation) corresponds to $L_1 = 1$. The solution below threshold ($P < 1$) is

$$V(\theta, -\theta, \omega) = 1 - 4P/\{(1 + P)^2 + \bar{\omega}^2\} \qquad (4.12)$$

a simple Lorentzian.

Figure 3 plots V for various r and P. $V \rightarrow 0$ is indicative of an EPR correlation. The coherent noise spike for $P > 1$ at $\bar{\omega} = 0$ is due to phase diffusion. Near threshold, both phase $\Delta\phi_+$ and intensity ΔI_- fluctuations are perfectly suppressed and $V(\theta, -\theta, \omega) \rightarrow 0$ near $\bar{\omega} = 0$. This is true in fact for all values of r and is also true below threshold as $P \rightarrow 1$. For very small r, near perfect suppression of phase fluctuations becomes possible well above threshold at higher frequencies $\bar{\omega} \approx \sqrt{2rP}$. Figure 3, curve b illustrates the appearance of such sidepeaks. The central dip is the intensity fluctuation spectrum. Particularly interesting is the situation depicted in curve c of an excellent pump ($r \rightarrow 0$). The bandwidth for perfect intensity fluctuation reduction extends out to $\bar{\omega} \approx 2$. This becomes broader than the frequencies $\bar{\omega} \approx \sqrt{2rP}$ showing perfect phase fluctuation reduction. Thus in summary we have perfect quantum correlation of signal-idler phase and intensity for fields of macroscopic photon number.

5. APPLICATIONS: SUB SHOT NOISE ABSORPTION MEASUREMENTS

The above calculations assumed there was no intracavity absorption and the fluctuations in the difference current exactly cancel at zero frequency. If there is an absorber inside the cavity this introduces a port for uncorrelated vacuum fluctuations to enter the cavity. Hence the fluctuations no longer cancel even if the absorption is equal for both signal and idler beams.

We shall denote the total losses (cavity plus internal) by κ_j and the cavity losses by γ_j ($\gamma_j \lesssim \kappa_j$). The internal equations remain unchanged but the relationship Eq.(3.1) between the external modes $b_j(t)$ and the internal ones $a_j(t)$ is given by the γ_j ($\gamma_j \lesssim \kappa_j$).

First we consider the case where the dampings are symmetric between the signal and idler, that is, $\gamma_1 = \gamma_2 = \gamma$, $\kappa_1 = \kappa_2 = \kappa$. The spectrum for the intensity fluctuations in the difference current normalized so the vacuum level is unity is,

$$\bar{S}_D(\omega) = \frac{\omega^2 + 4\kappa^2(1 - R_+)}{\omega^2 + 4\kappa^2} \qquad (5.1)$$

where $R_\pm = \frac{1}{2}\left(\frac{\gamma_1}{\kappa_1} \pm \frac{\gamma_2}{\kappa_2}\right)$

at $\omega = 0$:

$$\bar{S}_D(0) = \left(1 - \frac{\gamma}{\kappa}\right) \qquad (5.2)$$

Thus the effect of extra losses is to reduce the correlation between the modes and there is no longer perfect suppression of the noise as is shown in Fig.(4).

Returning now, to the more general case of arbitrary γ_i, κ_i; it has been suggested[12] that the correlations between the signal and idler modes of a parametric amplifier be used to enhance the sensitivity of absorption

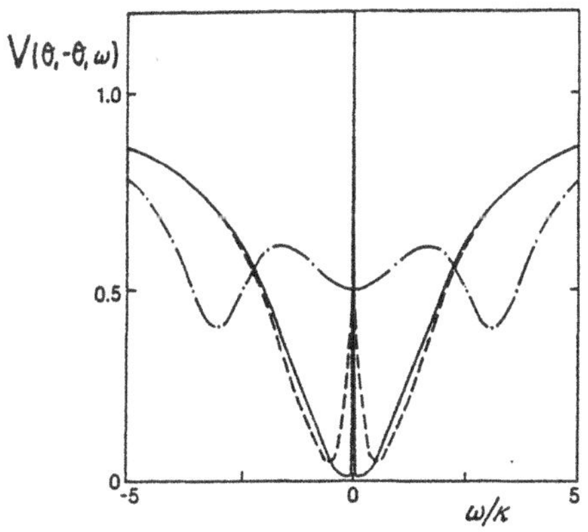

Figure 3. Plot of $V(\theta,-\theta,\omega)$ for the spectrum of fluctuations in the signal and idler quadrature amplitude difference, $X_1^{\theta} - X_2^{-\theta}$.

— Near threshold, $P = 1.01$, $\kappa_0/\kappa_1 = 0.01$, $I_1^0 = 10$
—·— Above threshold, $P = 50$, $\kappa_0/\kappa_1 = 0.1$
— — Above threshold, $P = 20$, $\kappa_0/\kappa_1 = 0.01$

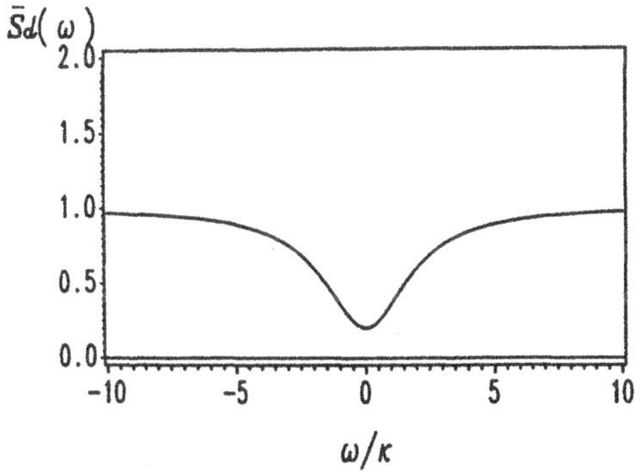

Figure 4. Effect of Intracavity Absorption: Plot of $\bar{S}_D(\omega)$

$\gamma_1 = \gamma_2 = \gamma$
$\kappa_1 = \kappa_2 = \kappa$, $\kappa_0/\kappa = 1$, $\gamma/\kappa = 0.7$

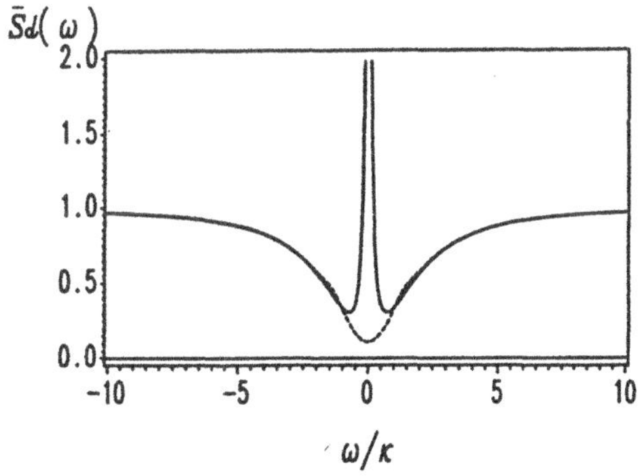

Figure 5. Effect of Asymmetrical Intracavity Absorption: Plot of $\bar{S}_D(\omega)$

$\frac{\kappa_0}{\kappa_+} = 1$, $\frac{\gamma_1}{\kappa_1} = 1$; $\frac{\gamma_2}{\kappa_2} = 0.8$; $\frac{\kappa_2}{\kappa_1} = 1.25$

p = 1.05——— ; p = 2— —

measurements.[24,25] We wish to suggest an absorption spectrometer where an absorber at the idler frequency is introduced into the cavity. In this case the spectrum of fluctuations at $\omega = 0$ is

$$\bar{S}_D(0) = 1 - R_+ + \frac{R_-^2}{(P-1)^2 R_+} \tag{5.3}$$

There is an extra noise term which is sensitive to the difference $R_- = \frac{1}{2}\left(\frac{\gamma_1}{\kappa_1} - \frac{\gamma_2}{\kappa_2}\right)$. In Fig.(5) we show the intensity spectrum for the difference current in the presence of an intracavity absorber at the idler frequency. We see that from a noise background of zero for no absorption, the presence of an absorber gives a marked increase in noise.

This may be used as a very sensitive detector for absorption. The level of absorption may be measured by inserting a calibrated variable absorber in one beam. The absorption is then varied until a null in the noise is obtained. This indicates the absorption in both arms is equal. This signal to noise rates in this case is

$$\frac{S}{N} = \frac{\bar{S}_D(0)\ \text{(unequal absorption)}}{\bar{S}_D(0)\ \text{(equal absorption)}}$$

$$= 1 + \frac{R_-^2}{(P-1)^2\ (1-R_+)R_+} \tag{5.4}$$

The feature of this intracavity scheme is that the sensitivity may be enhanced by operating near threshold. In the region of threshold there are critical fluctuations present which become manifest if there is an imbalance in the absorption in the two arms.

6. SQUEEZED PUMP LASERS

The usual laser produces coherent light with a linewidth caused by phase fluctuations which may be attributed to the pumping mechanism and the vacuum fluctuations entering the cavity. The quantum noise properties of the laser light are shot noise limited due to the Poissonian photon statistics. In order to reduce the laser fluctuations one may either choose to reduce the fluctuations in the pumping mechanism or reduce the vacuum fluctuations entering the cavity.

Yamamato *et al.* have treated a laser pumped with a source with reduced amplitude fluctuations. In this paper we describe a laser pumped by a squeezed field and analyse the effects the squeezed pump may have on the steady state operating characteristics, the laser linewidth and the quantum statistics. We compare this with a laser with a squeezed vacuum input to the cavity as discussed by Gea-Banacloche. We shall model the laser by adapting the methods developed by Haken[26] to include a squeezed pump and a squeezed input to the cavity. The laser model consists of N two level atoms interacting with a single cavity mode of the electromagnetic field. The cavity mode and the atoms are coupled to heat baths which model the pumping and decay mechanisms.

The atom field interaction is given by the dipole coupling Hamiltonian

$$H_{ATOMFIELD} = \hbar\omega_0 a^+ a + \frac{\hbar\omega}{2} S_z + ig\hbar(a^+ S_- - aS_+) \quad (6.1)$$

where a, a^+ are the boson operators for the field mode and S_+, S_-, S_z are collective atomic operators for N atoms, g is the dipole coupling constant.

The atoms are coupled to heat baths which model both the pumping mechanism and their spontaneous decay. The incoherent pumping of the atoms may be modelled by a bath of inverted harmonic oscillators denoted by Γ_p. The atom bath coupling is

$$H_{ATOM-BATH} = S_+\Gamma_p^+ + S_-\Gamma_p \quad (6.2)$$

Instead of the bath being in a thermal state, we assume the modes of the bath are squeezed. The correlation functions for the bath are taken as[27]

$$<\Gamma_p^+(t)\ \Gamma_p(t')> = (n_p + 1)\ \delta(t - t')$$

$$<\Gamma_p(t)\ \Gamma_p^+(t')> = n_p\ \delta(t - t')$$

$$<\Gamma_p(t)\ \Gamma_p(t')> = m_p^*\ e^{i\omega_0(t,t')}\ \delta(t - t')$$

$$<\Gamma_p^+(t)\ \Gamma_p^+(t')> = m_p\ e^{-i\omega_0(t,t')}\ \delta(t - t') \quad (6.3)$$

with $|m_p|^2 \lessapprox n_p(n_p + 1)$.

We call our model an incoherent "squeezed pump" because it yields phase dependent fluctuations of the atomic inversion, as we shall see later.

The above coupling models the pump as a squeezed vacuum which corresponds to super-Poissonian statistics. In order to model a field with sub-Poissonian statistics, or reduced amplitude fluctuations, one must in addition to the bath add a coherent driving field with amplitude E_p, which is assumed to be resonant with the atomic transition. Thus the atom-inverted bath coupling is modified to

$$H_{ATOM-BATH} = i\hbar g\ [S_+\ (\Gamma_p^+ + E_p^*\ e^{-i\omega_0 t}) + S_-\ (\Gamma_p + E_p\ e^{i\omega_0 t})] \quad (6.4)$$

within the rotating wave approximation.

Whether the pump has sub or super-Poissonian statistics will now depend on the phase of E_p with respect to the squeezing parameter m_p - for simplicity, the definition of E_p is chosen to correspond to E^* in the earlier sections. The cavity mode is coupled to a heat bath Γ_c which models the vacuum modes of the electromagnetic field external to the cavity.

$$H_{field-Bath} = i\hbar\kappa(a\Gamma_c^+ - a^+\Gamma_c) \quad (6.5)$$

If the fluctuations in the vacuum field entering the cavity are to be squeezed the correlation functions for the cavity bath are

$$<\Gamma_c^+(t)\ \Gamma_c(t')> = n_c\ \delta(t - t')$$

$$<\Gamma_c(t)\ \Gamma_c^+(t')> = (n_c + 1)\ \delta(t - t')$$

$$<\Gamma_c(t)\ \Gamma_c(t')> = m_c\ e^{-i\omega_0(t+t')}\ \delta(t - t')$$

$$<\Gamma_c^+(t)\ \Gamma_c^+(t')> = m_c^*\ e^{i\omega_0(t+t')}\ \delta(t - t') \quad (6.6)$$

with $|m_c|^2 \lessapprox n_c(n_c + 1)$.

The time dependence of the density operators for the atom field system after tracing over the heat bath observables is given by the master equation

$$\frac{\partial \rho}{\partial t} = \frac{1}{i\hbar}[\rho, H_0] + \left(\frac{\partial \rho}{\partial t}\right)_A + \left(\frac{\partial \rho}{\partial t}\right)_F \qquad (6.7)$$

(in a frame rotating at ω_0) where

$$\left(\frac{\partial \rho}{\partial t}\right)_A = [E_p S^+ - E_p^* S_-, \rho] + \frac{\gamma n_p}{2}(2S_-\rho S_+ - \rho S_- S_+ - S_- S_+ \rho)$$
$$+ \frac{\gamma(n_p + 1)}{2}(2S_+\rho S_- - \rho S_- S_+ - S_- S_+\rho) - \gamma m_p S_+\rho S_+ - \gamma m_p^* S_-\rho S_- \qquad (6.8)$$

with $\gamma = \omega_{12} - \omega_{21}$.

Here ω_{12} is the transition rate $|1\rangle \to |2\rangle$ due to the incoherent pumping, ω_{21} is the transition rate $|2\rangle \to |1\rangle$ due to non lasing transitions (spontaneous emission) and

$$\left(\frac{\partial \rho}{\partial t}\right)_F = \frac{\kappa}{2}\Big\{(1 + n_c)(2a\rho a^+ - a^+a\rho - \rho a^+ a) + n_c(2a^+\rho a - aa^+\rho - \rho aa^+)$$
$$- m_c(a^+\rho a^+ - a^+a^+\rho - \rho a^+a^+) - m_c^*(a\rho a - aa\rho - \rho aa)\Big\} \qquad (6.9)$$

where κ is the cavity decay rate.

The effect of the squeezed baths has been to add the terms with m_p and m_c to the master equation.

7. OPTICAL BLOCH EQUATIONS AND STEADY STATE SOLUTIONS

We shall first consider how the squeezed baths affect the steady state solutions. The equations for the average values of the atomic and field observables follow directly from the master equation

$$\dot{\alpha}_x = -\kappa\alpha_x + g v_x$$
$$\dot{\alpha}_y = -\kappa\alpha_y + g v_y$$
$$\dot{v}_x = -\gamma_\perp(1 + M_p)v_x + gD\beta_x \qquad (7.1)$$
$$\dot{v}_y = -\gamma_\perp(1 - M_p)v_y + gD\beta_y$$
$$\dot{D} = -\gamma_{\parallel}D + \gamma N - 4g(v_x\beta_x + v_y\beta_y)$$

where $\alpha = \alpha_x + i\alpha_y$ is the field amplitude, $\beta = \alpha + \frac{E_p^*}{g}$, v_x and v_y are the components of the atomic polarization and D is the atomic inversion. The atomic decay rates are

$$\gamma_{||} = \omega_{12} + \omega_{21}$$

$$\gamma_{\perp} = \tfrac{1}{2}\gamma_{||}$$

and $M_p = \dfrac{\gamma m_p}{\gamma}$; note that $|M_p|$ is bounded by one.

We see that the effect of squeezing the pump bath has been to modify the time scales of the two components of the atomic polarization.[28]

We shall consider the following cases: First we will briefly recapitulate our previous results on the effects of squeezing the pump alone, i.e. $m_p \neq 0$, $m_c = 0$, without a coherent driving field ($E_p = 0$). Then we will generalize this by adding a coherent part and finally we will investigate the effects of squeezing the cavity alone, i.e. $m_p = 0$, $m_c \neq 0$ (as has recently been done by Gea-Banacloche[18]).

i. Squeezed Pump ($m_p \in R_+$, $n_c = 0$)

(a) Pump in a squeezed vacuum state. For a squeezed vacuum pump the steady state solutions of Eqs.(7.1) are

$$\alpha = 0$$

$$\alpha = \pm\, i\, \sqrt{n_0\,[C - (1 - M_p)]} \quad \text{for } C > 1 - M_p$$

$$\alpha = \pm \sqrt{n_0\,[C - (1 + M_p)]} \quad \text{for } C > 1 + M_p \tag{7.2}$$

where C is the cooperativity parameter is defined by

$$C = \frac{g^2 N\gamma}{\gamma_{\perp}\kappa\gamma_{||}}$$

and n_0 the saturation photon number is

$$n_0 = \frac{\gamma_{||}\gamma_{\perp}}{4g^2}$$

The usual incoherently pumped laser has a solution which is independent of phase. The phase dependence in the noise of the squeezed pump light breaks this symmetry. As a consequence only fields which are either purely real or purely imaginary may exist. We require a stability analysis to reveal which of these solutions is stable if the reference phase choice $m_p \in R_+$ is made. Elimination of the steady state atomic variables yields the set of equations

$$\dot{\alpha}_x = \kappa\left(\frac{C}{R}(1 - M_p) - 1\right)\alpha_x \equiv A_x(\alpha_x, \alpha_y)$$

$$\dot{\alpha}_y = \kappa\left(\frac{C}{R}(1 + M_p) - 1\right)\alpha_y \equiv A_y(\alpha_x, \alpha_y) \tag{7.3}$$

with R standing for

$$R(\alpha_x,\alpha_y) = (1 - M_p^2) + \frac{\alpha_x^2}{n_0}(1 - M_p) + \frac{\alpha_y^2}{n_0}(1 + M_p) \qquad (7.4)$$

which yields the steady state solutions already given in Eq.(7.2). The stability of a solution requires all eigenvalues of the Jacobian corresponding to the linearization of Eq.(7.3) around the steady state to be negative. For the choice of phase M_p real and positive, we find the imaginary solutions

$$\alpha = \pm i\sqrt{n_0 (C - (1 - M_p))} \qquad (7.5)$$

are stable, if they exist, i.e. for $C > 1 - M_p$. Thus we have a stability exchange between the solution $|\alpha| = 0$ and the two imaginary solutions with $|\alpha|^2 = n_0[C - (1 - M_p)]$ occurring at $C = 1 - M_p$.

It is the steady states which are in phase with the low noise quadrature in the squeezed bath which are stable. This may be seen from the potential solution. A potential may be defined by

$$\frac{\partial\Phi(\alpha_x,\alpha_y)}{\partial\alpha_x} = - A_x(\alpha_x,\alpha_y)$$
$$\frac{\partial\Phi(\alpha_x,\alpha_y)}{\partial\alpha_y} = - A_y(\alpha_x,\alpha_y) \qquad (7.6)$$

(A_x and A_y from Eq.(7.3)).

The potential conditions

$$\frac{\partial A_x}{\partial\alpha_y} = \frac{\partial A_y}{\partial\alpha_x} \qquad (7.7)$$

may be shown to hold globally. Thus a potential exists and is found by integration of Eq.(7.6) to be

$$\Phi(\alpha_x,\alpha_y) = \frac{\kappa}{2}\left(\alpha_x^2 + \alpha_y^2 - C\,n_0\,\ln R(\alpha_x,\alpha_y)\right) \qquad (7.8)$$

We see from Fig.6 that for $M_p \neq 0$ the phase symmetry of the usual laser is broken. The squeezed bath acts to imprint a particular phase onto the steady state as is seen from the increasingly deep valleys along $\psi = \pm\pi/2$ and increasingly high ridges at $\psi = 0, \pi$ as M_p is increased. We note the difference with the symmetry breaking which occurs in the laser with injected signal[29] and also in the micromaser, where only one phase is stable.

(b) Pump in an arbitrary squeezed state. We include a coherent driving field coupled to the atoms to model a bath in a squeezed state with a non-zero coherent part. Viewing Eq.(7.1), we see that the inclusion of such a driving field results in a replacement of α by
$\beta = \alpha + \frac{E_p^*}{g}$ in the equations of motion of the atomic polarization and inversion, but leaving the equations of motion of the field unchanged.

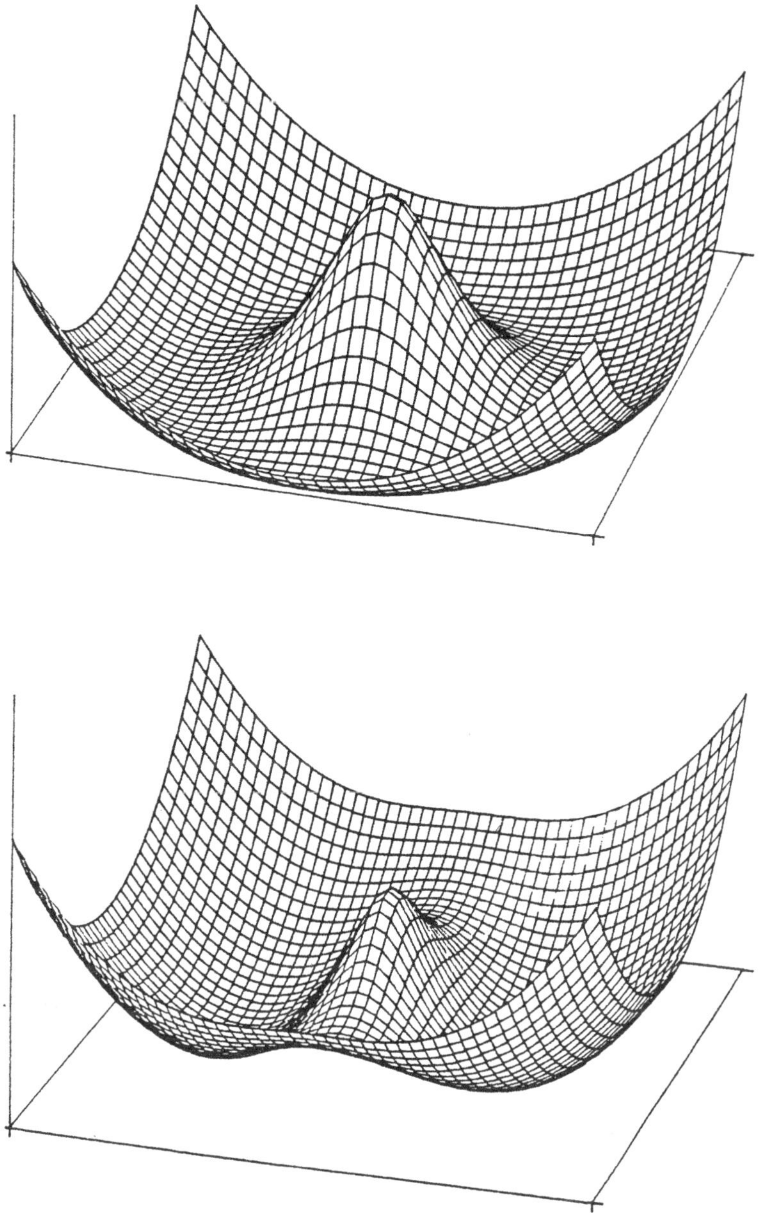

Figure 6. The potential function $\Phi(\alpha_x, \alpha_y)$ of the laser above threshold

(a) Ordinary laser with phase diffusion $M_p = 0$, $C = 2$.
(b) Squeezed pump laser $M_p = 0.5$, $C = 2$.

It is clearly visible how the potential surface is distorted to form two local minima along the imaginary axis leading to a phase locking phenomena.

Consequently, Eq.(5.3) is replaced by

$$\dot{\alpha}_x = \kappa[-\alpha_x + \frac{C(1 - M_p)}{R(\beta_x,\beta_y)}\beta_x]$$

$$\dot{\alpha}_y = \kappa[-\alpha_y + \frac{C(1 + M_p)}{R(\beta_x,\beta_y)}\beta_y] \tag{7.9}$$

Hence the expression for the potential Φ is formally unchanged, except for the fact that the saturation denominator R (see Eq.(7.4)) has to be evaluated at β instead of at α.

However, the determination of the steady state solutions turns out to be more complicated, in as much as it involves the solution of a cubic polynomial.

In the following, we wish to restrict ourselves to the case of either purely real or purely imaginary amplitudes of the coherent part. It is well known that, for the choice $M_p \in R_+$, the former case gives rise to super-Poissonian statistics in the bath, whereas the latter leads to sub-Poissonian statistics in the bath. Henceforth, we will refer to these two cases, as bunched and antibunched input, respectively.

For the case of an antibunched input, i.e. $\varepsilon_p = i|\frac{E_p}{g}|$ and thus $\beta_x = \alpha_x$, $\beta_y = \alpha_y - \varepsilon_p$, two steady state solutions are easily found:

$$\alpha_x^a = \pm\sqrt{n_o(C - C_a)},\quad \alpha_y^a = -|\varepsilon_p|\frac{1 + M_p}{M_p} \tag{7.10}$$

for $C > C_a = 1 + M_p + \frac{|\varepsilon_p|^2}{n_o}\frac{1 - M_p^2}{4M_p^2}$

The remaining three roots satisfy

$$\alpha_x^a = 0 = \beta_x^a \tag{7.11}$$

and a cubic equation in α_y:

$$\alpha_y\left((1 - M_p) + \frac{\beta_y^2}{n_o}\right) = C\beta_y \tag{7.12}$$

It is straightforward to prove that Eq.(7.12) possesses one negative and two positive solutions - provided three real solutions exist.

A stability analysis shows that the solutions (7.10) are unstable where they exist, i.e. for $C > C_a$. (Note that these solutions turn into the pair of unstable solutions of the previous section for $|E_p| \to 0$.) Examining the stability of the roots of the cubic numerically, we found that the negative root and the larger positive root are stable for $C > C_a$, whereas for $C < C_a$ the negative one is always stable and the larger positive one can be stable, depending on the parameters E_p and M_p. This may be illustrated by the potential function (Eq.(7.8) with α replaced by β) for the antibunched input depicted in Fig.(7a).

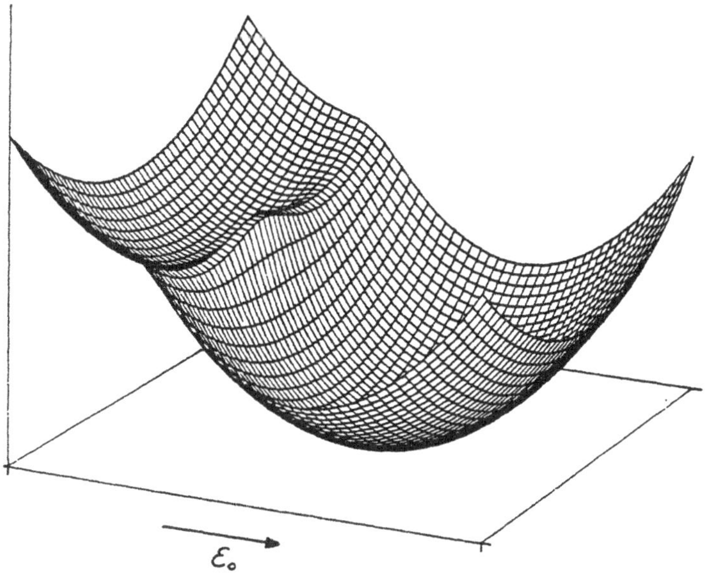

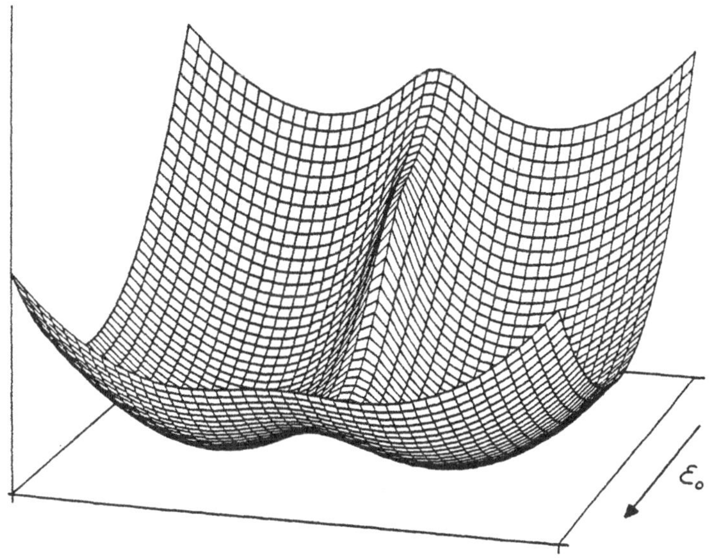

Figure 7. The potential function of the squeezed pump above threshold

(a) antibunched input $M = 0.9$, $C = 10$, $|\varepsilon| = 4$
(b) bunched input $M = 0.9$, $C = 10$, $|\varepsilon| = 4$

The corresponding results for the bunched input $\varepsilon_p = |\frac{E_p}{g}|$, are easily obtained by replacing $M_p \to -M_p$ and interchanging $\alpha_y \longleftrightarrow \alpha_x$ and $\beta_y \longleftrightarrow \beta_x$ in the expressions derived for the antibunched input. In particular, we find that the two steady state solutions similar to Eq. (7.10) are always stable for

$$C > C_b = 1 - M_p + \frac{|\varepsilon_p|^2}{n_0} \frac{1 - M_p^2}{4M_p^2},$$

whereas for $C < C_b$, one of the roots of the cubic is stable (see Fig. (7 b)).

Summarizing, we have found that there exist two stable steady state solutions (for $C > C_a$) along the imaginary axis in the antibunched input - which is not surprising, since such an input would lock the laser phase to $\pi/2$, while the squeezed vacuum in the bath leads to a locking of the phase to $\pm\,\pi/2$, as we have seen in the previous section. Note, that consequently the potential well in the direction $\pi/2$ is deeper than the one in the direction $-\pi/2$, as is clearly visible in Fig. (7a).

For the bunched input, however, there is a trade-off between a tendency to lock to $\phi = 0$ due to the driving field and a tendency to lock to $\phi = \pm\,\pi/2$ due to the squeezed bath, so we get two stable steady states (for $C > C_b$) in the first and fourth quadrant, which lie symmetrically about the real axis; this can be seen as two (equally deep) valleys in the potential surface Fig. (7b). For increasingly large $|E_p|$ and fixed $(1 - M_p)$, these two solutions approach each other by wandering symmetrically towards the real axis.

ii. Squeezed Cavity ($m_c \in R_+$, $m_p = 0$)

It is easily derived from the master Eq. (4.9) that the corresponding Bloch equations for α, α^*, v, v^* and D do not differ from the set of equations describing an ordinary laser. However, the noise in the bath is phase-dependent, giving rise to new noise correlations in the corresponding Fokker-Planck equation or stochastic differential (Ito) equation.

From what we have seen before, for the case of a squeezed bath coupled to the atoms, one might suspect that the anisotropy of the noise here similarly destroys the phase symmetry of the laser; that is that a "locking" of the phase might occur, due to phase dependent correlations of the noise in the laser. This "phase-locking" of course, would be hidden in the semiclassical equations for α and α^*, since it stems from an anisotropy in the noise - and not from a distortion of the potential function (such as in section (7.1)).

Adiabatically eliminating the atoms, one may derive from the master equation a two-dimensional Fokker-Planck equation in the variables α and α^* (via the usual truncation procedure of Haken's laser theory[26]). Converting this Fokker-Planck equation (or the corresponding Ito stochastic differential equation) to intensity and phase variables involves the diffusion matrix elements, i.e. the noise correlations. Thus it may be hoped that choosing the variables ϕ and I, given by

$$I = \alpha^* \alpha\,.$$

$$\phi = \frac{1}{2i} \ln \frac{\alpha^*}{\alpha} \qquad (7.13)$$

or $\alpha = \sqrt{I}\, e^{-i\phi}$

$$\alpha^* = \sqrt{I}\, e^{i\phi} \tag{7.14}$$

one may gain more insight into the situation. (It is well known that the semiclassical solutions need not stay the same under a nonlinear coordinate transform). Carrying this conversion out, we get

$$\dot{I} = -2\kappa I \left[1 - \frac{C}{1 + \frac{I}{n_o}} \right] + 2\kappa n_c + \frac{g^2 N}{\gamma_\perp^{\,2}}\, \omega_{12} \tag{7.15}$$

$$\dot{\phi} = \frac{\kappa m_c}{I} \sin 2\phi \tag{7.16}$$

The last expression in the equation for I stems from the spontaneous emission noise in the laser. n_c is the number of photons in the bath coupled to the lasing mode, $m_c = \sqrt{n_c(n_c + 1)}$ stands for the squeezing parameter in the bath. The important thing is that $\dot{\phi} \neq 0$ for non-zero m_c, in contrast to an ordinary laser, thus revealing the "steady state phases" to be $\phi = 0$ (or π) and $\phi = \pi/2$ (or $-\pi/2$).

Viewing Eq.(7.16), it is obvious that $\phi = \pm\, \pi/2$ corresponds to stable points, whereas $\phi = 0, \pi$ are unstable if m_c is chosen real and positive in agreement with previous results by Gea-Banacloche.[18] Note the resemblance to the situation encountered for the squeezed pump!

One might wish to derive a potential function for the semiclassical equations (7.15). However, from the dependence of the r.h.s. of the phase equation on the laser intensity for $m_c \neq 0$, while the intensity equation is phase <u>in</u>dependent, it is clear that the potential conditions are not satisfied. This, again, reflects the fact, that this "phase-locking" is not due to some minimum in the potential surface, but is entirely due to the anisotropy in the noise.

The empty cavity limit is easily recovered from Eq.(7.15) by setting $g = 0$ (thus $c = 0$). The drift term in the phase equation is seen to be still the same in this simplest possible case, where we have a constant diffusion matrix, i.e. an Ornstein-Uhlenbeck process. (This is so, although $\alpha = 0$ in the steady state in an empty cavity and hence ϕ cannot be interpreted as a proper phase.)

8. ROTATING WAVE VAN DER POL EQUATION WITH PHASE DEPENDENT GAIN

We shall consider the laser with a squeezed pump and the usual vacuum entering the cavity ($n_c = m_c = 0$).

The quantum statistics of the laser output may be calculated by converting the master equation to a Fokker Planck equation and then deriving stochastic differential equations.

The resulting stochastic differential equations are

$$\dot{\alpha} = -\kappa\alpha + gv + \Gamma_\alpha$$

$$\dot{v} = -\gamma_{\perp} v - \gamma m_p v^* + gD\alpha + \Gamma_v \qquad (8.1)$$

$$\dot{D} = -\gamma_{\parallel} D + \gamma N - 2g(v^*\alpha + v\alpha^*) + \Gamma_D$$

where the correlation functions are

$$< \Gamma_\alpha(t)\, \Gamma_\alpha^*(t')> = 2\kappa\, n_c\, \delta(t - t')$$

$$<\Gamma_v(t)\, \Gamma_v(t')> = [\gamma m_p (N + D) + 2gv\alpha]\, \delta(t - t')$$

$$<\Gamma_v(t)\, \Gamma_v^*(t')> = \omega_{12}\, N\, \delta(t - t')$$

$$<\Gamma_v(t)\, \Gamma_D(t')> = - 2\omega_{12}\, v\, \delta(t - t') \qquad (8.2)$$

$$<\Gamma_D(t)\, \Gamma_D(t')> = [2(\gamma_{\parallel} N - \gamma D) - 4g(\alpha^* v + v^*\alpha)]\, \delta(t - t')$$

To make a comparison with the usual laser we shall derive the Fokker Planck equation for the laser mode in the limit that the atoms may be adiabatically eliminated, that is,

$$\gamma_{\perp}(1 \pm M_p) >> \kappa$$

In this section we shall assume that α and α^* are the complex conjugate of each other because in our case of small M_p which implies that spontaneous emission is the dominant noise source, the diffusion matrix is nearly positive definite somewhat above threshold.

In the region close to threshold where saturation terms may be neglected, we obtain

$$\dot{\alpha} = G(\alpha, \alpha^*)\alpha - \alpha_{nl} + F_\alpha \qquad (8.3)$$

with

$$G(\alpha, \alpha^*) = \kappa\left[-1 + \frac{C}{(1 - M_p^2)}\left(1 - \frac{\alpha^*}{\alpha} M_p\right)\right]$$

and

$$\alpha_{nl} = \kappa C \frac{\left(1 - \frac{\alpha^*}{\alpha} M_p\right)}{(1 - M_p^2)^2}\left[\frac{|\alpha|^2}{n_0} - M_p\left(\frac{\alpha^2}{2n_0} + \frac{\alpha^{*2}}{2n_0}\right)\right]\alpha$$

This is a rotating wave van der Pol equation with phase dependent gain. Setting $\alpha = \sqrt{I}\, e^{-i\psi}$, we may write $G(\alpha,\alpha^*)$ as

$$G(\alpha,\alpha^*) = \kappa\left[-1 + \frac{C}{(1 - M_p^2)}\left[1 - e^{2i\psi} M_p\right]\right] \qquad (8.4)$$

It is clear that the gain is largest for $\psi = \pm \pi/2$, hence a steady state field with $\psi = \pm \pi/2$ is built up above threshold.

Converting the Langevin Eq.(6.3) to intensity and phase variables we can derive the phase diffusion rate evaluated about the stable steady state solution $\bar{I}$ with phase $\bar{\psi} = \pm \pi/2$, we find

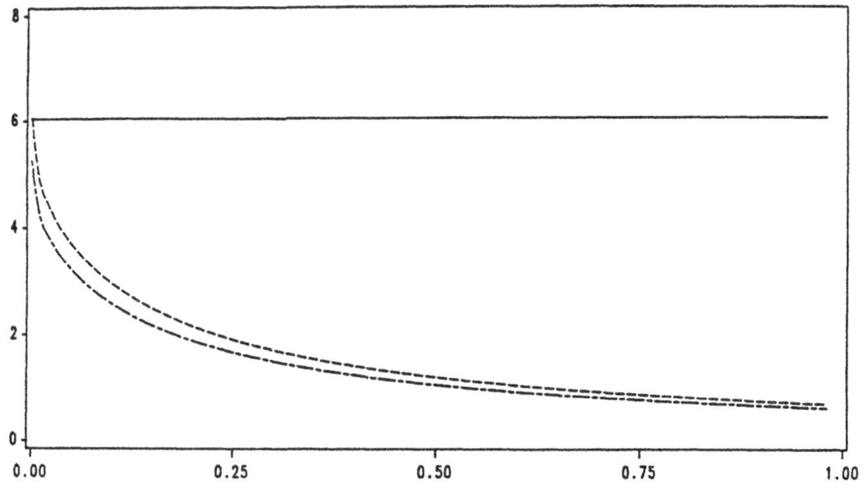

Figure 8. Phase diffusion coefficient $D_{\psi\psi}$ for the laser at 10% above threshold

— — — exact solution
---- rotating wave van der Pol solution

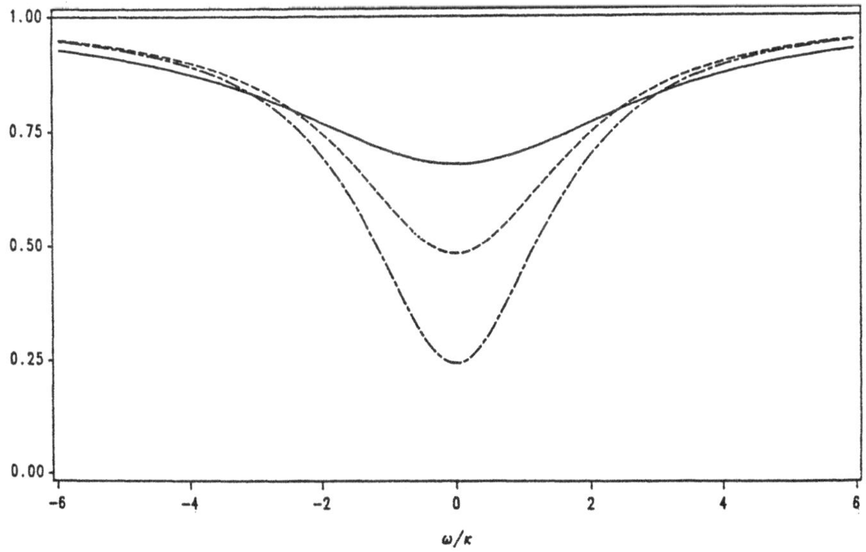

Figure 9. Variance of the quadrature amplitude Y with antibunched input for fixed output intensity

——— $M_p = 0.471$
---- $M_p = 0.700$
— — — $M_p = 0.934$

$$D_{\psi\psi}\ (\bar{I},\bar{\psi}) = \frac{1}{2\bar{I}} \left[\frac{2C\kappa}{(1 + M_p)^2} \left(1 + n_p + m_p \left(1 + \frac{1}{2n_p + 1}\right)\right) \right] \tag{8.5}$$

For $M_p = 0$, this expression reduces to the usual phase diffusion rate due to spontaneous emission noise, which may be written as

$$D_{\psi\psi} = \frac{1}{2\bar{I}} \frac{g^2 \bar{N}_2}{2\gamma_\perp} \tag{8.6}$$

$\bar{N}_2$ being the number of excited atoms in the steady state.

In Fig.(8) the phase diffusion coefficient $D_{\psi\psi}$ is plotted against the squeezing parameter n_p (for m_p taken maximal $m_p = \sqrt{n_p(n_p + 1)}$). The results from the rotating wave van der Pol oscillator equation are compared with the results from the full equations (8.1). In Fig.(8a) the laser is ten percent above threshold and in Fig.(8b) the laser is operating two times above threshold.

We see that the effect of the squeezed pump is to reduce the phase diffusion coefficient significantly below that of the usual laser. A similar effect has been found by Gea-Banacloche[18] for the laser with a squeezed vacuum entering the cavity. In this case a reduction of the phase diffusion coefficient by a factor of two may be achieved, although this occurs about an unstable state.

We now consider the variances and squeezing in the squeezed pump laser. The quantum fluctuations of the output light may be calculated from the full set of equations (8.1). The details of this calculation may be found in references (30 and 31). The squeezing spectra in two quadratures $V_{xx}(\omega)$ and $V_{yy}(\omega)$ are calculated corresponding to phase and amplitude noise respectively. The vacuum level is $V(\omega) = 1$. There is no reduction in the phase noise below the vacuum level, however the amplitude fluctuations may drop below the vacuum level leading to sub-Poissonian statistics, similar to the model described by Yamamoto *et al.*[14] This effect is shown in Fig.(9) where the atoms are pumped with an antibunched input.

REFERENCES

1. D.F. Walls, Nature, 306, 141 (1983).
2. Special Issue on Squeezed States of the Electromagnetic field, Journal of the Optical Society of America, 4, No.10, (1987). Editors H.J. Kimble and D.F. Walls.
3. Special Issue on Squeezed Light, Journal of Modern Optics, 34, No. 6,7, (1987). Editors R. Loudon and P.L. Knight
4. R.E. Slusher, L.W. Hollberg, B. Yurke, J.C. Mertz and J.F. Valley, Phys.Rev.Lett. 55, 2409 (1985).
5. M.W. Maeda, P. Kumar and J.H. Shapiro, Optics Lett. 12, 161 (1987).
6. R.M. Shelby, M.D. Levenson, S.H. Perlmutter, R.G. de Voe and D.F. Walls, Phys.Rev.Lett., 57, 691 (1986).
7. L. Wu, H.J. Kimble, J.L. Hall and H. Wu, Phys.Rev.Lett, 57, 2520 (1986).
8. R. Short and L. Mandel, Phys.Rev.Lett. 51, 384 (1983).
9. R. Brown, E. Jakeman, R. Pike, J. Rarity and P. Tapster, Europhys.Lett. 2, 279 (1986).
10. D. Burnham and D. Weinberg, Phys.Rev.Lett. 25, 84 (1970).
11. S. Friberg, C. Hong and L. Mandel, Phys.Rev.Lett. 54, 2011 (1985).
12. S. Reynaud, C. Fabre and E. Giacobino in ref(3) p1520.

13. S. Reynaud, Europhysics Letters, 4, 427 (1987).
14. Y. Yamamoto, S. Machida and S. Nilsson, Phys. Rev. A34, 4025 (1986).
15. S. Machida, Y. Yamamoto and Y. Itaya, Phys.Rev.Lett. 58, 1000 (1987).
16. P. Filipowicz, J. Javanainen and P. Meystre, Phys.Rev. A34, 3077 (1986).
17. J. Krause, M.O. Scully and H. Walther, Phys.Rev. A34, 2032 (1986).
18. J. Gea-Banacloche, Phys.Rev.Lett. 59, 543 (1987).
19. K.J. McNeil and C.W. Gardiner, Phys.Rev. 28A 1560 (1983).
20. P.D. Drummond and C.W. Gardiner, J.Phys. A13, 2353 (1980).
21. R. Graham and H. Haken, Z.Physik, 210, 276 (1968).
22. M.J. Collett and C.W. Gardiner, Phys.Rev. A30, 1386 (1984).
23. A.S. Lane, M.D. Reid and D.F Walls (to be published).
24. N.C. Wong and J. Hall, JOSA B2, 1510 (1985).
25. M. Gehrtz, G. Bjorklund and E. Whittaker, JOSA, B2, 1510 (1985).
26. H. Haken "Laser Theory", Springer (1984).
27. M.J. Collett and C.W. Gardiner, Phys.Rev. A31, 3761 (1985).
28. C.W. Gardiner, Phys.Rev.Lett. 56, 1917 (1986).
29. A. Heidmann, R.J. Horowicz, S. Reynaud, E. Giacobino, C. Fabre and G. Camy. Phys.Rev.Lett. 59, 2555 (1987).
30. M.A. Marte and D.F.Walls, Phys.Rev.A37, 1235 (1988).
31. M.A. Marte, H. Ritsch and D.F. Walls (to be published).

LASER STABILIZATION USING SQUEEZED LIGHT*

Carlton M. Caves

Center for Laser Studies
University of Southern California
Los Angeles, California 90089-1112 USA

I. INTRODUCTION

Many lasers do not approach ideal operation. They have technical noise—both in intensity and frequency—which exceeds—often far exceeds—the noise predicted by ideal laser theory. Should other considerations nonetheless prompt one to use such a noisy laser, the noise can be controlled by using negative feedback to stabilize the laser's intensity and/or frequency. Here I investigate the use of squeezed light to enhance the performance of such feedback schemes.

Feedback suppresses noise from the laser but imposes on light extracted from the feedback loop noise that arises within the loop. Whether such a tradeoff is desirable depends on how noisy the laser is and how well the feedback scheme operates. Often it is desirable because the feedback loop, being free of the messy and nonlinear gain medium, offers a more pristine environment than the laser cavity. Indeed, I assume here that technical noise within the feedback loop can be eliminated. Then, from a quantum-mechanical point of view, noise within the feedback loop arises from two sources: (i) fluctuations associated with losses, e.g., inefficiencies, absorption, and scattering, and (ii) vacuum fluctuations that sneak into the feedback loop through one or more open ports. The strategy investigated here is to reduce losses and to cover open ports with squeezed vacuum light—i.e., squeezed light with no mean field. The squeezed light is produced by tapping a small portion of the laser light to pump one or more "squeezers"—nonlinear optical devices that generate squeezed light.

Negative feedback schemes can be analyzed quantum mechanically using an approach developed by Shapiro *et al.*[1] It involves linearizing the fluctuations about the lock point of the feedback loop. I do not give detailed calculations here; they have been or will be presented elsewhere. The focus is on describing the feedback schemes, identifying the important noise sources, and illustrating how squeezed light can be used to reduce the noise.

The performance of a feedback scheme is summarized here by giving a spectral density for the intensity or frequency (or phase) of the light extracted from the

* Supported in part by the US Office of Naval Research [Contract No. N00014-88-K-0042]

feedback loop. All the results are predicated on the following assumptions: (i) The RF bandwidth B is small compared to the inverse of the propagation time around the feedback loop. (ii) The feedback gain is large enough over the bandwidth B that laser noise in the extracted light can be ignored. Thus the spectra display only feedback-imposed noise. (iii) The squeezers operate in the classical-pump limit, in which pump quantum fluctuations can be ignored. This does not imply, of course, ideal minimum-uncertainty squeezing. (iv) The squeezers are fast enough that phase jitter between the squeezed light and the laser light can be ignored. (v) Single-pass losses at optical elements are small enough to be ignored. In optical cavities used for frequency stabilization, however, multiple-pass losses are included.

In Sec. II I consider a simple scheme to stabilize intensity, and in Sec. III I turn to a frequency-stabilization scheme, based on the fringe-side technique.

II. INTENSITY STABILIZATION

A scheme for stabilizing intensity is sketched in Fig. 1.[2] It relies on negative feedback from a conventional photodetector to stabilize the intensity to a fixed reference level. Useful, intensity-stabilized light is removed from the feedback loop at an "extraction beamsplitter," which has transmissivity χ. In Fig. 1 the element that controls intensity is a variable attenuator (outside the laser), represented by a beamsplitter with variable transmissivity $T(t)$. This sort of intensity control is used, following Shapiro *et al.*,[1] for ease in the mathematical analysis, but it is not an essential feature. In particular, the consequent reduction in mean power can be avoided by using a more realistic element to control the intensity of the laser.

The laser produces light with frequency Ω and photon intensity $\mathcal{P}_L$. [All intensities are measured in photon units—i.e., $\mathcal{P}_L = (\text{energy/sec})/\hbar\Omega$.] The laser light first traverses the variable beamsplitter, after which it encounters a beamsplitter with transmissivity $1-\xi$, which removes a (small) fraction ξ of the intensity to pump the squeezer. The squeezer takes in vacuum through an isolator (depicted in Fig. 1 as a three-port circulator) and produces squeezed vacuum light (labeled a_2). The remaining laser light (labeled a_1) is combined with the squeezed light at the extraction beamsplitter (transmissivity χ). One of the outputs of the extraction beamsplitter (labeled b_1) is directed onto a photodetector, which has quantum efficiency η and which produces a photocurrent $I(t)$. The feedback loop is closed by choosing the transmissivity of the variable beamsplitter to be $T(t) = T_0[1 - \varsigma\delta I(t)]$, where T_0 is the fiducial transmissivity of the variable beamsplitter, ς characterizes the strength of the feedback, and $\delta I(t)$ is the difference between $I(t)$ and a fixed reference level. The other output of the extraction beamsplitter (labeled b_2) is the extracted light.

A heuristic quantum-mechanical analysis illustrates how this scheme works when the feedback is perfect and the photodetector has unity quantum efficiency ($\eta = 1$). Suppose the a_1-light has an amplitude fluctuation δ_1 and the in-phase quadrature of the squeezed light has a fluctuation δ_2, both these fluctuations being measured in vacuum units. The variance $\langle\delta_2^2\rangle \equiv R$ characterizes the amount of squeezing in the a_2-light relative to vacuum ($R = 1$) in units of power. The fluctuations δ_1 and δ_2 are not independent because the perfect feedback cancels precisely the amplitude fluctuation,

$$\Delta_1 = \delta_1\sqrt{\chi} - \delta_2\sqrt{1-\chi} = 0, \tag{2.1}$$

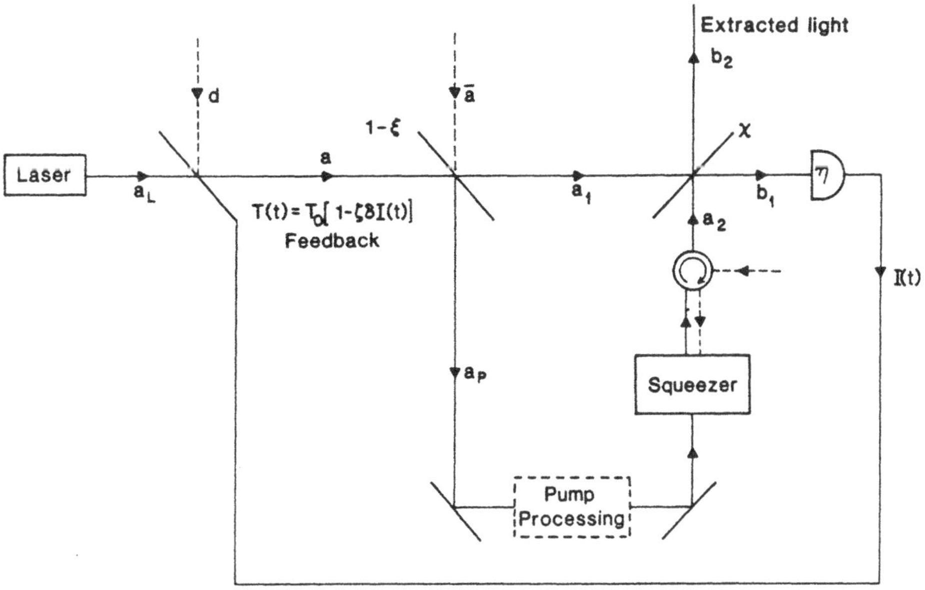

Figure 1. Scheme for laser intensity stabilization using squeezed light

seen by the photodetector. The resulting amplitude fluctuation in the a_1-light,

$$\delta_1 = \delta_2\sqrt{\frac{1-\chi}{\chi}} \qquad \Rightarrow \qquad \langle\delta_1^2\rangle = R\frac{1-\chi}{\chi}, \tag{2.2}$$

means that the extracted light has an amplitude fluctuation

$$\Delta_2 = \delta_1\sqrt{1-\chi} + \delta_2\sqrt{\chi} = \delta_2/\sqrt{\chi}. \tag{2.3}$$

The crucial sign change between Eq. (2.1) and Eq. (2.3) is enforced by energy conservation (or unitarity). The variance of the amplitude fluctuations in the extracted light becomes

$$\langle\Delta_2^2\rangle = \langle\delta_2^2\rangle/\chi = R/\chi; \tag{2.4}$$

thus the extracted light has squeezed amplitude fluctuations—sub-vacuum amplitude fluctuations—if $R < \chi$. These squeezed fluctuations lead to sub-shot-noise photon statistics if the mean intensity carried by the extracted light is sufficiently high.

This heuristic analysis reveals the following. Feedback cancels precisely the amplitude fluctuations that come from the laser, both within the feedback loop (e.g., at the photodetector) and in the extracted light. At the same time, however, by canceling within the feedback loop the amplitude fluctuations that arise from the in-phase quadrature of the a_2-light, feedback inevitably "anti-corrects" those same amplitude fluctuations as they appear in the extracted light. In other words, the feedback imposes on the extracted light amplitude noise that arises within the feedback loop; the source of this noise is the in-phase quadrature of the light that enters the a_2-port of the extraction beamsplitter. The point of the squeezing is to reduce the size of this feedback-imposed noise.

In a semiclassical analysis the feedback-imposed noise arises within the feedback loop as "shot noise" at the photodetector. Thus a semiclassical analysis does not suggest any way to reduce this noise. The virtue of the quantum-mechanical analysis is that it traces the feedback-imposed noise to its source and reveals squeezing as a way to enhance the intensity stability of the extracted light.

The idea[2] to combine negative feedback with the use of squeezed light was stimulated by work of Shapiro *et al.*,[1] who gave an authoritative account of the connection between semiclassical and quantum-mechanical analyses of negative feedback and the presence of sub-shot-noise statistics within and outside the feedback loop. Shapiro *et al.*[1] were interested in understanding an experiment performed by Machida and Yamamoto,[3] in which negative feedback from a conventional photodetector was used to stabilize a laser's intensity, and a proposed variant,[4] in which the conventional photodetector would be replaced by a Kerr-effect quantum nondemolition photon counter.[5] With a conventional photodetector the Machida-Yamamoto[3] experiment yielded sub-shot-noise photon statistics within the feedback loop [Eq. (2.1)], but super-shot-noise statistics in out-of-loop light extracted at a beamsplitter [Eq. (2.4) with $R = 1$ (vacuum a_2-input)]. This behavior has been analyzed and understood both semiclassically[1,6,7] and quantum mechanically.[1,3,4,6] Use of a quantum nondemolition photon counter[4] would allow extraction of sub-shot-noise light. The squeezed-light scheme described here provides a different method for extracting sub-shot-noise light from a feedback loop. It is worth mentioning that sub-shot-noise light has been produced by pump-amplitude noise suppression in a semiconductor laser.[8,9] Another potential source is photon twinning, with feedback from one twin, in a nondegenerate parametric oscillator.[10]

The detailed analysis[2] of the squeezed-light feedback scheme shown in Fig. 1 is nearly as simple as the heuristic analysis given above. What are the results? The extracted light has mean photon intensity

$$\mathcal{P}_{\text{out}} = (1-\chi)(1-\xi)T_0\mathcal{P}_L. \tag{2.5}$$

Fluctuations in the amplitude quadrature of the extracted light are characterized by a spectral density $R_{b_2,1}(\epsilon)$, which is a function of RF frequency ϵ and which measures the fluctuations relative to vacuum noise. In the limit of large feedback the detailed calculation yields

$$R_{b_2,1}(\epsilon) = \frac{R_{a_2,1}(\epsilon)}{\chi} + \frac{1-\eta}{\eta}\frac{1-\chi}{\chi} = \frac{(1-\eta)(1-\chi)+\eta R_{a_2,1}(\epsilon)}{\eta\chi}, \tag{2.6}$$

where $R_{a_2,1}(\epsilon)$ is the spectral density of the in-phase (squeezed) quadrature of the a_2-light [cf. Eq. (2.4)]. Provided the extracted intensity $\mathcal{P}_{\text{out}}$ is large enough, the spectral density of intensity fluctuations in the extracted light is given by

$$S_{\mathcal{P}_{\text{out}}}(\epsilon) = 2\mathcal{P}_{\text{out}}R_{b_2,1}(\epsilon). \tag{2.7}$$

The factor $2\mathcal{P}_{\text{out}}$ is the shot-noise-limit, so the extracted light beats the shot-noise limit at RF frequency ϵ if $R_{b_2,1}(\epsilon) < 1$ (squeezed amplitude fluctuations)—i.e., if

$$R_{a_2,1}(\epsilon) < \frac{\eta+\chi-1}{\eta}, \tag{2.8}$$

a condition that can be met so long as $\eta + \chi > 1$.

It is perhaps worth spelling out how large the extracted intensity $\mathcal{P}_{\rm out}$ must be in order for Eq. (2.7) to be valid. Left out of Eq. (2.7) are the fluctuations in the phase quadrature of the extracted light, which carry real power and give rise to fluctuations in intensity. The phase quadrature of the extracted light has a spectral density (relative to vacuum) $R_{b_2,2}(\epsilon) = 1 - \chi + \chi R_{a_2,2}(\epsilon)$, where $R_{a_2,2}(\epsilon)$ is the spectral density of the out-of-phase (amplified) quadrature of the a_2-light. The characteristic size of the resulting intensity fluctuations is $R_{b_2,2}^2 B_{\rm sq}$, where $R_{b_2,2}$ is a typical value of $R_{b_2,2}(\epsilon)$ and $B_{\rm sq}$ is the bandwidth over which the a_2-light is squeezed. Thus the condition for the validity of Eq. (2.7) is

$$\mathcal{P}_{\rm out} R_{b_2,1} >> R_{b_2,2}^2 B_{\rm sq} \qquad \Rightarrow \qquad \mathcal{P}_{\rm out} >> (R_{b_2,2}^2 / R_{b_2,1}) B_{\rm sq}. \tag{2.9}$$

More enlightening than the intensity spectral density $S_{\mathcal{P}_{\rm out}}(\epsilon)$ is a dimensionless figure of merit,

$$\mathcal{F}(\epsilon) \equiv \frac{S_{\mathcal{P}_{\rm out}}(\epsilon)/\mathcal{P}_{\rm out}^2}{2\mathcal{P}_L/\mathcal{P}_L^2} = \frac{R_{b_2,1}(\epsilon)}{(1-\chi)(1-\xi)T_0} = \frac{(1-\eta)(1-\chi) + \eta R_{a_2,1}(\epsilon)}{\eta\chi(1-\chi)(1-\xi)T_0}, \tag{2.10}$$

which gives the intensity noise-to-signal spectrum $S_{\mathcal{P}_{\rm out}}(\epsilon)/\mathcal{P}_{\rm out}^2$ of the extracted light, normalized to the noise-to-signal spectrum $2\mathcal{P}_L/\mathcal{P}_L^2 = 2/\mathcal{P}_L$ of a shot-noise-limited source that carries the full laser power $\mathcal{P}_L$. If $\mathcal{F}(\epsilon) < 1$, the extracted light has better noise-to-signal at RF frequency ϵ than does a shot-noise-limited source with photon intensity $\mathcal{P}_L$, even though the extracted light carries less than the full laser power $\mathcal{P}_L$. The condition $\mathcal{F}(\epsilon) < 1$ is equivalent to

$$R_{a_2,1}(\epsilon) < \frac{1-\chi}{\eta}\left[\eta + \eta\chi(1-\xi)T_0 - 1\right], \tag{2.11}$$

a condition more stringent than that in Eq. (2.8), but which can be met if $\eta + \eta\chi(1 - \xi)T_0 > 1$. In assessing Eq. (2.11), one should bear in mind that the important restrictions are set by η and χ; the fraction of light ξ needed to pump the squeezer can be small, and the reduction in power by the factor T_0 can be avoided by using a more realistic control element.

III. FREQUENCY STABILIZATION

I turn now to a squeezed-light scheme for stabilizing frequency. Frequency stabilization is not so straightforward as intensity stabilization, nor have the schemes I have investigated been as promising. Here I illustrate the use of squeezed light in a variant of the well-developed fringe-side scheme[11,12] for stabilizing frequency to a resonance of an optical cavity (a preliminary account is given in Ref. 13). The fringe-side scheme has a serious drawback: improved frequency resolution comes at the expense of decreased bandwidth. An RF modulation technique[14] for stabilizing frequency to an optical cavity overcomes this drawback, but the modulation introduces problems (to be discussed elsewhere) in adapting the technique to the use of squeezed light.

A variant of the fringe-side scheme is depicted in Fig. 2. Differences from the usual fringe-side scheme are intended to reduce the number of open ports that must be covered by squeezed light. The frequency reference is a resonance of an optical cavity (ultimately, the length of the cavity). Light from a laser is incident on

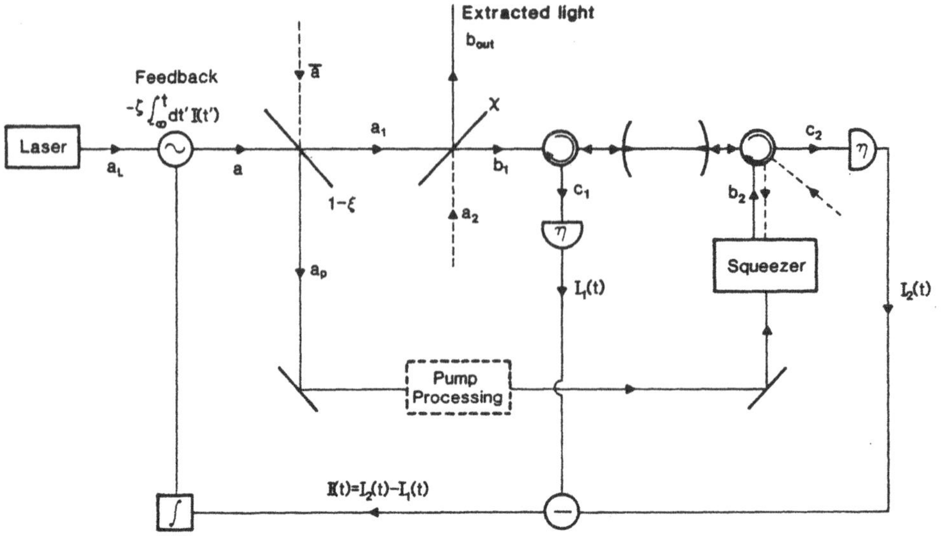

Figure 2. Fringe-side laser frequency stabilization using squeezed light

the front face of the cavity. The light reflected and transmitted from the cavity is directed onto identical photodetectors, whose output photocurrents are differenced. The differenced photocurrent is integrated and fedback to an electro-optic device that changes the phase of the laser light. The feedback loop locks at a frequency on the side of the cavity resonance, where equal intensities are reflected from and transmitted through the cavity. A fluctuation in frequency away from this lock frequency leads to opposite intensity changes in the reflected and transmitted light and thus to a signal in the differenced photocurrent. In contrast, an intensity fluctuation gives rise to equal intensity changes in the reflected and transmitted light and does not contribute to the differenced photocurrent. The choice of a phase shifter as the control element is made for ease in the mathematical analysis. A more realistic control element would control directly the frequency of the laser and avoid integration of the differenced photocurrent. Useful, frequency-stabilized light is removed at an extraction beamsplitter, which has transmissivity χ.

The fringe-side scheme is torn between two irreconcilable desires. To make the cavity a better frequency reference, one narrows the cavity resonance by reducing the transmissivity of the cavity mirrors, but doing so increases the time delay around the feedback loop and thus limits the bandwidth over which the frequency can be stabilized. Throughout the following, in fact, I assume that the RF frequency ϵ is small compared to the half-width β of the cavity resonance.

A crucial feature of the fringe-side scheme is that the important open port is not the second input port of the extraction beamsplitter, but rather the far end of the optical cavity. [This conclusion depends on the assumption that χ is not too close to unity; see remarks following Eq. (3.7).] This feature can be understood by a heuristic argument, which ignores mirror losses and photodetector inefficiencies. At the lock frequency of the feedback loop, the fringe-side scheme is perfectly balanced and formally equivalent to a balanced homodyne detector;[15,16] noise in the differenced photocurrent arises solely from fluctuations in one quadrature of the light incident on the far end of the optical cavity. Just away from the lock frequency—at RF frequency $\epsilon << \beta$—the scheme is nearly balanced. Fluctuations in the phase quadrature of the light incident on the front face of the cavity—whether due to the laser phase fluctuations that one is trying to correct or due to fluctuations in one quadrature of the light that enters the extraction beamsplitter's second input port—appear in the differenced photocurrent reduced by a factor $\epsilon/2\beta$. In contrast, fluctuations in the appropriate quadrature of the light incident on the far end of the cavity make nearly the same contribution to the differenced photocurrent as at the lock point. These fluctuations in one quadrature of the far-end light are thus the dominant source of noise within the feedback loop.

The feedback loop imposes these far-end fluctuations on the phase quadrature of the extracted light; their effect can be reduced by covering the far end of the optical cavity with appropriate squeezed light. The heuristic argument yields one quantitative conclusion: the feedback-imposed noise in the phase quadrature of the extracted light has a spectral density proportional to $(2\beta/\epsilon)^2$. Such an ϵ^{-2} spectrum is characteristic of phase diffusion; hence the point of squeezing in this case is not to produce sub-vacuum (sub-shot-noise) phase fluctuations, but rather to reduce the phase diffusion.

A more detailed description of the fringe-side scheme (Fig. 2) runs as follows. The laser produces light with photon intensity $\mathcal{P}_L$ and nominal frequency Ω equal to the lock frequency of the feedback loop. This laser light is processed by the electro-optic phase shifter and then encounters a beamsplitter with transmissivity $1-\xi$, which removes a (small) fraction ξ of the intensity to pump the squeezer. The remaining laser light (labeled a_1 in Fig. 2) is directed onto the extraction beamsplitter (transmissivity χ), the other side of which is exposed to vacuum. One output of the extraction beamsplitter (labeled b_{out}) is the extracted light. The other output (labeled b_1) illuminates the front face of the stabilization cavity. The squeezer takes in vacuum through an isolator (depicted as a four-port circulator) and produces squeezed vacuum light (labeled b_2) that illuminates the far end of the stabilization cavity. The outputs of the optical cavity (labeled c_1 and c_2) are separated from the inputs by isolators and directed onto identical photodetectors, which have quantum efficiency η and which produce photocurrents $I_1(t)$ and $I_2(t)$. The photocurrents are differenced to produce a current $I(t) = I_1(t) + I_2(t)$, and the feedback loop is closed by choosing the phase shift produced by the electro-optic phase shifter to be $-\varsigma\int_{-\infty}^{t} I(t')\,dt'$ [frequency shift $-\varsigma I(t)$], where ς characterizes the strength of the feedback.

The detailed analysis includes intrinsic multipass losses in the optical cavity due to absorption and scattering at the cavity mirrors. The mirrors, assumed to be identical, are characterized by reflectivity $\mathcal{R}$, transmissivity $\mathcal{T}$, and intrinsic loss coefficient $\mathcal{L}$. Energy conservation requires, of course, that $\mathcal{R}+\mathcal{T}+\mathcal{L}=1$. It is convenient to use the cavity length ℓ to convert these dimensionless numbers into quantities with units of frequency:

$$\gamma \equiv (c/4\ell)\mathcal{T}, \tag{3.1a}$$

$$\mu \equiv (c/4\ell)\mathcal{L}. \tag{3.1b}$$

The half-width of the cavity resonance is given by

$$\beta = 2(\gamma+\mu). \tag{3.2}$$

The lock point of the feedback loop is on the side of the cavity resonance at a frequency

$$\Omega = \omega_0 + 2(\gamma^2 - \mu^2)^{1/2}, \qquad \gamma > \mu, \tag{3.3}$$

where ω_0 is the resonant frequency. At the lock frequency the cavity reflects and transmits equal photon intensities, each being a fraction $\gamma/2(\gamma+\mu)$ of the photon intensity incident on the cavity,

$$\mathcal{P} = \chi(1-\xi)\mathcal{P}_L. \tag{3.4}$$

The power dissipated in the cavity (in photon units) is

$$\mathcal{P}_{\text{diss}} = \frac{\mu}{\gamma+\mu}\mathcal{P}. \tag{3.5}$$

Notice that when $\mu = 0$, the lock frequency becomes $\Omega = \omega_0 + \beta$.

The detailed analysis of the fringe-side scheme (to be presented elsewhere) is straightforward, but tedious. What are the results? The extracted light has mean photon intensity

$$\mathcal{P}_{\text{out}} = (1-\chi)(1-\xi)\mathcal{P}_L = \frac{1-\chi}{\chi}\mathcal{P}. \tag{3.6}$$

In the limit of large feedback the phase quadrature of the extracted light has a spectral density (relative to vacuum noise)

$$R_{b_{\text{out}},2}(\epsilon) = \left(\frac{4\gamma}{\epsilon}\right)^2 \frac{1-\chi}{\chi}\left[R_{b_2,2}(\epsilon) + \frac{\mu}{\gamma-\mu} + \frac{\gamma}{\gamma-\mu}\frac{1-\eta}{\eta}\right], \qquad \epsilon << 4\gamma, \tag{3.7}$$

where $R_{b_2,2}(\epsilon)$ is the spectral density of the appropriate (squeezed) quadrature of the b_2-light. In Eq. (3.7) the first term in large brackets is due to fluctuations in one quadrature of the light incident on the far end of the cavity, the second term comes from fluctuations associated with cavity losses, and the last term arises from photodetector inefficiencies. Equation (3.7) neglects the (flat) contribution to $R_{b_{\text{out}},2}(\epsilon)$ from vacuum fluctuations entering the second input port of the extraction beamsplitter. This neglect requires that χ not be too close to unity; the precise validity condition is $\chi R_{b_{\text{out}},2}(\epsilon) >> 1$.

If the extracted intensity $\mathcal{P}_{\text{out}}$ is sufficiently large, the spectral density of phase fluctuations in the extracted light is given by

$$S_\Phi(\epsilon) = \frac{1}{2\mathcal{P}_{\text{out}}}R_{b_{\text{out}},2}(\epsilon) = \frac{1}{2\mathcal{P}}\left(\frac{4\gamma}{\epsilon}\right)^2\left[R_{b_2,2}(\epsilon) + \frac{\mu}{\gamma-\mu} + \frac{\gamma}{\gamma-\mu}\frac{1-\eta}{\eta}\right]. \tag{3.8}$$

The factor $1/2\mathcal{P}_{\text{out}}$ is the shot-noise limit for phase fluctuations. Thus, as noted earlier, the point of squeezing in the fringe-side scheme is not to produce sub-shot-noise phase fluctuations—i.e., squeezed phase fluctuations $[R_{b_{\text{out}},2}(\epsilon) < 1]$—but rather to reduce the size of the phase diffusion, which has a spectrum proportional to ϵ^{-2}.

Since phase diffusion corresponds to a flat frequency spectrum, it is useful to rewrite the above results in terms of the spectral density of frequency fluctuations in the extracted light, $S_\omega(\epsilon) = \epsilon^2 S_\Phi(\epsilon)$. At the same time it is convenient to introduce the quantity

$$Q(\epsilon) \equiv 1 - \eta + \eta R_{b_2,2}(\epsilon), \tag{3.9}$$

which takes into account the deleterious effect of sub-unity quantum efficiency. In terms of $Q(\epsilon)$, the frequency spectral density becomes

$$S_\omega(\epsilon) = \frac{(4\gamma)^2}{2\eta \mathcal{P}} \left(Q(\epsilon) + \frac{\mu}{\gamma - \mu} \right). \tag{3.10}$$

One way to improve the frequency stability is simply to narrow the cavity resonance—i.e., to decrease γ—until $\gamma \sim \mu$—i.e., until transmission losses from the cavity are about equal to the intrinsic losses. With this strategy squeezing is essentially useless, because S_ω is dominated by fluctuations associated with intrinsic losses in the cavity. This strategy has a serious drawback, however, because it narrows the bandwidth over which the frequency is stabilized. If one desires a bandwidth much larger than μ, then one chooses the transmissivity of the cavity mirrors so that $\gamma >> \mu$. In this case squeezing *is* useful: if $1 - \eta \lesssim \mu/\gamma$, squeezing can reduce S_ω by a factor $\sim \mu/\gamma$ to a value $S_\omega \sim 16\gamma\mu/\eta\mathcal{P}$. The frequency stability can always be further improved by increasing the power $\mathcal{P}$ available to the feedback loop, but this improvement might be limited by the amount of power $\mathcal{P}_{\rm diss} \simeq (\mu/\gamma)\mathcal{P}$ [Eq. (3.5)] that can be dissipated in the cavity mirrors. In this case squeezing allows one to achieve a frequency spectral density $S_\omega \sim 16\mu^2/\eta\mathcal{P}_{\rm diss}$.

It is interesting to compare the squeezed-light fringe-side scheme with a configuration investigated by Gea-Banacloche,[17] who found reduced phase diffusion when the vacuum incident on a laser cavity is replaced by squeezed light. To facilitate comparison with Gea-Banacloche's results, it is useful to isolate the effects of squeezing in the fringe-side scheme by specializing to the case $\mu = 0$ (no cavity losses) and $\eta = 1$ (perfect photodetectors) and further to assume a flat squeezing spectrum, $R_{b_2,2}(\epsilon) = R =$ constant, over the bandwidth of interest. The frequency spectrum (3.10) for the fringe-side scheme can then be written in terms of a phase-diffusion rate

$$D = \tfrac{1}{2} S_\omega = \frac{(4\gamma)^2}{4\mathcal{P}} R = \frac{(2\beta)^2}{4\chi(1-\xi)\mathcal{P}_L} R. \tag{3.11}$$

For a fair comparison with Eq. (3.11), consider an ideal lasing medium inside a one-ended optical cavity with half-width β. Assume that the laser produces photon intensity $\mathcal{P}_L$, which is related to the mean number n_0 of photons in the lasing cavity by $\mathcal{P}_L = 2\beta n_0$, 2β being the photon loss rate from the cavity. In the absence of intrinsic losses within the lasing cavity, Gea-Banacloche[17] finds a phase-diffusion rate

$$D_{\rm G-B} = \frac{(2\beta)^2}{4\mathcal{P}_L}(1 + \overline{R}) \tag{3.12}$$

when the end of the laser cavity is illuminated by squeezed light for which $\overline{R}$ measures the degree of squeezing relative to vacuum. If $\overline{R} = 1$, then $D_{\rm G-B} = (2\beta)^2/2\mathcal{P}_L$ becomes the Schawlow-Townes linewidth. Squeezing the light incident on the lasing cavity can reduce the Schawlow-Townes linewidth by a factor of two; the remaining part of $D_{\mathrm{G-B}}$ is due to noise sources within the lasing medium. In comparison, the phase-diffusion rate (3.11) for the fringe-side scheme does worse because of the factor $\chi(1-\xi)$, which expresses the fact that the feedback loop works with less than the total laser power, but it gets greater benefit from squeezing because there are no additional noise sources (when $\mu = 0$ and $\eta = 1$).

REFERENCES

1. J. H. Shapiro, G. Saplakoglu, S.-T. Ho, P. Kumar, B. E. A. Saleh, and M. C. Teich, *J. Opt. Soc. Am. B* **4**, 1604 (1987).
2. C. M. Caves, *Opt. Lett.* **12**, 971 (1987).
3. S. Machida and Y. Yamamoto, *Opt. Commun.* **57**, 290 (1986).
4. Y. Yamamoto, N. Imoto, and S. Machida, *Phys. Rev. A* **33**, 3243 (1986).
5. N. Imoto, H. A. Haus, and Y. Yamamoto, *Phys. Rev. A* **32**, 2287 (1985).
6. H. A. Haus and Y. Yamamoto, *Phys. Rev. A* **34**, 270 (1986).
7. J. H. Shapiro, M. C. Teich, B. E. A. Saleh, P. Kumar, and G. Saplakoglu, *Phys. Rev. Lett.* **56**, 1136 (1986).
8. S. Machida, Y. Yamamoto, and Y. Itaya, *Phys. Rev. Lett.* **58**, 1000 (1987).
9. S. Machida and Y. Yamamoto, *Phys. Rev. Lett.* **60**, 792 (1988).
10. A. Heidmann, R. J. Horowicz, S. Reynaud, E. Giacobino, C. Fabre, and G. Camy *Phys. Rev. Lett.* **59**, 2555 (1987).
11. R. L. Barger, J. S. Sorem, and J. L. Hall, *Appl. Phys. Lett.* **22**, 573 (1973).
12. J. Helmcke, S. A. Lee, and J. L. Hall, *Appl. Opt.* **21**, 1686 (1982).
13. C. M. Caves, in *Laser Spectroscopy VIII*, edited by W. Persson and S. Svanberg (Springer, Berlin, 1987), p. 146.
14. R. W. P. Drever, J. L. Hall, F. V. Kowalski, J. Hough, G. M. Ford, A. J. Munley, and H. Ward, *Appl. Phys. B* **31**, 97 (1983).
15. H. P. Yuen and V. W. S. Chan, *Opt. Lett.* **8**, 177 (1983).
16. B. L. Schumaker, *Opt. Lett.* **9**, 189 (1984).
17. J. Gea-Banacloche, *Phys. Rev. Lett.* **59**, 543 (1987).

PULSED SQUEEZED LIGHT

R. E. Slusher, A. LaPorta, P. Grangier and B. Yurke

AT&T Bell Laboratories
Murray Hill, New Jersey 07974

INTRODUCTION

Generation of squeezed light is at present limited to narrow regions of the optical spectrum.(1–3) For the parametric amplifiers used to generate squeezed light, this limit is imposed by the following properties of the nonlinear material used for the parametric process: 1) small nonlinear response, 2) linear losses, 3) insufficient laser pump power and 4) optical damage at high pump powers. Optical cavities have been used(1,2) to enhance the nonlinear interaction for continuous wave CW pumping but this limits the bandwidth and also increases the total linear loss.

We have used a pulsed laser pump to generate squeezed light without the aid of an optical cavity.(4) Pulsed pumping can help to extend the spectral range and bandwidth by increasing the pump power for short intervals. In this pumping mode, damage to the nonlinear crystal is less likely, since peak power is increased without an increase in average power. Linear losses can be discreased since shorter interaction lengths are required. In general total losses must be very low, i.e. $\alpha L << 1$ where α is the linear absorption coefficient and L is the interaction length. Finally, many pulsed laser systems are now available and they operate in ranges where large optical nonlinearities are expected.

Both the generation process and detection process for pulsed squeezed light require careful analysis.(5) The phase relationships for parametrically generated light are the key to obtaining large squeezing. Good squeezed light is like good comedy, it's all in the timing. In the pulsed case the increase and decrease in parametric gain during the pulse must be considered in predicting the level and bandwidth of the squeezing generated. Section II of this paper will review this analysis and show that large squeezing over broad bandwidths is predicted. The pulses of squeezed light generated can be detected by pulsed homodyne detectors if a pulse-train is used. This detection process is described in Section III. The experimental results to date are described in Section IV. Finally, applications of pulsed squeezed light are described in Section V.

---

* Institute d'Optique Theorique, Batiment 503 BP 43, France 91406

THEORY OF GENERATION BY PARAMETRIC DOWNCONVERSION

We will describe pulsed squeezed light generation by parametric down-conversion[5] although other processes such as four-wave mixing would yield similar results. A nonlinear crystal with no inversion symmetry can couple a pump field $E_p(x,t)$ to signal $E_s(x,t)$ fields through the second-order polarizability as described by the wave equation

$$\frac{\partial^2 E_s}{\partial t^2} - v^2 \frac{\partial^2 E_s}{\partial x^2} = \kappa E_p E_s \tag{1}$$

strength of the nonlinearity. The high intensity pump field can be approximated as a classical field

$$E_p(x,t) = A(t-x/v)\sin[2\omega_0(t-x/v)+\phi] \tag{2}$$

where ϕ is the optical phase at $(t-x/v)=0$. The pump pulse is characterized by the amplitude envelope function $A(t-x/v)$ which is assumed to vary slowly relative to an optical period over a time of the order of τ, i.e., $\tau >> 1/\omega_0$ where ω_0 is the optical frequency about which the signal fields will be centered. Our detection technique requires that the pump pulses are sequenced in a train of identical pulses with period T where $T >> \tau$. The signal field builds up from the incident vacuum field at the entrance face of the nonlinear medium at $x=0$ throughout the length of the medium, $x=L$. After exiting the medium the squeezed signal field propagates in free space to a homodyne detector at $x=\ell$. The squeezed field at the detector is then calculated to be

$$E_s(\ell,t) = \{E_s^{(+)}(0,t)\cosh[\beta(t)] - E_s^{(-)}(0,t)e^{-i\phi}\sinh[\beta(t)]\}e^{-i\omega_0 t} + \text{h.c.} \tag{3}$$

where

$$E_s^{(+)}(0,t) = \mathcal{E}_0 \int_\Omega a(\omega)e^{-i\omega t}d\omega \tag{4}$$

and the nonlinear phase shift is

$$\beta(t) = \frac{\kappa A(t-x/v)L}{8\pi(\lambda/n_0)}. \tag{5}$$

The field operators $E_s^{(+)}$ and $E_s^{(-)}$ are hermetian conjugates and $a(\omega)$ is the creation operator for a mode at frequency ω. Squeezing extends over a range of frequencies Ω centered at ω_0; i.e., the optical frequencies of the squeezed light are $\omega+\omega_0$ and extend over a range

$$-\Omega < \omega < \Omega \tag{6}$$

where

$$2\omega_0 >> \Omega \tag{7}$$

and we assume that

$$\Omega > \frac{2\pi}{\tau}, \frac{2\pi}{T} \tag{8}$$

Perfect phase matching is assumed in this analysis. In practice the upper bound on Ω is set by the frequency limit for phase matching $1/\tau_{pm}$. Typical frequencies involved in the experiments are

$$\frac{1}{T} \sim 0.2\text{GHz} < \frac{1}{\tau} \sim 10\text{GHz} < \frac{\Omega}{2\pi} < 1/\tau_{pm} \sim 100\text{GHz} \tag{9}$$

Equation (3) expresses the canonical transformation of the vacuum field required for squeezed light of the form[6]

$$b = \mu a + \nu a^\dagger \tag{10}$$

For large squeezing the peak value of the nonlinear phase shift $\beta(0)$ must be of the order of unity. Note that for the pulsed case $\beta(t)$ varies with the amplitude of the pump field A(t) and has frequency components over a range of $1/\tau$. However, the range of frequencies that are squeezed is limited only by the upper bound on Ω, i.e. the phase matching frequency limit for parametric downconversion. This limiting frequency increases with decreasing length L of the nonlinear medium. This length is determined by the pump field amplitude A(t) and the requirement that the peak β value be near unity. For large pump fields the frequency range for the squeezed noise can be very large. This frequency range will eventually reach a limit when $\Omega/\omega_0 \sim 1$. In this broadband limit the degree of squeezing is limited by Ω/ω_0 when referenced to unity for the unsqueezed vacuum.[6] In the present experiments $\Omega/\omega_0 \sim 10^{-4}$ so that this fundamental limit does not play a significant role.

PULSED DETECTION AND THE MEASURED SPECTRUM

The basic idea for detection of pulsed squeezed light is to use a pulsed local oscillator for a balanced homodyne detector. This strobes on the sensitivity of the homodyne detector to the squeezed light pulses at a time near the peaks of the squeezed pulses. Experimentally the local oscillator pulse is a small portion of a mode-locked pulse-train at the central squeezing frequency which is delayed to coincide with the squeezed pulse-train at the homodyne detector. The major portion of the initial mode-locked laser pulse-train drives a frequency doubler which forms the pump pulse train at optical frequency $2\omega_o$.

A pulse train is essential for the detection scheme. The fourier transform of a pulse train has strong frequency components only at integral multiples of the inverse of the period between pulses 1/T. At frequencies between these peaks the frequency components of the local oscillator fields are very small compared to the broadband noise associated with the shot noise and sub-shot noise squeezed signal fields. By using a balanced detector the peaks at multiples of 1/T can be reduced by a factor of 10 to 100. However at this reduced levels they are still far above (~ 80 dB) the shot-noise level. The balanced detector also has the advantage of reducing low frequency ($>> 1/T$) intensity fluctuations in the local oscillator or signal light.

The large amplitude local oscillator field can be assumed to be classical

$$E_{LO}(t) = F(t)\cos(\omega_0 t + \theta) \tag{11}$$

where F(t) is not an operator and represents the amplitude envelope of the LO pulse train. Each pulse in the LO is characterized by a duration τ_{LO}. The current produced at the homodyne detector at frequencies ω is proportional to the product of $E_s(t)$ and $E_{LO}(t)$. Details of the calculation of the noise spectrum $S(\omega)$ of this current are found in reference 5. It is assumed that the electronic response time of the detector τ_D is much longer than the optical period. In the experiment τ_D is near 20 ns, slightly larger than $T = 5$ ns.

A schematic diagram of the expected noise spectrum is shown in Fig. 1. The shot-noise spectrum is normalized to unity. This level is interpreted as the vacuum fluctuation level of the signal field E_s and is constant as a function of frequency as well as the relative phase ϕ_{LO} between E_s and E_{LO}. Squeezing is the reduction of the noise below unity at the optimum relative phase between E_s and E_{LO}. The predicted noise reduction extends uniformly out to the phase mismatch limit at $1/\tau_{pm}$. Unbalanced local oscillator frequencies at multiples of 1/T extend out to frequencies of the order of $1/\tau$.

The level of noise reduction depends on the peak value of the nonlinear phase shift β as well as a convolution in time of the local oscillator and squeezed pulses,[5]

$$S(\omega)_m = \frac{\int_{-T/2}^{T/2} F^2(t) e^{-2\beta(t)} dt}{\int_{-T/2}^{T/2} F^2(t) dt} \tag{12}$$

where the subscript m indicates the minimum noise obtained by optimizing the relative phase of ϕ_{LO}. As an example of the effect of varying the relative values of τ and τ_{LO} consider pulses which have a Gaussian time dependence

$$F(t) = F(0)\exp(-t^2/\tau_{LO}^2) \tag{13}$$

In the present experiments a pulse-train identical to the LO pulse train is doubled to form the pump field. In this case the nonlinear phase shift, proportional to the pump field has the form

$$\beta(t) = \beta(0)\exp(-2t^2/\tau_{LO}^2) \tag{14}$$

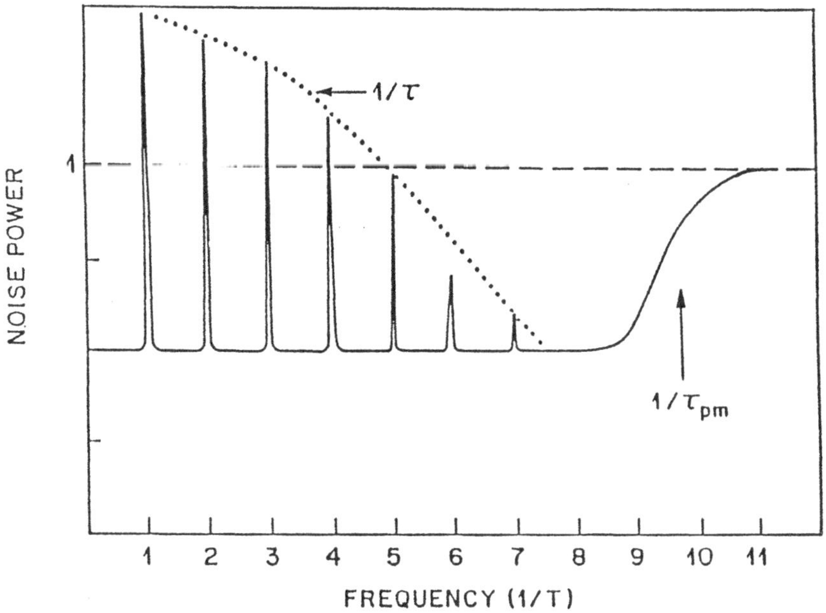

Fig. 1 Predicted noise squeezing spectrum for a train of pulses each with duration τ and separated by a period T. The noise power is normalized with respect to the shot-noise or vacuum fluctuation level indicated by the dashed line. Noise reduction below the shot-noise level by a factor determined by the parametric gain and system losses extends uniformly out to the frequency limit imposed by phase matching $1/\tau_{pm}$. Only spikes remain at integral multiples of 1/T out to frequencies of the order of $1/\tau$ which are due to imbalance in the homodyne detector.

The calculated squeezed noise power is shown in Fig. 2 as a function of $\beta(0)$, the peak nonlinear phase shift. Note that the measured noise reduction is slightly less than the equivalent CW experiment since the LO pulse in this case is slightly longer than the squeezed pulse and thus samples non-optimally squeezed portions of the pulse. Also shown in Fig. 2 is the measured noise for $\tau = \tau_{LO}$. If $\tau_{LO} << \tau$ the squeezed field can be measured at essentially its peak value by centering the short LO pulse on the squeezed pulse. By delaying the LO pulse relative to the squeezed pulse, a short LO pulse could map out the squeezing noise reduction as a function of time during the squeezed pulse. This might be a valuable diagnostic technique for studying pulsed squeezed light generation.

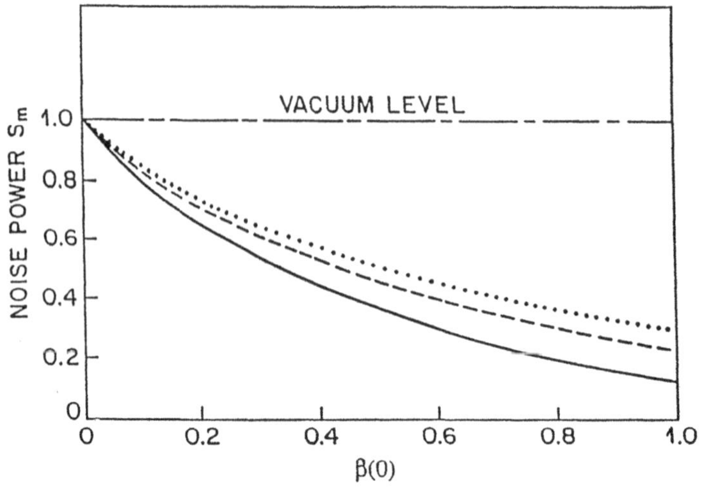

Fig. 2 The optimum noise reduction below the shot-noise level (dashed/solid line) is shown as a function of peak nonlinear phase shift. The solid line corresponds to ideal CW homodyne detection. The dashed and dotted lines correspond to pulsed homodyne detection of squeezed pulses with Gaussian shapes of duration τ using local oscillator pulse durations of $\tau_{LO} = \tau$ and $\tau_{LO} = \sqrt{2}\tau$ respectively. Experiments described here correspond to the dotted curve calculation.

EXPERIMENTAL RESULTS

Both the pump and local oscillator pulse-trains where obtained from a mode-locked Nd: YAG laser operating at a wavelength of 1.064 μm. Its average output power was 10W, the pulse repetition rate was 200 MHz and the pulse duration was 100 ps. The experimental arrangement is shown in Fig. 3. The nonlinear material used to double this pulse-train was a KTP crystal 5 mm in length. Typical conversion efficiency was 10% resulting in a green pulse-train with pulse widths near 70 ps, an average power of 1W and peak power near 2MW/cm^2. A prism is used to separate the green and IR light. The green light is then focussed into the paramp to form a focal region with a beam waist radius $w_0 \sim 20\ \mu$m over a length of approximately 1cm. Two KTP crystals parametrically downconvert the pump to the strongly correlated signal and idler beams. The crystals are antireflection coated at both 0.532 and 1.064μm in order to minimize losses of pump and squeezed light. Two KTP crystals, each 5 mm in length, are used to eliminate walkoff of the parametrically downconverted signal and idler beams. Type II phase matching is used, i.e. the pump and signal beams are polarized at 90° to the z axis of the crystals and the idler is polarized along the z axis. In this case the angular variation of the refractive index in the plane of the signal polarization causes an angular displacement of the signal beam relative to the idler beam. This results in a 17 μm walkoff of the signal from the idler at the output face of each 5 mm KTP crystal. Since the squeezed light results only from an exact superposition of the

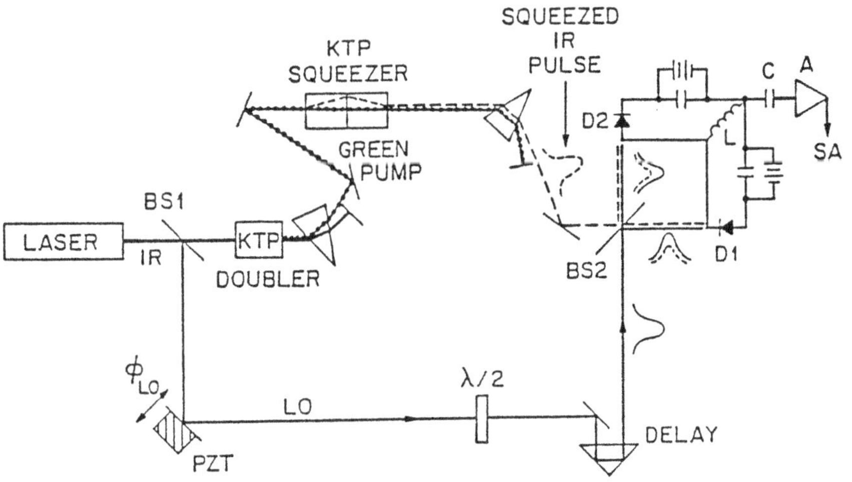

Fig. 3 A schematic diagram of the experimental apparatus used to generate and detect pulsed squeezed light. A mode-locked Nd:YAG laser is doubled by a KTP crystal to form a green pump pulse-train for the KTP squeezer used to generate squeezed light. A 1mW portion of the initial IR pulse train is phase shifted and delayed to overlap the squeezed pulses and serve as the local oscillator for the balanced homodyne detectors D1 and D2. The $\lambda/2$ plate in the LO path is rotated to match the polarization of the LO and squeezed pulses.

signal and idler beams, polarized at 45° with respect to the z-axis, this walkoff must be compensated. This could be done with additional optical elements, however increased parametric gain and walkoff compensation can be obtained simply by using a second crystal with the same orientation of the z-axis but a rotation in the x-y plane in the opposite sense with respect to the first crystal. For example, if the first crystal is at a +23.75° phase match angle, the second crystal should be cut at −23.75° for walkoff compensation. It is important to note that simply turning over a crystal cut at +23.75° to serve as the second crystal does not suffice since this inverts the z axis resulting in a net null in the parametric gain. In the initial experiments this mistake was made and the null parametric gain was avoided by orienting the two crystals at a small angle. As a final step in the generation process, the IR squeezed light is separated from the green pump light by a second prism and directed to the balanced homodyne detector.

A small 1mW portion of the initial laser pulse-train is split off to form the local oscillator for the balanced homodyne detector. The phase, polarization and delay of the LO can be independently set relative to the squeezed light beam. The LO and squeezed light are combined at the 50/50 beamsplitter BS2. A second small portion of the initial laser pulse-train is split off and frequency shifted by 60 MHz with an acousto-optic modulator. This probe beam is focussed to match the pumped region of the parametric downconverter and serves to measure and optimize the parametric gain and mode-matching between the squeezed and LO beams.

The spectrum of the photocurrent from a single photodiode detector is shown in Fig. 4. The detector is a InGaAs photodiode (EPITAXX model ETX300) with a 300 μm diameter active area and the photocurrent is amplified by a Trontech model 40F amplifier. The photocurrent spectrum consists of broadband shot-noise (approximately 3 dB above the amplifier noise at 50 MHz) and strong peaks at multiples of 200 MHz, i.e., the pulse-train frequency 1/T. In Fig. 4 these peaks decrease with increasing frequency due to the photodiode (~100 MHz) and amplifier (~50 MHz) frequency response. These peaks begin to saturate the amplifier at photocurrents near 0.1 mA where the shot-noise is not yet dominating the amplifier input noise. An improvement in the ratio of shot-noise to amplifier noise can be obtained by using the tuned balanced arrangement shown in Fig. 5. A low Q (~3) tuned circuit is formed by L and C with a resonant frequency near 60 MHz. Bias voltages V_A and V_B are bipassed by large capacitors C_b. The photocurrent response shown in Fig. 6 is obtained with this network with photocurrents near 0.4 mA in each detector. Between 50 and 60 MHz the shot-noise is nearly 9 dB above the amplifier noise which is adequate for measuring squeezed noise reduction. Balancing of the coherent current component reduces the peaks at multiples of 200 MHz by nearly 20 dB so that the amplifier operates below saturation. The measured shot-noise level is compared and found to be equal to similar noise produced by a CW beam of the same power as the average pulse-train power. This level is calibrated to be at the theoretically predicted shot-noise level by careful measurements of the

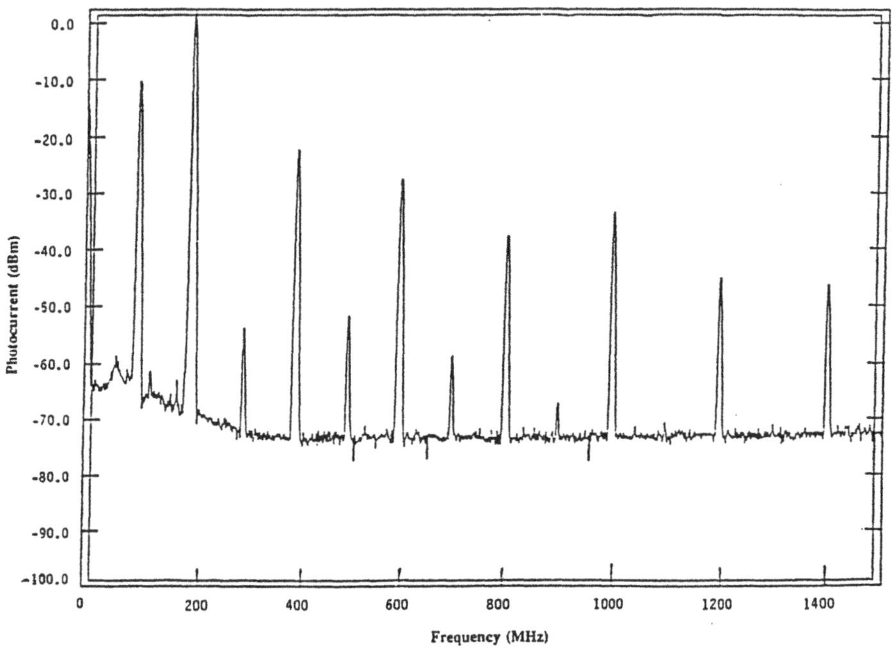

Fig. 4 A spectrum of photocurrent ($i_o = 0.1$ mA) is shown for a single photodiode excited by a mode-locked laser pulse-train with a pulse repetition rate of 200 MHz. The current is displayed vertically at 10 dB/division. The peaks at intervals of 200 MHz are due to the mode-locked pulse train. Noise between the peaks is a combination of shot-noise and amplifier noise. Peaks at multiples of 100 MHz are due to rf pickup and mixing. Both detector and amplifier response cause the rolloff in peak levels with frequency.

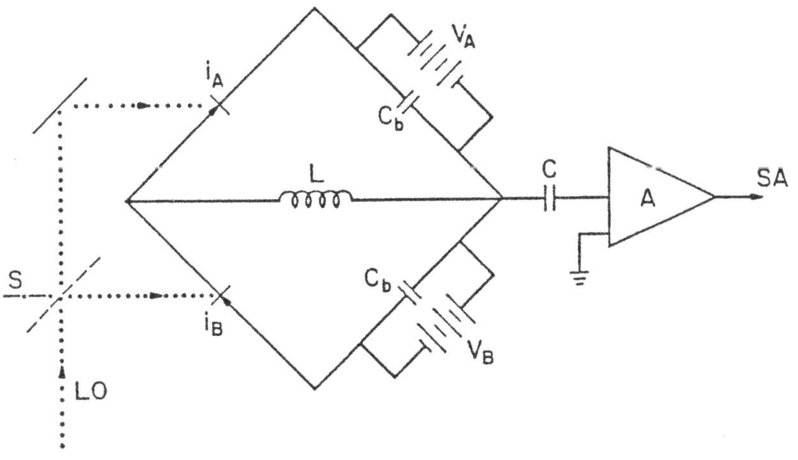

Fig. 5 Balanced homodyne detector for measurement of pulsed squeezing. The resonant circuit formed by the LC circuit enhances shot noise between peaks in the photocurrent spectrum at multiples of 200 MHz and reduces these large current peaks by balancing the coherent portion of the photocurrent.

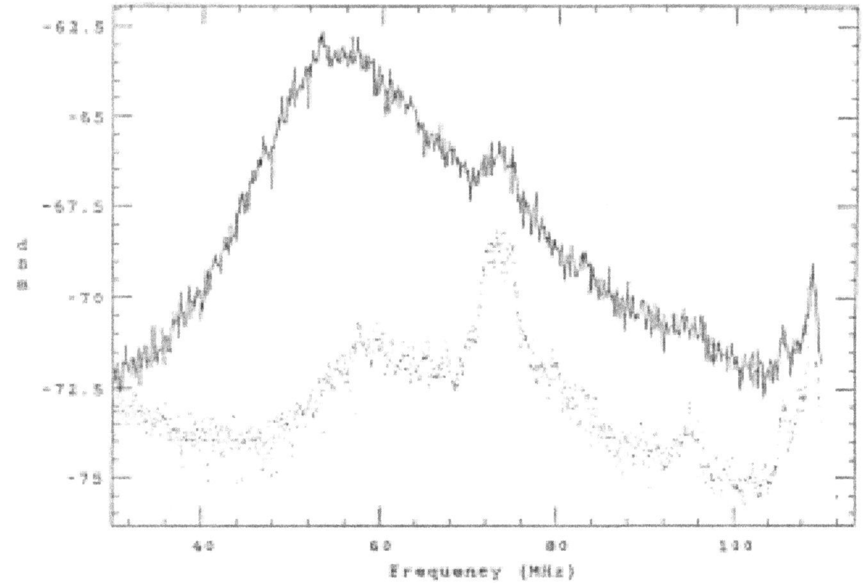

Fig. 6 Noise spectra measured using the balanced homodyne detector in Fig. 5 with (solid line) and without (dotted line) equal photocurrents of 0.4 mA at each detector. The dotted data is a measure of the amplifier noise with the resonant balancing circuit at the input. Peaks in the noise near 70 and 110 MHz are caused by electrical pickup.

LO power and system gain. It is a pleasant surprise to find the theoretical shot-noise level for the pulse-train from a "standard" mode-locked Nd:YAG laser at frequencies to within a megahertz of the large peaks at multiples of 200 MHz.

Squeezed noise reduction below the shot-noise level obtained with this arrangement is shown in Fig. 7. As the LO phase ϕ_{LO} is varied, the noise level varies between 0.6 dB below the vacuum level and 0.8 dB above with a period of π. The average photocurrent remains unchanged as the noise level varies with ϕ_{LO}, eliminating any explanation on the noise variation based on optical interference effects. Squeezing is observed from 30 to 170 MHz. This frequency range should extend out to 100 GHz if it were not limited by the detector/amplifier bandwidth.

The maximum and minimum measured squeezed noise levels $P_{S\pm}$ are consistent with the measured gain and losses in the system,

$$P_{S\pm} = [\eta 10^{\pm G/10} + 1 - \eta] P_V + P_A \tag{15}$$

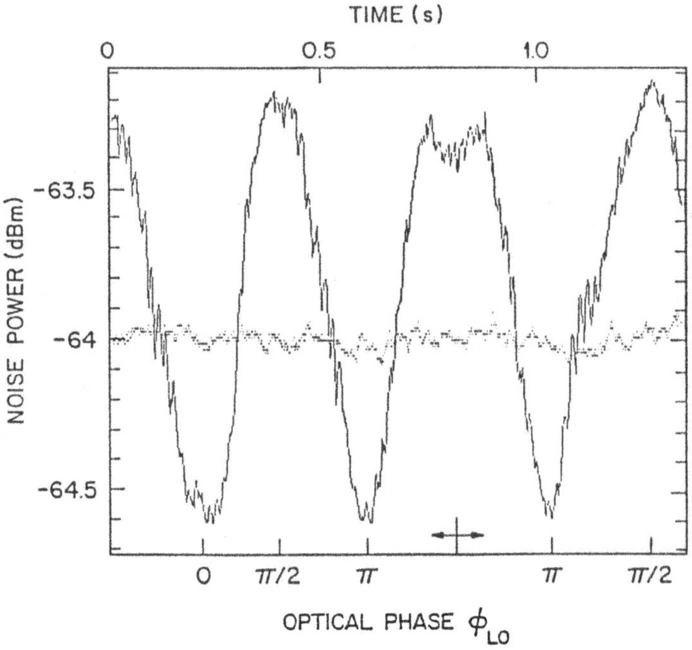

Fig. 7 Noise power for shot-noise (dotted line) and pulsed squeezed light (solid line) as a function of homodyne detector local oscillator phase ϕ_{LO} and time. Noise at $\phi_{LO} = \pi/2$, $3\pi/2$, etc., drops as much as 0.6 dB below the combined shot-noise and amplifier level. The center frequency is 53 MHz, the radio frequency bandwidth is 1 MHz and the video averaging bandwidth is 10 Hz. Phase jitter of the LO relative to the squeezed pulse of $\pm 20°$ at a frequency of 56 Hz (laser cooling water turbulence excited optical table resonance) is evident in the solid data and degrades the observed noise reduction. The double arrow indicates a reversal in the direction of the ϕ_{LO} sweep.

where the loss is characterized by an effective quantum efficiency

$$\eta = \eta_d \eta_m^2 \eta_o = 0.58 \tag{16}$$

G = 1.35 dB is the measured parametric gain, P_V is the shot-noise or vacuum fluctuation power and P_A is the amplifier input noise. The system losses include the detector efficiency $\eta_d = 0.9$ the mode-matching efficiency $\eta_m = 0.85$ and losses in the optics characterized by an efficiency $\eta_o = 0.9$. At this point in the experiments the squeezing is limited by the pump power. Preliminary results show that increasing the pump power results in gains of 6 to 10 dB which should produce large squeezed noise reduction. With these gains the squeezed noise reduction will be primarily limited by the net system quantum efficiency, η.

These experimental results show that pulsed generation and detection of squeezed light agrees with the theoretical predictions and is relatively simple to obtain. A wide range of nonlinear materials and pumps should be useful for generation of quantum light states with these techniques and should extend the spectral regions where squeezed light can be generated. Table I listed some promising $\chi^{(3)}$ materials where four-wave-mixing paramps may generate squeezed light. Since pump, signal and idler frequencies are nearly equal for four-wave-mixing, phase matching is easier to achieve and a doubler is not required. However $\chi^{(3)}$ typically results in smaller gain in many crystals and resonant enhancement near an exciton or biexciton feature is often required. A more detailed analysis of the possibilities for both $\chi^{(3)}$ and $\chi^{(2)}$ nonlinearities is in progress.

Table I Nonlinear optical properties of materials promising of parametrically generated squeezed light. A peak pump intensity of I(0) = 10 MW/cm^2 is assumed except for CuCl and ZnSe where saturation limits the pump level to 0.1 MW/cm^2. The wavelengths listed are only to indicate the spectral region of interest. Actual wavelengths depend on the temperature and dimensionality of the material. The peak nonlinear phase shift and the nonlinear index are related by $\beta(0) = 2\pi L n_2 I(0)/\lambda$. PDA stands for polydiacetylene.

Material	$n_2(cm^2/W)$	L(μm)	Wavelength (μm)	$\beta(0)$
GaAs				
resonant	10^{-6}	1	0.87	60
nonresonant	$10^{-13} - 10^{-10}$	1000	1.0 - 0.88	0.01 - 10
PDA				
resonant	10^{-9}	100	0.66	10
nonresonant	$10^{-12} - 10^{-10}$	1000	1.0 - 0.68	0.1 - 10
CuCl	10^{-5}	1	0.389	10
ZnSe	10^{-5}	1	0.44	10
(resonant biexciton)				

APPLICATIONS TO POLARIZATION INTERFEROMETRY

It has recently been demonstrated[7,8] that the sensitivity for interferometric measurement of optical phase can be enhanced beyond the previous shot-noise limit by using squeezed light at the normally unused "dark" input port of the interferometer. In a wave description of the basic interferometer, it is the vacuum fluctuations entering this dark port which interfere with the coherent field at the illuminated input port resulting in a limit to the sensitivity for phase measurement. By combining squeezed noise reduction and pulsed techniques we have been designing squeeze enhanced polarization interferometers which achieve ultrahigh sensitivity *and* picosecond time resolution.

An example of a pulsed squeeze enhanced polarization interferometer is shown in Fig. 8. One input port is illuminated with linearly polarized pulsed light, equivalent to the LO in the previous discussion. This light is combined at the first polarization beam splitter with time coincident pulses of squeezed vacuum with polarization at 90° to the LO. If there is no sample in the interferometer the $\lambda/2$ plate rotates both the LO and squeezed beam polarizations by 45° so that equal portions of each are incident on the balanced detectors D1 and D2 through the second polarization beam splitter. In this case the noise levels and squeezed noise reduction are identical to those described in sections II through IV. If a sample is introduced which rotates the LO and squeezed polarizations, an imbalance between the detector currents is measured. At LO levels near 1 mW and averaging bandwidths of 30 kHz the shot-noise limited sensitivity of this interferometer is several μrad of polarization rotation. This sensitivity is enhanced beyond this limit by the degree of squeezing at the dark port, hopefully in the range from 1 to 10 dB.

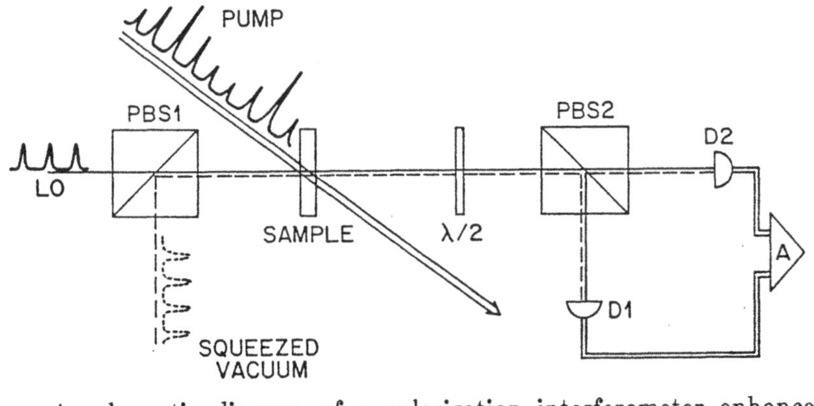

Fig. 8 A schematic diagram of a polarization interferometer enhanced by pulsed squeezed light entering the dark port. Polarization beamsplitter PBS1 combines the orthogonally polarized LO and squeezed pulses, the $\lambda/2$ rotates both polarizations by 45° and beamsplitter PBS2 directs equal light intensities to each detector in the balanced network. Rotation of the polarization in the sample induced by the pump pulses is measured as an imbalance between the detectors D1 and D2.

Our first experiments are directed at studies of the nonlinear response of thin semiconductor layers, e.g. quantum multiwells. A measure of $\chi^{(3)}$ is easily obtained along with its time response by pumping an optical birefringence in the crystal with a train of pump pulses at a small angle to the LO beam. For example, a 1 MW/cm^2 pump beam at 1.06 μm in GaAs is expected to result in a rotation of polarization near 0.5 μrad/μm due to the electronic $\chi^{(3)}$ process. At present the value of $\chi^{(3)}$ for this wavelength is inferred from third harmonic generation[9] to be 2×10^{-10} esu. At sensitivities near the shot-noise level, the $\chi^{(3)}$ value can be measured accurately so that pump power dependence and the effects of lower dimensionality can be studied in detail. In order to eliminate technical noise we will modulate the pump amplitude at frequencies near 50 MHz by using acoustooptic modulators. As shown in section IV the noise levels in this frequency range are limited at the basic shot-noise or reduced squeezed noise levels.

Another interesting application of sensitive polarization interferometry[10] is shown in Fig. 9. In the case the "sample" is a small piece of electrooptic material of the order of 10 to 50 μm in diameter and 20 μm in length. It has a high reflectance coating at the bottom face to reflect the light back through the crystal into the second polarization beam splitter. This crystal forms a sensitive monitor of electric fields since fields across this probing tip rotate the polarization of the light fields. Probes of this type are now being used to monitor the electric fields near VLSI electronic circuits.[10] Since the picosecond duration LO pulses can be synchronized and delayed relative to periodic electric fields, one can obtain the time dependence of the field patterns. If operated

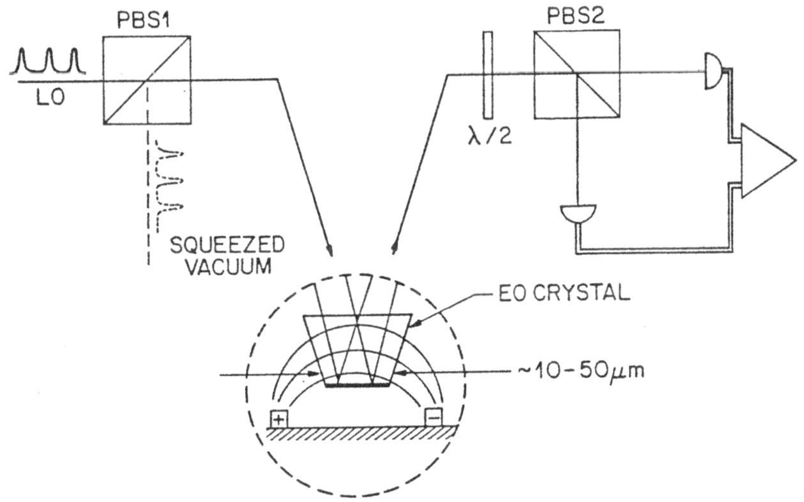

Fig. 9 A polarization interferometer used to measure small, rapidly varying electric fields is shown schematically. Polarization beamsplitters PBS1 and PBS2 form the interferometer and function along with the $\lambda/2$ plate as described in Fig. 8. Polarization rotation is measured resulting from electric field induced birefringence in a small electrooptic crystal. Since the LO and squeezed light at the input ports are pulsed, the time dependence of repetitive electric fields can be measured. Squeezed light enhancement near the shot-noise level results in electric field sensitivities in the range below 0.1V/cm for measurement averaging times near 1ms.

with squeezed noise enhancement near the shot-noise level, the sensitivity of these electric field probes is below 0.1V/cm with 1ms averages over a pulse-train with average power levels near 1mW.

SUMMARY

Experiments have demonstrated that short duration pulses of squeezed light can be generated and detected in the form of pulse-trains. High peak powers in the pulsed pump laser generate the large nonlinear phase shifts in short parametric interaction lengths required to produce large squeezed noise reduction. The squeezed light pulse-trains generated by parametric downconversion can be detected using balanced homodyne detection with a pulsed local oscillator. Each squeezed pulse can be very short in duration corresponding to a large bandwidth limited by the phase matching length of the nonlinear crystal.

Applications of pulsed squeezed vacuum to polarization interferometry can enhance the sensitivity of measurement of polarization rotation with picosecond time resolution. Studies of thin layered nonlinear materials are now in progress using shot-noise limited polarization interferometry. Optimized nonlinear materials in combination with pulsed techniques may allow squeezed light and other interesting quantum light states to be generated and applied over large ranges of the infrared and visible spectrum.

Finally, in some experiments, short pulses may be dictated by the physics to be studied. For example, radiation from relativistic electrons in magnetic undulators have been predicted to generate quantum light states.(11) Electron sources for these experiments are typically in the form of pulse trains. Squeezing generated by solitons in optical fibers(12) may have interesting time dependences on short time scales. There are also quantum light experiments where pulsed techniques may be important in order to temporally separate fields in order to obtain clearly defined correlation effects. These experiments may include interfere in phase space(13) and EPR experiments.(14,15)

REFERENCES

1. R. E. Slusher, L. W. Holberg, B. Yurke, J. C. Mertz and J. F. Valley, Phys. Rev. Lett. *55*, 2409 (1985); and the collection of articles in "Squeezed States of the Electromagnetic Field," H. J. Kimble and D. F. Walls, eds., JOSA B, October 1987.
2. Ling- An Wu, H. J. Kimble, J. L. Hall and H. Wu, Phys. Rev. Lett. *57*, 2520 (1986).
3. R. M. Shelby, M. D. Levenson, S. H. Perlmutter, R. G. DeVoe and D. F. Walls, Phys. Rev. Lett. *57*, 691 (1986).
4. R. E. Slusher, P. Grangier, A. La Porta, B. Yurke and M. J. Potasek, Phys. Rev. Lett. *59*, 2566 (1987).
5. B. Yurke, P. Grangier, R. E. Slusher and M. J. Potasek, Phys. Rev. A *35*, 3586 (1987).
6. B. Yurke, Phys. Rev. A *32*, 300, 311 (1985).

7. Min Xiao, Ling- An Wu, and H. J. Kimble, Phys. Rev. Lett. *59*, 278 (1987).

8. P. Grangier, R. E. Slusher, B. Yurke and A. La Porta, Phys. Rev. Lett. *59*, 2153 ((1987).

9. W. K. Burns and N. Bloembergen, Phys. Rev. B *4*, 3437 (1971).

10. J. A. Valdmanis, Elec. Lett. *23*, 1308 (1987).

11. W. Becker and J. K. McIver, Phys. Rev. A *27*, 1030 (1983).

12. S. J. Carter, P. D. Drummond, M. D. Reid and R. M. Shelby, Phys. Rev. Lett. *58*, 1841 (1987).

13. W. Schleich and J. A. Wheeler, Nature *326*, 574 (1987).

14. P. Grangier, M. J. Potasek and B. Yurke, Phys. Rev. A, to be published.

15. M. D. Reid, submitted for publication.

QUANTUM FLUCTUATIONS IN OPTICAL MEASUREMENTS

H. J. Kimble

Department of Physics
University of Texas at Austin
Austin, Texas 78712

A principal motivation for the investigation of squeezed states of light has been the potential that such states offer for the improvement of the precision of optical measurements beyond the limit set by the zero-point or vacuum fluctuations of the electromagnetic field. In the past two years, this expectation has been heightened by the successful generation of squeezed states of light by several groups working with a variety of physical systems (1). If we set aside for the moment the numerous scientific and technological questions associated with the optimization of the squeezing produced in these initial experiments, it is an exciting prospect to attempt to employ these sources to implement the various measurement schemes with squeezed light that have been discussed in the theoretical literature for many years.

One of the first applications of squeezed light (and certainly one of the most important in terms of the development of this field of research) is to interferometry. Caves first proposed that the photon counting error in a conventional two-beam interferometer could be reduced by injecting squeezed light into the normally open or unused input of the interferometer (2). We have recently implemented Caves' proposal with a Mach-Zehnder interferometer whose inputs are driven by the usual coherent state beam from a frequency-stabilized laser and by the squeezed vacuum state generated by a subthreshold optical parametric oscillator (3,4). We have demonstrated an improvement in sensitivity for the detection of phase modulation of 3dB relative to the shot-noise or vacuum-state limit. The observed enhancement is in good agreement with theoretical prediction, when an acounting is made for the various linear losses (measured in absolute terms) in our system. In fact it is the linear losses (such as the nonunity quantum efficiency of the photodiodes) and not the degree of squeezing available from our source that currently limits the enhancement in signal-to-noise ratio when squeezed light is employed. Similar results in polarization interferometry have been obtained by Grangier, et al. (5).

A second application of squeezed light to precision measurement that we have explored involves the detection of amplitude modulation. In this experiment the squeezed vacuum from the subthreshold optical parametric oscillator is combined with a laser to produce a squeezed signal field of nonzero mean amplitude. The signal field is passed through an acoustooptic modulator whose transmission is weakly modulated. The resulting amplitude modulation encoded on the signal beam is detected in a balanced homodyne detector. As in the measurements of phase modulation in the absence of squeezed light, the minimum detectable modulation index is limited by the fluctuations inherent in a coherent state that give rise to shot-noise. However, with the injection of squeezed light, we observe an increase in signal-to-noise ratio for the detection of amplitude modulation of 2.5dB above the shot-noise or vacuum state limit (6). The increase is again limited by the various passive losses of the system and not by the degree of squeezing generated by our source.

While these demonstrations offer rather modest gains in sensitivity considering their technical complexity, they are nonetheless the first applications of squeezed states to improve the precision of measurement beyond the vacuum-state limit. As the degree of squeezing is improved, as the system losses are reduced, and as new measurement strategies are invented, the impact of this research could be wide-ranging across a spectrum of applications in measurement science.

This work was supported by the Office of Naval Research, by the Venture Research Unit of British Petroleum, and by the Joint Services Electronics Program.

<u>References</u>

1. "Squeezed States of the Electromagnetic Field", Feature Issue of J. Opt. Soc. Am. B<u>4</u> (October, 1987).
2. C. M. Caves, Phys. Rev. D<u>23</u>, 1693 (1981).
3. L. A. Wu, H. J. Kimble, J. L. Hall, and H. Wu, Phys. Rev. Lett.<u>57</u>, 2520 (1986).
4. Min Xiao, L. A. Wu, and H. J. Kimble, Phys. Rev. Lett. <u>59</u>, 2781(1987).
5. P. Grangier, R. E. Slusher, B. Yurke, A. LaPorta, Phys. Rev. Lett. <u>59</u>, 2566 (1987).
6. Min Xiao, L. A. Wu, and H. J. Kimble, Optics Lett. <u>13</u>, (June,1988).

SQUEEZING THERMAL MICROWAVE RADIATION

B. Yurke and P. G. Kaminsky

AT&T Bell Laboratories
Murray Hill, New Jersey 07974

E. A. Whittaker

Stevens Institute of Technology
Hoboken, New Jersey 07030

A. D. Smith, A. H. Silver, and R. W. Simon

TRW Space & Technology Group
Redondo Beach, CA 90278

ABSTRACT

We have demonstrated squeezing of 4.2 K thermal noise using a Josephson-parametric amplifier operated at 19.4 GHz. A 42% reduction in the equilibrium noise was observed. When operated at 0.1K, the amplifier exhibited an excess noise of 0.28 K referred to the amplifier's input. This excess noise is less than the vacuum fluctuation noise at the input. The device can be operated in a nonchaotic mode with a signal gain of 9 dB.

I. INTRODUCTION

A number of laboratories [1-6] have successfully generated squeezed light at optical frequencies. Because squeezed light promises to be useful in increasing the sensitivity of precision instruments such as interferometers [7-8] and in exploring new physical phenomenon [9] it is desirable to have squeezed state sources which cover a wide range of the electromagnetic spectrum. We are currently engaged in an effort to generate squeezed microwaves in the K band region (18-26.5 GHz) of the electromagnetic spectrum. We have successfully squeezed 4.2 K equilibrium noise by 42% using a Josephson parametric amplifier [10-11] operated at 19.4 GHz. When operated at 0.1 K this amplifier exhibits an excess noise of 0.28 K referred to the input. This is less than the vacuum fluctuation noise $\hbar\omega/2k = 0.47$ K. Consequently, the amplifier is quieter than a phase insensitive linear amplifier can in principle be [12], that is, the amplifier has less internal noise than the internal noise of a maser due to spontaneous emission. Further, the amplifier can be operated in a nonchaotic mode with a parametric gain of 12 dB (the actual gain including losses into and out of the device being 9 dB). The amplifier thus can serve as the first stage for an extremely sensitive microwave detector which could be used to explore the quantum statistics of microwave radiation emitted by Rydberg atom masers [13,14] or nonlinear electronic circuits [10]. Here the results of our observations will be reported. First, however, some simple theory will be introduced to explain how the device works and how the measurements were performed. Also a description of the device and the detection instrumentation will be given.

II. DEVICE DESIGN

In order to generate quadrature-squeezed electromagnetic radiation one requires a phase sensitive amplifier which will amplify one amplitude component of the signal and deamplify the other component. A degenerate parametric amplifier is an example of such a device. In a degenerate parametric amplifier a time dependent reactance oscillating at twice the signal frequency does parametric work on the signal [15]. In the case of a Josephson parametric amplifier the nonlinear inductance of a Josephson-junction is modulated by a time dependent current in order to provide the time dependent reactance. Although a Josephson junction is not usually thought of as a nonlinear inductor it is quite straightforward to determine the inductance from the Josephson equations [16]

$$I = I_c \sin\phi \tag{1}$$

and

$$\frac{d\phi}{dt} = \frac{2eV}{\hbar} \tag{2}$$

where I is the current flowing through the junction, V is the voltage across the junction, I_c is the critical current, and ϕ is the junction phase. Using the definition for an inductance L,

$$V = \frac{d}{dt}(LI)\,, \tag{3}$$

it is a simple exercise to show from (1) and (2) that

$$L = L_J \frac{\sin^{-1}(I/I_c)}{I/I_c} \tag{4}$$

where L_J is the characteristic Josephson inductance

$$L_J = \frac{\hbar}{2e\,I_c}\,. \tag{5}$$

The circuit diagram for the parametric amplifier we employed is depicted in Fig. 1. The inductors L_s are small and can be neglected when considering the rf performance of the device. The device thus essentially consists of two Josepson junctions J_d, two capacitors C_d, and two inductors L_d all in parallel. The advantage of employing two Josephson junctions derives from the fact that the two junctions and inductors L_s form a dc superconducting quantum interference device (SQUID). Such a device acts as a single Josephson junction whose critical current is a function of the magnetic flux passing through the SQUID loop [16]. This flux is controlled by adjusting the current flowing through the inductors L_s via the $+I_c$ and $-I_c$ control lines. The circuit of Fig. 1 thus simplifies to the circuit depicted in Fig. 2 where the critical current is now a function of θ which is controlled by flux coupled into the dc SQUID loop and $L = L_{d/2}$ and $C = 2\,C_d$. The effective inductor L in parallel with the effective junction J can be viewed as an rf SQUID

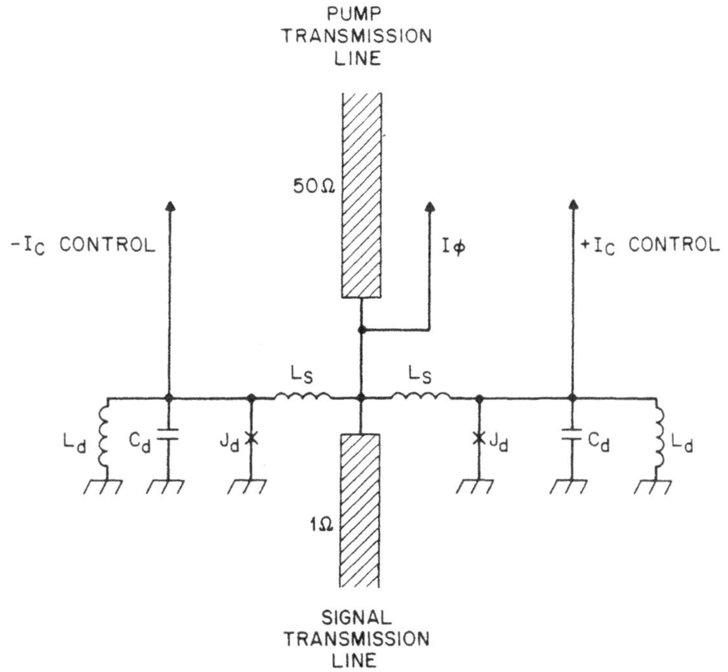

Fig. 1. The schematic diagram for the Josephson parametric amplifier employed in the experiment. The amplifier features two Josephson junctions in a dc SQUID arrangement. This SQUID acts like a single Josephson junction whose critical current I_c can be controlled, via the flux coupled into the SQUID loop through the inductors L_s, by applying current to the control lines $+ I_c$ and $-I_c$.

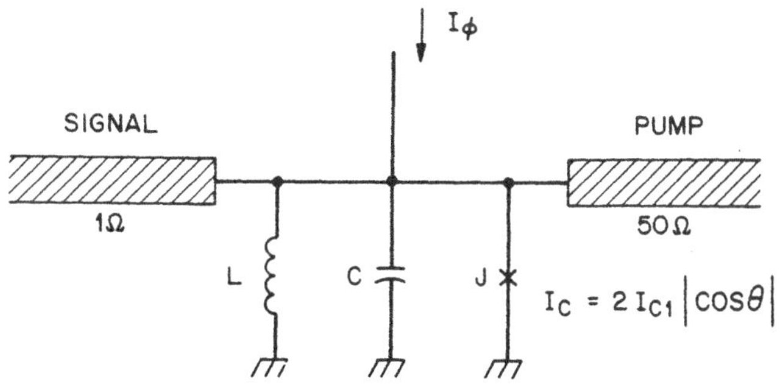

Fig. 2. The circuit of Fig. 1 can be simplified to the one depicted here. The parallel combination of the inductor and Josephson junction forms an rf SQUID loop. The device can thus be viewed as an rf SQUID parametric amplifier.

[16]. The relation between the current I_ϕ and the phase is given by

$$I_\phi = \frac{\hbar}{2eL}[\beta \sin\phi + \phi] \qquad (6)$$

where $\beta = L/L_J$. The junction phase ϕ is a single valued function of I_ϕ when $\beta \leqslant 1$. Hence, in order to avoid operational instabilities one wants β to be less than one. On the other hand, if β is too small, the inductance L shorts the nonlinear inductance making it incapable of supplying sufficient parametric work to significantly amplify a signal. Consequently, the device was designed with $\beta \approx 1$.

Josephson parametric amplifiers have generally been plagued with excess noise [17,18], however, recently several groups [19,20] have succeeded in obtaining low noise performance using rf SQUID parametric amplifiers. Our device is a second generation design of one of these devices [20].

The amplifier was fabricated on a 1cm×1cm silicon substrate. The junctions, having a 50 μm^2 area were fabricated for a critical current of 200 A/cm^2 using a niobium/aluminum oxide technology [21]. The inductors L_s consist of a 0.5 pH length of transmission line between the junctions. The rf SQUID inductance L/2 was 2.9 pH. The capacitors C were designed so that the total parallel capacitance (including the junction capacitance) would be 18.1 pF. The bias lines consisted of coplanar waveguides with impedances of 50 Ω in order not to load the device whose characteristic impedance is 1 Ω. In order to transform the 1 Ω impedance to 50 Ω a 4-stage impedance transformer consisting of various lengths of strip line and coplanar waveguide as described in Table 1 was also fabricated on the substrate. Microwave power was coupled onto and off of the chip through SMA connectors followed by coax-to-waveguide transitions. The 3dB passband of the amplifier was measured to be 1 GHz. The device is operated in the negative resistance reflection mode [22] where the amplified (or deamplified) signal propagates out of the same port that the signal enters (the 1 Ω transmission line of Fig. 1). Pump current is supplied to the junctions through a 50 Ω transmission line.

Table 1

Impedance	Section	Angle
50 ohm	end	
35 ohm	coplanar waveguide	90 degrees
1 ohm	microstrip	3.5 degrees
50 ohm	coplanar waveguide	13 degrees
3.7 ohm	microstrip	90 degrees
1.4 ohm	microstrip	90 degrees
1 ohm	end	

Table 1 lists the successive lengths of coplanar waveguide and microstrip line that were used to construct the 50 Ω to 1 Ω impedance transformer.

III. NOISE THEORY

The experimental measurements reported here were made at nonzero if frequencies, and hence a proper analysis of the experiment requires a two-mode or wideband analysis [23-25]. Such an analysis has been presented elsewhere [26]. Here for simplicity a single-mode analysis is presented to motivate the wideband formulas Eq. (15)-(19) used in the comparison of our data with theory. The parametric amplifier will be modeled by an ideal parametric amplifier, labeled JPA in Fig. 3, and the loss will be modeled by the beam splitter BS whose transmission is $\eta^{1/2}$. A signal entering a_{in} will pass through the beam splitter twice as it returns via b_{out} and hence the reflection loss is η. This quantity can be measured by comparing the reflection from a Josephson parametric amplifier with the pump off to that of a short. The ideal parametric amplifier performs the now familar mode transformation

$$b = G^{1/2}a + (G-1)^{1/2}a^{\dagger} \tag{7}$$

where G is the signal power gain and G−1 is the conjugate power gain. The beam splitter BS can be taken to perform the mode transformation

$$a = \eta^{1/4}\, a_{in} + (1-\eta^{1/2})^{1/2}\, c_{in}$$

$$b_{out} = \eta^{1/4} b + (1-\eta^{1/2})^{1/2}\, c_{out}\ . \tag{8}$$

A homodyne detector measures the operator [27-29]

$$X_{\phi} = b_{out} e^{-i\phi} + b^{\dagger}_{out}\, e^{i\phi}$$

where ϕ is the local oscillator phase. The boson operators a, b, a_{in}, b_{out}, c_{in}, and c_{out} all satisfy commutation relations of the form

$$[a, a^{\dagger}] = 1\ . \tag{10}$$

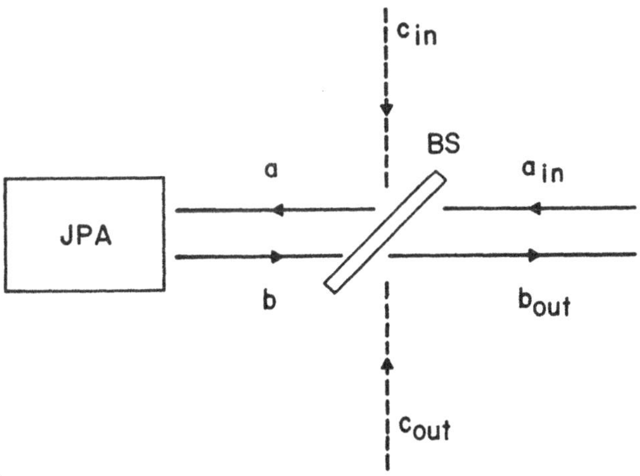

Fig. 3. Here a parametric amplifier is modeled as an ideal lossless parametric amplifier JPA with a beam splitter BS at the input to provide the loss.

Keeping in mind that the input a_{in} and the noise modes c_{in} and c_{out} are uncorrelated, and that for noise

$$<a_{in}\, a_{in}> \; = \; <c_{in}\, c_{in}> \; = \; <c_{out}\, c_{out}> \; = 0 \tag{11}$$

one can show that the noise power $P \equiv <X_{\phi}^{2}>$ observed by the homodyne detector is

$$P = \eta F(\phi)\; P_{in} + (1-\eta^{1/2})\,[\eta^{1/2}\, F(\phi) + 1]\, P_{n} \tag{12}$$

where

$$P_{in} = <a_{in}^{\dagger}\, a_{m}> + <a_{in}\, a_{in}^{\dagger}> \tag{13}$$

is the noise that a homodyne detector would see looking directly at the input and the loss noise sources are taken to have the same noise power P_n,

$$P_{n} = <c_{in}^{\dagger}\, c_{in}> + <c_{in}\, c_{in}^{\dagger}> \; = \; <c_{out}^{\dagger}\, c_{out}> + c_{out}\, c_{out}^{\dagger}>\,, \tag{14}$$

when observed with a homodyne detector and

$$F(\phi) = 2G-1 + 2G^{1/2}(G-1)^{1/2}\cos 2\phi\ . \tag{15}$$

The wideband generalization of (12) is

$$S(\nu) = \eta F(\phi) S_{in}(\nu) + (1-\eta^{1/2})\; [\eta^{1/2} F(\phi) + 1]\; S_{loss}\, (\nu) \tag{16}$$

where $S(\nu)$, $S_{in}(\nu)$ and $S_{loss}(\nu)$ are the power per unit bandwidth seen by an ideal homodyne detector at if frequency ν. If the loss and the signal source are both at the same temperature, then

$$S_{loss}(\nu) = S_{in}(\nu) = \frac{h\nu_0}{2}\coth\left(\frac{h\nu_0}{2kT}\right) \tag{17}$$

where it has been assumed that the if frequency ν is much less than the carrier frequency ν_0, $\nu << \nu_0$.

In the real experiment, there is loss η_d between the detector and the output port of the homodyne detector. The homodyne detector employed in the experiment is a conventional diode mixer with a noise temperature T_d of 660K. Consequently, one must look for extremely small changes on top of a large noise floor. This was accomplished by using a lock-in detection technique where the pump is turned on and off and the noise with the pump on is compared with the noise when the pump is off. The spectrum analyzer, which was set to measure noise in dB's, using this detection method reports

$$\Delta S = 10 \log_{10}\left(\frac{1+\eta_d S_{on}(\nu)/kT_d}{1+\eta_d S_{off}(\nu)/kT_d}\right) \tag{18}$$

where $S_{on}(\nu)$ is given by Eq. (16) and $S_{off}(\nu)$ is given by Eq. (16) with $G = 1$. Eqs. (15)-(18) constitute the noise theory that we will use to analyze the data. For a large classical signal where $\eta_d S_{off}(\nu)/kT_d >> 1$, Eq. (18) reduces to

$$\Delta S = 10 \log_{10}(F(\phi))\ . \tag{19}$$

Hence, by injecting a strong classical signal at one of the two sideband frequencies $\nu_0+\nu$ and $\nu_0-\nu$ into the parametric amplifier one can measure $F(\phi)$ directly. Consequently, all the quantities G, ϕ, η, ν, ν_0, T, T_d, and η_d that enter into the noise theory can be independently measured.

IV. INSTRUMENTATION

The instrumentation used in the data taking is depicted in Fig. 4. Breaks in the figure indicate which components are at room temperature (300 K) and which components are in the cold part (4.2 K) of the cryostat. The components in the cold part of the cryostat are in fact heat sunk to a dilution refrigerator and some components such as the Josephson parametric amplifier JPA and the cold termination T can be cooled to 0.1 K. The transition between the room temperature components and the low temperature components is made by stainless steel waveguides (WR-42 for the signal waveguides; WR-22 for the pump) 1.5 m long.

A 19.4 GHz source provides both the local oscillator for the mixer M and the pump for the Josephson parametric amplifier. On its way from the source to the pump port of the JPA the power passes first through a 3 dB coupler and then through an isolator I4 which isolates the mixer from reflections off of components in the pump line. Similarly, I5 isolates the pump from the mixer. The relative phase between the mixer LO and the pump is adjusted via a current controlled phase shifter ϕ. The pump is turned on and off via a diode switch S2. The pump power is then doubled via doubler D and finally attenuated to the desired level with

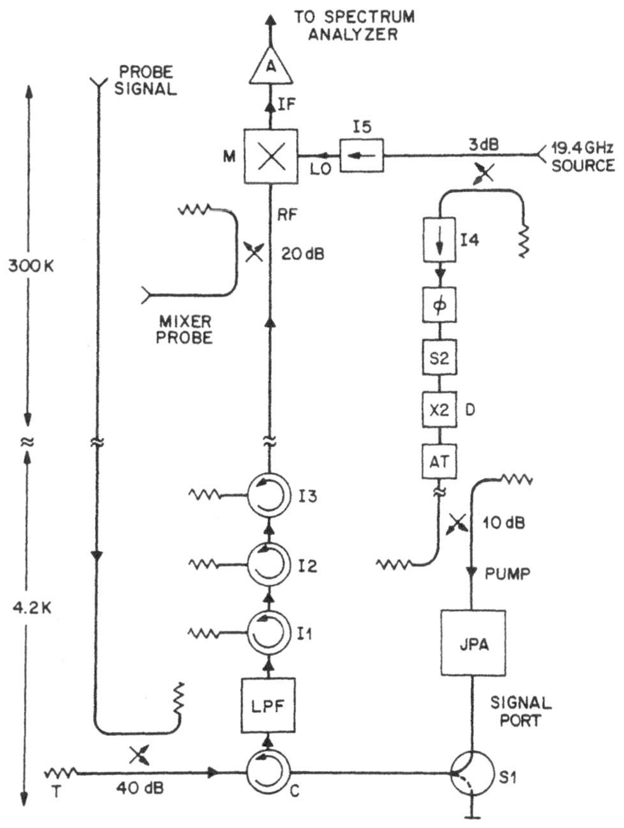

Fig. 4. A schematic of the instrumentation employed in data taking. See text for description.

the attenuator AT. Upon entering the cold part of the cryostat the pump power is attenuated an additional 10 dB via a directional coupler in order to reduce thermal noise.

The source of the noise to be squeezed is the variable temperature cold termination T of a design described by McGrath et al. [30]. A probe signal is combined with the noise through a 40 dB directional coupler. The circulator C directs the probe signal and noise to the signal port of the JPA. A waveguide switch S1 allows one to replace the JPA with a short in order to make accurate reflection loss measurements [31]. The squeezed signal from the JPA is directed by circulator C through a low-pass filter (to block pump power from reaching the mixer) and then through a series of three cryogenic isolators [32], I1, I2, and I3 which block thermal noise from reaching the JPA through the mixer waveguide. The total isolation provided by I1, I2, and I3 was measured to be 68 dB. For diagnostic purposes, a signal can be injected into the rf port of the mixer via a 20 dB coupler. In this manner one can check for possible detector saturation. The mixer M is a conventional diode mixer [33]. Upon amplification by amplifier A the mixer if is delivered to a spectrum analyzer for measurement. All the noise measurements reported here were taken with an HP 8566B spectrum analyzer. A resolution bandwidth of 1 MHz, a video bandwidth of 10 Hz, a frequency span of 1 MHz, and a setting of 1 dB/div were employed. The data taking was automated with a computer which also controlled the phase shifter ϕ and the temperature of the cold termination T when data as a function of temperature was taken. The average noise level of a given spectrum analyzer trace was recorded by the computer and typically 150 pairs of pump on/pump off readings were averaged per setting of oscillator phase ϕ or temperature T.

V. RESULTS

For the 4.2K thermal noise squeezing experiments the cryostat was filled with 1 Torr of ^{4}He exchange gas in order to keep all the cryogenic microwave components at 4.2K. Classical amplification and deamplification as a function of ϕ, observed under these conditions, is depicted in Fig. 5. The classical probe was offset of a frequency $\nu = 55$ MHz from the 19.4 GHz carrier frequency. The open circles are the lock-in zero obtained by adjusting the pump attenuator AT for maximum attenuation. The filled circles are the measured $\Delta S(\nu)$. The smooth curve is a fit of Eq. (15) and (19) to the data. This provides a measurement of $F(\phi)$ and in particular G was found to be 1.48. By comparing the reflection off the parametric amplifier with the pump off to that of a short (using waveguide switch S1) the reflection loss η was measured to be 0.38 (4.2 dB loss). Figure 6 shows the experimentally measured noise power level $\Delta S(\nu)$ as a function of ϕ. This data was taken at the same time as the data of Fig. 5, i.e., the spectrum analyzer was automated first to make a signal power measurement at $\nu = 55$ MHz then to make a noise power measurement at $\nu = 70$ MHz. Again the open circles indicate the lock-in zero. The solid circles are the data. The detector system noise temperature was measured by terminating the input port of the mixer first with a room temperature termination and then with a 77K termination. The detector system noise temperature was found to be $T_d = 660 \pm 20$K. Also in a separate experiment the total attenuation from the output port of the parametric amplifier to the rf port of the mixer was measured to be $\eta_d = 0.28 \pm .02$ or 5.5 dB. Hence, all the numbers that enter into the noise theory Eq. (15)-(18) have been independently measured. The solid line of Fig. 6 is a comparison of this theory with the data. We stress that there are no adjustable parameters in the theory. From the observed maximum noise reduction $\Delta S = -3.7 \times 10^{-3}$ dB, a drop ΔT

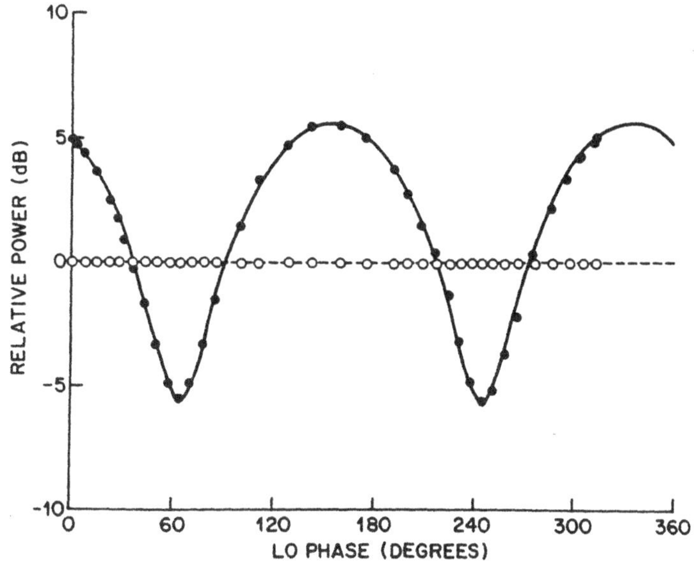

Fig. 5. Classical parametric gain as a function of the relative phase between the local oscillator and the pump (filled circles). The open circles are a zero check. The solid curve is a theory fit to the data.

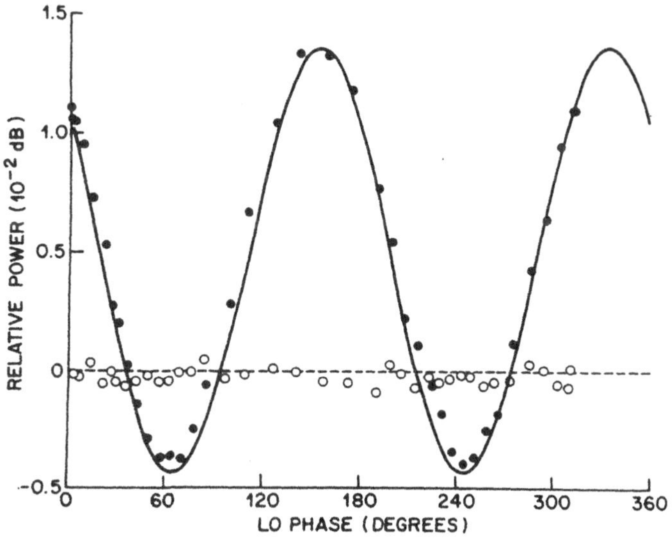

Fig. 6. Noise data (filled circles) taken at the same time as the classical gain data of Fig. 1. The open circles establish the lock-in zero. The solid line is the theoretical prediction with no adjustable parameters. A 1.8 K drop or squeezing below the 4.2 K equilibrium noise is observed.

$$\Delta T = \frac{T_d}{\eta_d}\left(1 - 10^{\frac{\Delta S}{10}}\right) \tag{20}$$

of 1.8 K below the 4.2 K equilibrium noise floor can be inferred. This corresponds to a 42% squeezing of the equilibrium noise.

In order to rule out detector saturation an experiment was performed in which a probe signal with $\nu = 55.6$ MHz was injected directly into the input port of the mixer to monitor the detector system gain and at the same time the noise coming from the parametric amplifier was measured at 70 MHz. The difference between the power of the probe signal when the pump is on to that when the pump is off is plotted as a function of ϕ in Fig. 7 (open circles). It was found that the detector system gain remained constant to within $\pm 5 \times 10^{-4}$ dB. In contrast the noise $\Delta S(\nu)$ exhibited 3×10^{-3} dB squeezing. Hence, detector saturation has at most a 10% effect on the data.

By pumping out the ^{4}He exchange gas and operating the dilution refrigerator the parametric amplifier and cold termination were cooled to 0.1 K. By adjusting the current through a resistor on the cold termination its temperature T_{in} could then be varied. Since now the temperature of the parametric amplifier losses T_{loss} (presumed to be 0.1 K) is different from the temperature T_{in} of the noise entering the input port one no longer has equality of $S_{loss}(\nu)$ and $S_{in}(\nu)$ in Eq. (17),

$$S_{loss}(\nu) = \frac{h\nu_0}{2} \coth\left[\frac{h\nu_0}{2kT_{loss}}\right] \tag{21}$$

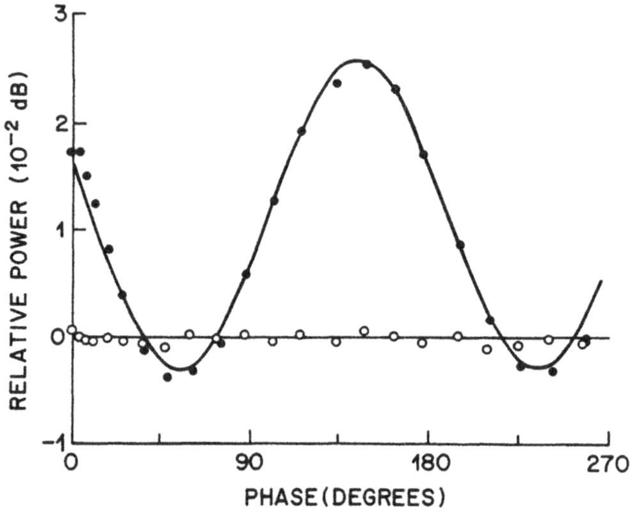

Fig. 7. Monitoring the detector gain (open circles) and the noise (filled circles) as a function of local oscillator phase. The detector system gain remains constant indicating that no saturation effects are present while the noise data shows sizable squeezing.

and

$$S_{in}(\nu) = \frac{h\nu_0}{2} \coth\left(\frac{h\nu_0}{2kT_{in}}\right). \tag{22}$$

Fig. 8 depicts noise data taken as a function of termination temperature T_{in} for the case when ϕ was adjusted for maximum amplification. In this case G was measured to be 1.84 and the reflection loss was measured to be 0.50 (3 dB). The expected ΔS from the theory Eq. (15), (16), (18), (21) and (22) is plotted as the solid line in Fig. 8. Note that again there are no adjustable parameters. The data (filled circles) follows the theory for T_{in} above 0.7 K. Below this temperature the data points deviate from the expected curve indicating excess noise. The expected ΔS is 1.92×10^{-3} dB. The measured ΔS at 0.1 K is $2.82 \pm .4 \times 10^{-3}$ dB indicating an excess noise of $9.0 \pm .4 \times 10^{-4}$ dB ($.44 \pm .02$K in temperature units). The origin of this noise has not yet been identified; the effect could be due to the cold termination decoupling from the dilution refrigerator at low temperatures.

Taking the view that this noise is generated internally in the amplifier, the noise can be referred to the amplifier's input by dividing by $\eta(F(0) - 1)$. One then obtains 0.28 K for the amplifier's noise temperature. Since the vacuum fluctuation noise at 19.4 GHz is $h\nu/2\,k = 0.47$ K, the amplifier is quieter than a linear phase insensitive amplifier is allowed to be.

Taking the view that this noise is not generated in the amplifier but coupled into the cryostat from the outside, this noise contributes 6% to the 4.2 K equilibrium noise. Since the signal measured is the difference between the

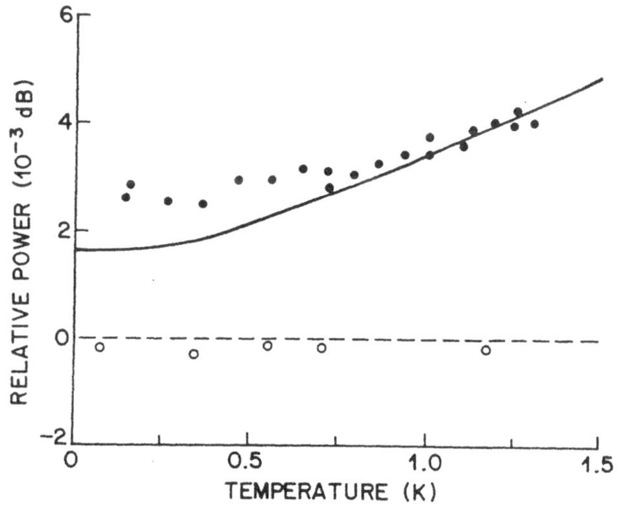

Fig. 8. An attempt to use the Josephson parametric amplifier with the LO phase adjusted for maximum gain to measure the $\frac{1}{2}h\nu\coth(h\nu/2\,kT)$ function by varying the temperature of the cold load. Below 0.7 K the data points deviate from the theoretical prediction indicating either the presence of excess noise or that the cold termination has lost thermal contact with the dilution refrigerator.

pump-on and the pump-off noise levels it has thus been established that the y-axis zero of Fig. 6 is the 4.2 K equilibrium noise floor to within 6%. The open circles of Fig. 8 again indicate the lock-in zero.

Holding the amplifier and cold termination at 0.1 K, the phase ϕ can be varied in order to look for squeezing. Such data is depicted in Fig. 9. The dotted line indicates the lock-in zero. The data points (filled circles) do fall below the lock-in zero for certain ranges of ϕ indicating noise deamplification. As in the 4.2 K squeezing G and η were measured using a classical probe. In this case $G = 1.84$ and $\eta = 0.50$. From the variable temperature cold load measurements discussed above the excess noise level and vacuum noise level can be inferred. These are labeled in Fig. 9. As can be seen the noise does not drop below the vacuum noise level, but comes close. The solid line is the theoretical prediction based on the classically determined $F(\phi)$ assuming there is no excess noise.

At 0.1 K we have also explored the onset of chaos as the pump power is increased. Plotted in Fig. 10 is the noise to gain ratio (circles) as a function of attenuation of the pump attenuator AT. The parametric gain (triangles) is also plotted. The actual gain which includes the reflection loss is 3 dB less. The filled circle data was taken by successively increasing the pump power (decreasing the attenuation). The noise to gain ratio remains fixed until a parametric gain of 12 dB (9 dB actual gain) is achieved. Then the noise to gain ratio rapidly increases as chaos sets in. In an attempt to look for hysteresis data was also taken by successively increasing the pump attenuation (open circles). This data follows the closed circle data within the scatter of the data.

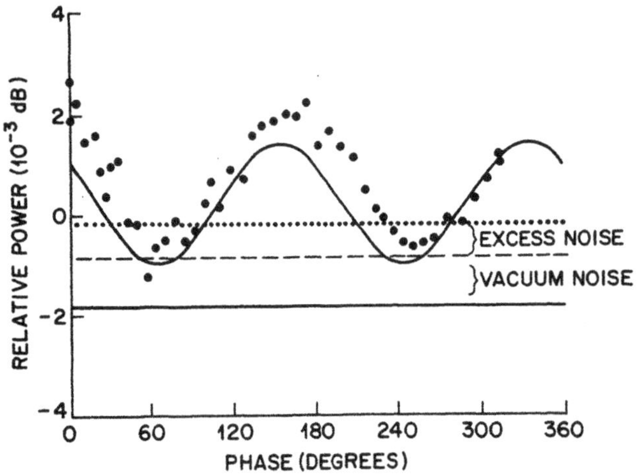

Fig. 9. The observed noise squeezing as a function of local oscillator phase when the parametric amplifier and cold termination are at 0.1 K. The dotted line indicates the lock-in zero. The data (filled circles) exhibits noise deamplfication, however due to the presence of excess noise the noise is not deamplified below the vacuum noise floor.

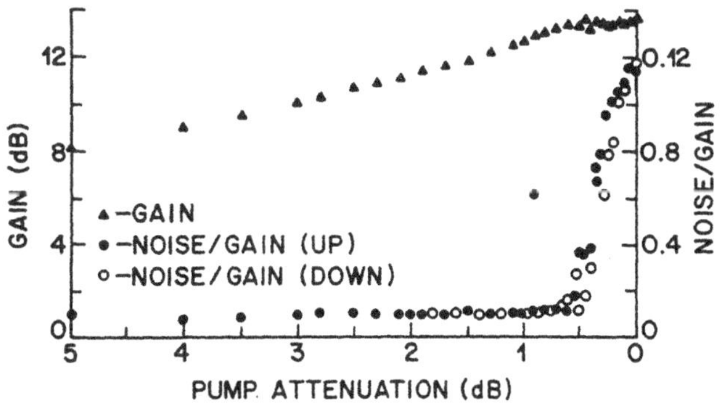

Fig. 10. The noise to gain ratio and the parametric gain plotted as a function of pump attenuator setting. The noise to gain ratio remains flat until a parametric gain of 12 dB is achieved. The noise then rapidly rises due to the onset of chaos. This data was taken with the parametric amplifier and cold termination at 0.1 K.

VI. CONCLUSIONS

We have demonstrated squeezing of 4.2 K thermal noise using a Josephson-parametric amplifier operated at 19.4 GHz. A 42% reduction in the equilibrium noise was observed. Detector saturation effects contribute no more than 10% to this number. Further, it has been established that the 42% noise reduction is with respect to a noise floor that is equal to the 4.2 K equilibrium noise floor to within 6%.

When operated at 0.1 K the amplifier exhibits an excess noise of 0.28 K which is less than the vacuum fluctuation noise level of 0.47 K. The amplifier is thus less noisy than a linear phase insensitive amplifier could possibly be. Further, the amplifier does not become choatic until a signal gain of 9 dB has been achieved.

REFERENCES

[1] R. E. Slusher, L. W. Hollberg, B. Yurke, J. C. Mertz, and J. F. Valley, Phys. Rev. Lett. *55*, 2409 (1985).

[2] R. M. Shelby, M. D. Levenson, S. H. Perlmutter, R. G. De Voe, and D. F. Walls, Phys. Rev. Lett. *57*, 691 (1986).

[3] L. Wu, H. J. Kimble, J. L. Hall, and H. Wu, Phys. Rev. Lett. *57*, 2520 (1986).

[4] M. W. Maeda, P. Kumar, and J. H. Shapiro, Opt. Lett. *3*, 161 (1986).

[5] S. Machida, Y. Yamamoto, and Y. Itaya, Phys. Rev. Lett. *58*, 1000 (1987).

[6] A. Heidmann, R. J. Horowicz, S. Reynaud, E. Giacobino, and C. Fabre, Phys. Rev. Lett. *59*, 2555 (1987).

[7] M. Xiao, L. Wu, and H. J. Kimble, Phys. Rev. Lett. *59*, 278 (1987).

[8] P. Grangier, R. E. Slusher, B. Yurke, and A. LaPorta, Phys. Rev. Lett. *59*, 2153 (1987).

[9] C. W. Gardiner, Phys. Rev. Lett. *56*, 1917 (1986).

[10] B. Yurke, L. W. Rupp, and P. G. Kaminsky, IEEE Trans. Mag. *MAG-23*, 458 (1987).

[11] B. Yurke, P. G. Kaminsky, R. E. Miller, E. A. Whittaker, A. D. Smith, A. H. Silver, and R. W. Simon, manuscript in preparation.

[12] C. M. Caves, Phys. Rev. D*26*, 1817 (1982).

[13] D. Meschede, H. Walther, and G. Müller, Phys. Rev. Lett. *54*, 551 (1985).

[14] M. Brune, J. M. Raimond, P. Goy, L. Davidovich, and S. Haroche, Phys. Rev. Lett. *59*, 1899 (1987).

[15] H. Takahasi, *Advances in Communication Systems*, edited by A. V. Balakrishnan (Academic, New York, 1965).

[16] A. Barone and G. Paternò, *Physics and Applications of the Josephson Effect* (John Wiley, New York, 1982).

[17] P. L. Richards, *SQUID 76 Superconducting Quantum Interference Devices and Their Applications*, edited by H. D. Hallbohm and H. Lubbig (Walter de Gruyter, Berlin, 1977) pp. 323-338.

[18] N. F. Pedersen, *SQUID 80 Superconducting Quantum Interference Devices and Their Applications*, edited by H. D. Hallbohm and H. Lubbig, (Walter de Gruyter, Berlin, 1980) pp. 739-762.

[19] N. Calander, T. Claeson, and S. Rudner, J. Appl. Phys. *53*, 5093 (1982).

[20] A. D. Smith, R. D. Sandell, J. F. Burch, and A. H. Silver, IEEE Trans-Mag. *MAG-21*, 1022 (1985).

[21] M. Gurvitch, M. A. Washington, and H. A. Huggins, Appl. Phys. Lett. *42*, 472 (1983).

[22] L. A. Blackwell and K. L. Kotzebue, *Semiconductor-Diode Parametric Amplifiers* (Prentice-Hall, Englewood Cliffs, 1961).

[23] B. Yurke and J. S. Denker, Phys. Rev. A*29*, 1419 (1984).

[24] C. M. Caves and B. L. Schumaker, Phys. Rev. A*31*, 3068 (1985).

[25] B. L. Schumaker and C. M. Caves, Phys. Rev. A*31*, 3093 (1985).

[26] B. Yurke, J. Opt. Soc. Am. B *4*, 1551 (1987).

[27] H. P. Yuen and J. H. Shapiro, IEEE Trans. Inf. Theory *IT-26*, 78 (1980).

[28] B. L. Schumaker, Opt. Lett. *9*, 189 (1984).

[29] B. Yurke, Phys. Rev. A*32*, 311 (1985).

[30] W. R. McGrath, A. V. Räisänen, and P. L. Richards, International Journal of Infrared and Millimeter Waves *7*, (1986).

[31] For a description of the switch see P. G. Kaminsky, L. W. Rupp, and B. Yurke, Rev. Sci. Instrum. *58*, 894 (1987).

[32] The circulators, KYG 2122K, are manufactured by Passive Microwave Technology, 1 Remmet Ave., Canoga Park, CA 91304.

[33] The mixer, model number CK-1, was manufactured by Honeywell Spacekom, 214 East Gutierrez Street, Santa Barbara, CA 93101.

CLASSICAL AND QUANTUM STATISTICS IN PARTITION OF HIGHLY DEGENERATE LIGHT

F. De Martini

Dipartimento di Fisica, Università di Roma I, Roma, Italy 00185

1) INTRODUCTION

The statistical properties of photons have been found to be well accounted for by the methods of Quantum Electrodynamics. In spite of this satisfactory situation, recently the attention of some theoretical physicists has been attracted by an interesting paradox which shows, in the context of light beam-splitting, the existence of a surprising conflict between the expectations of QED and those dictated by the indistinguishability property of the photon [1,2]. Let us present the problem first. Assume that a quasi-monochromatic light beam belonging to a single field mode k, in a chaotic state and with an average number of photons per mode $<\hat{n}_k> \equiv \bar{n}$, excites an optical beam-splitter (BS) or a two way interferometer (IF). Label by i=1,2 the two output modes. Each input state may be given in the form: $|n> = \sum_{l(0-n)} p_{nl} \, |l,n-l>$, where l is the photon number in state say i=1, and $P_{nl} = |p_{nl}|^2$ are the two-way partition probabilities. The field density operator is given by: $\rho = \sum_{n(0-\infty)} P_n \, |n><n| = \sum_{n(0-\infty)} P_n \sum_{l,h(0-n)} p_{nl} \, p_{nh} |l,n-l><n-h,h|$, and $P_n = \bar{n}^n/(1+\bar{n})^{(1+n)}$ (Ref.2,3). Consider now the QED theory of the process. The input state $|n>$ is obtained by successive applications to the vacuum-state of the operator $b^{\dagger}_k = \sum_i w_i a^{\dagger}_i$, $a^{\dagger}_i$ expressing the field on mode i : $|n> = ((b^{\dagger}_k)^n/\sqrt{n!}) \, |0>$. This procedure leads to $P_{nl} = W^l_1 W_2^{(n-l)} \, n! \, / \, (l!(n-l)!)$, i.e., the Bernoulli partition law. The weight-factors $W_i \equiv |w_i|^2$, $\sum_i W_i = 1$, are assumed in the context of QED to be constant values depending on macroscopic boundary conditions as slit-apertures, BS-reflectances, scattering cross-sections etc. In this case the above partition corresponds to the Maxwell-Boltzmann (M-B) distribution, the well known "*classical statistics*"[4]. It is indeed surprising that QED, when applied to the $|n>$ - state field representation, the "nonclassical" one because of complete phase non-definition, leads to an utmost classical result: one that, among other physical consequences, implies complete absence of inter-particle quantum correlations in partition. As an example, the above argument may be cast in a simple form if we consider an input pure state with two photons per mode, n=2. In this case we can devise an experiment, similar to the (b) one reported later in this paper, in which two phototubes inserted on the output arms of a lossless BS detect the photon number combinations (l, 2-l) through the output states: $|0,2>$, $|1,1>$,

|2,0>. If we evaluate the detection probabilities by QED we obtain $P_{2l} \equiv |<2\,|\,1,2\text{-}l>|^2 = 1/4$ for l=0,2 and $P_{2l}=1/2$ for l=1. This *double counting* of state | 1,1> implies "classical" distinguishability of the two photons travelling through BS. The Bose-Einstein (B-E) two-way partition law, which accounts for "quantum indistinguishability', $P_{nl}=1/(n+1)$, would lead instead to $P_{2l}=1/3$, viz., to a *single counting* of | 1,1>. This paradox appears to refer to a general behaviour of particles and it shows up in all experiments involving field-interference effects, as for instance in optical interferometry and diffraction. Since in these processes the interference follows a field-partition effect taking place in the same instrument, the "classical" partition given by QED leads to a maximum "contrast" of the fringing pattern, i.e., for IF's, to a value of the fringe "Visibility":
$V=(I_{max}-I_{min})/(I_{max}+I_{min})=1$, where $I=<\hat{I}>$ is the average detected light intensity [5]. This can be explained by saying that the "classical" absence of quantum correlations in photon beam-splitting leads to a perfect field cancellation at the fringe "minima". On the contrary, if a "stimulated-emission" - like, inter-photon correlation takes place in beam-splitting, according to (B-E) partition law, the field-unbalance and the consequent imperfect field cancellation, leads to a decreasing value of $V(\bar{n}) < 1$ for increasing $\bar{n}>1$, as shown in Fig.2. Interestingly enough, no interferometry experiments with $\bar{n} >1$ have been carried out so far and the theoretical results given above have not yet been tested experimentally[5]. Consider also that (B-E) and Fermi-Dirac (F-D) statistical effects appear to be observable experimentally when "microscopic" particle-particle interaction processes take place, i.e., in the context of the "microscopic" domain of "quantum many-body" phenomena[4]. Owing to these considerations, we do believe that the above paradox, far from being a curious accident, hides very general and fundamental properties of particles and awaits clarification. In the present paper, we shall start with the presentation of a test of the BS theory performed by means of two, first- and second-order coherence, experiments. These ones, that may be thought of as involving selectively the two complementary aspects of the photon, are carried out with a highly-degenerate light, viz.$\bar{n} >> 1$, a condition that is essential for any reliable investigation on statistics. Afterwards, we shall propose what we believe to be the only consistent solution to the paradox: a solution that shall have a price. Before dealing with our test note that, according to QED, if photons belong to a superposition of coherent states $|\,\alpha>$, viz., to a laser beam, again a "classical" (and here well understandable) absence of inter-photon quantum correlation in the splitting process is expected[2].

2) EXPERIMENTAL

(a) - *Michelson Interferometry* -. A chaotic superposition of | n>-states with large $\bar{n}$ was obtained by incoherently scattering a 0.7 W, λ = 5145 A, Ar-laser beam (Spectra-Physics 2025-03) with a 3 MHz linewidth obtained by an intracavity etalon.The beam, with a TEM00 mode structure, was focused by a 2 cm f.l.lens on a ground-glass disk driven at 50 rps by a stable motor. The tangential speed of the disk at the focus was v=20 m/sec., the diameter of the focal spot was d=20 μm so that the coherence-time of the chaotic field was $t_c=(d/v)=1\mu sec$[3,6]. The angular extent of the single-mode under investigation was determined by d and by the diameter (1,5 mm) of the end-face of an optical-fiber placed at a distance $L=(L_1+L_2)=(1.2+0.5)$ m from the disk in the k-direction: this corresponded to a "degree of spatial coherence" >0.98 for the k-mode exciting the IF[5]. The k-direction was taken at 15° from the laser beam-axis to avoid unwanted collection of scattered

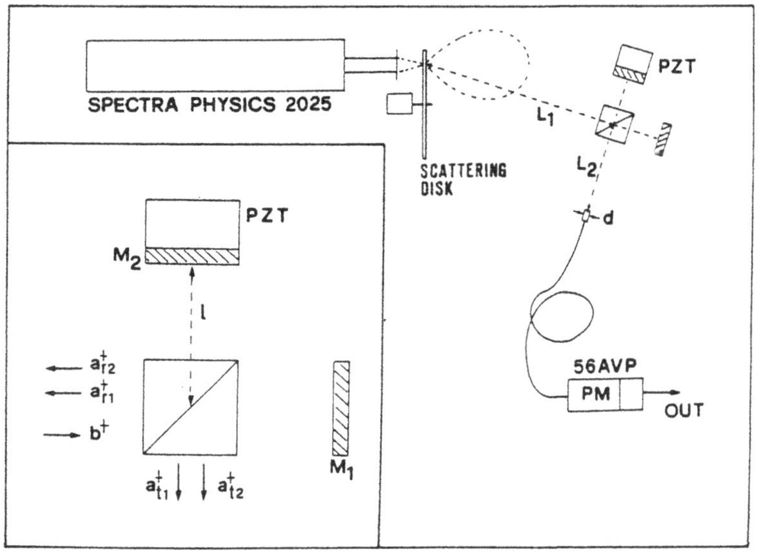

Fig.1 - Apparatus for Michelson Interferometry with chaotic light. For (HB-T) Interferometry mirrors M_i are replaced by two photon-counters.

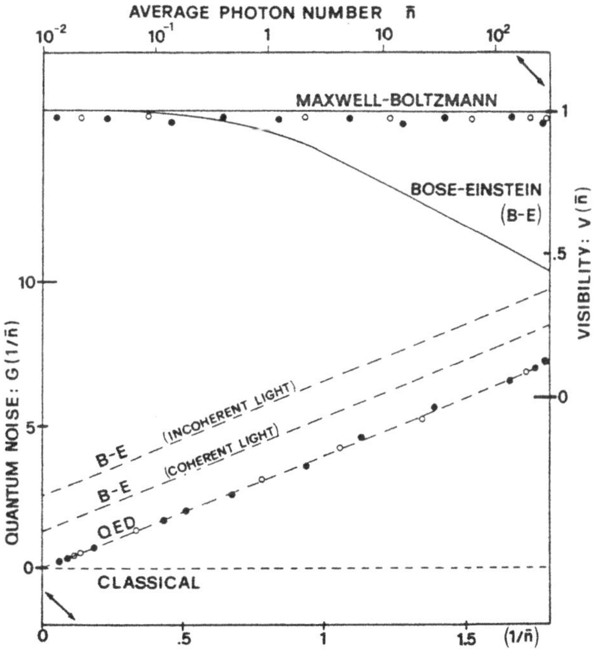

Fig.2 - Michelson "fringe visibility" $V(\bar{n})$ as function of the average photon number per mode. (solid-line, upper-right scales). "Beam-Splitter Quantum Noise Function" $G(1/\bar{n})$ (dotted-lines, lower-left scales). The plots refer to Maxwell-Boltzmann and Bose-Einstein statistics for coherent (o) and incoherent (•) light.a=0.99. Scales for $\bar{n}$ and $1/\bar{n}$ are not mutually consistent.

coherent light (Fig.1). Light was transferred through a polarizer and the fiber to a 56AVP photomultiplier (PM) DC-connected to a milliampermeter with 1 sec. integration time. Neutral-Density filters (ND) were inserted in the light path between the scattering focal spot and the BS of IF in order to vary $\bar{n}$ at BS. The ND's were exchanged between two stacks placed, along k, on the input and output arms of IF, so that the detected light intensity was independent of $\bar{n}$ at BS and nearly constant during the experiment. A (25-9093) Ealing Michelson-IF equipped with an (.5/.5) Halle-cube BS (TWK40) and two metal mirrors M_i was placed between the scattering source and the fiber-end. The probability of "photon survival" (i.e.non-absorption) in a single BS-process was measured to be: $a > 0.97$. The two equal round-trip paths between BS and M_i were 2.l=10 cm. and one of M_i was driven by a PZT-Transducer at a frequency of 0.05 Hz to scan over the IF-fringes of the detected I. The k-field polarization was taken parallel to the BS surface and IF was aligned in a condition of "optical contact"[5]. The plots of "Visibility" (i.e., "degree of first-order coherence") versus $\bar{n}$ are drawn in Fig.2. Since a double-splitting action takes place on each arm of the Michelson IF and since, out of the pairs of interfering output fields a_{ri}, a_{ti}, we investigated the transmitted ones, the quantum expression of the Visibility is[3]:

$$V(\bar{n}) = (2/\bar{n})\cdot \mathrm{Tr}\,[\rho(a^\dagger_{t1}\, a_{t2}+a^\dagger_{t2}\, a_{t1})]\,.$$

By inserting the expression of ρ, evaluated for double-splitting, in the one of the partition probabilities P_{nl} given by QED, we obtain $V(\bar{n})=1$ for coherent as well as for incoherent light. This expresses absence of quantum photon-correlation, as mentioned. If we evaluate P_{nl} according to the (B-E) partition law we find, for our IF:

$$V(\bar{n})=(4/\bar{n}) \sum_{n(0-\infty)} (P_n\, a^{2n+3}/(n+1)) \sum_{j(1-n)} \sqrt{j} \sum_{i(0-(n-j))} \sqrt{i+1} \sum_{l(i-(n-j))} \sqrt{1/(l+1)(l+2)(n-l)(n+1-l)}\,.$$

The plots of Fig.2 show that the crucial problem related to the realization of "classical" versus (B-E) statistics in light-splitting can be settled experimentally *only* by means of an highly-degenerate photon beam (viz., $\bar{n}>>1$), as expected. In our experimental conditions the equivalent temperature of the quasi-Planckian photon distribution of the chaotic light was: $T=(\bar{n}h\nu/k)=2.83\ 10^5\ K^o$, for $\bar{n}=10$. This corresponds to the temperature of the solar "Corona".

(b) - *Hanbury Brown-Twiss Interferometry* -. The experimental setup of Fig.1 has been transformed to allow for BS photon beam-splitting investigation by a second-order coherence technique[2,7]. The experiment was done with the Ar-laser and repeated with an ultra-stable, TEM00 mode, λ=6328 A, 1 mW, He-Ne-laser (Spectra-Physics 117A). The coherence time could be varied by changes of v up to t_c=0.1 sec. The IF mirrors M_i were replaced with micrometric holders of two, 1 mm. diameter, optical-fibers transferring photons to the cathodes of two low-noise-selected, 56AVP PM's. The k-mode spatial coherence definition, optical-fiber aligment, filtering methods were obtained according to the standard procedures for this kind of experiment[2]. Single-photon counting measurements were taken by a two-channel computer-interfaced counter, developed in our Laboratory, by which ~10^8 photocounts/(channel*second) could be processed.

A quite significant quantity, here measured for the first time, is the "*Beam Splitter Quantum-Noise Function*": $G \equiv \langle \hat{g} \rangle$, defined in terms of the operator:

$$\hat{g} \equiv ((\hat{n}_1/\langle\hat{n}_1\rangle)-(\hat{n}_2/\langle\hat{n}_2\rangle))^2, \text{ where: } \hat{n}_i \equiv a^\dagger_i\, a_i.$$

G is the average mean-square difference between the relative number of transmitted and reflected photons on the output channels i=1,2 (Ref.7). Three cases are relevant:

1) Classical BS : $G_{cl} = 0$

2) Quantum (QED) BS : $G_{qed} = 4/\bar{n}$ (for a symmetrical BS)

3) Quantum (B-E) BS : $G_{be} = G_{qed}+c$ (c=8/3 (chaotic-), c=4/3 (coherent-light)).

Note that the function G is related to the "degrees of second-order coherence", $g_{ij}(0)$, measured in Hanbury Brown-Twiss (HB-T) interferometry, through a normal-ordering procedure[2]. In fact, if $G' \equiv (g_{11}+g_{22}-2\ g_{12})$ we obtain: $G'=\langle:\hat{g}:\rangle$, $G=\langle\hat{g}\rangle$. The experimental conditions corresponding to the data reported in Fig.2 were: t_c=6 msec., counter gate δt=80μsec., sample number/datum=50.000.

3) CLASSICAL AND QUANTUM STATISTICS FOR DISTINGUISHABLE PARTICLES

The results of the experiments (a,b) clearly support the validity of QED. In spite of this satisfactory result, the paradox is yet to be solved and a question arises: "Why and in what circumstances should the photon, an intrinsically indistinguishable "quantum" particle, behave as it were an intrinsically distinguishable "classical" one?". We are able to give a consistent answer to this highly notrivial question, and then to resolve the paradox.

The most important propositions of our conjecture are the following:

1) The "Indistinguishability Principle" (IP) should be *rejected* as it is violated in our experiments. However this does not imply any lack of validity of the well known algebraic wavefunction-symmetrization method of Quantum Statistical Theory (QST) (or anti-symmetrization, if Pauli principle applies, viz., for F-D particles)[4]. Let us comment on the above statement as follows:

(a) - The key motivation for rejection of (IP) resides in the impossibility for intrinsically indistinguishabile particles to satisfy formally the (M-B) partition law. "Indistinguishability" as an intrinsic particle property was introduced in the first formulation of QST precisely in order to legitimate "a posteriori" the use of the quantum statistics. In this respect, the following statements expressed by A.Einstein in 1924-1925 may be enlightening[8]:

"The differences between the Boltzmann and the (B-E) counting express indirectly a certain hypothesis on a mutual influence of the molecules which, for the time being,is of quite misterious nature.." and,

"The statistical method of Herr Bose and myself is by no means beyond doubt, but seems only *a posteriori* justified by its success for the case of radiation..".

We interpret the long lifetime of the "indistinguishability" concept as an example of the well known historical persistence of the scientific "paradigms" (T.S.Kuhn, 9, 10). In fact, in the last decades the physical and epistemological difficulties inherent this concept, together with the reluctance to abandon it, has led to a series of semantic reinterpretations that, in some cases, turned out to be misleading and even logically inconsistent. This is precisely the case of the semantics involving "Indistinguishable" versus "Identical" particles, or the current identification of "Bose-Einstein Statistics" and "Indistinguishability Principle" with the mentioned formal algebraic method of wavefunction-symmetrization[4].

(b) - According to the wavefunction-symmetrization method, the overall wave-function $U(q_1,q_2,..q_i,..)$ of an ensemble of weakly-interacting, indistinguishable elements with coordinates q_i is given in terms of products of permuted wavefunctions $u_k(q_i)$ corresponding to the dynamics of the single isolated elements (Tolman, 4, pag.366). The current application of this method, which is found to be physically correct, is logically inconsistent with the indistinguishability hypothesis itself. In fact this hypothesis implies that these particles, when are part of a many-body ensemble, cannot be identified by their coordinates nor by any other label, even in the context of a provisional assumption[11].

2) The "classical" partition law, viz., the M-B statistics, is satisfied by *any* particle of *any* type (i.e.photons, electrons, tennis balls, etc.) if the splitting (viz., scattering) is provided by a device that complies with Bohr's definition of "Measurement Apparatus", i.e., one which "absorbs" without reaction-motions the momentum transferred by the impinging particles[12]. This is exactly what happens in the case of our experiments as well as in any optical experiment involving a massive BS or splitting-device[13].

Bose-Einstein or Fermi-Dirac, "quantum" statistical effects show up in scattering events involving *only* "microscopic" objects (i.e., particles). The "microscopic object" is one that, among other properties, is free to move under particle-momentum transfer. For this reason, "many-body" QST applies in the "microscopic world" and elementary "quantum-scattering" effects are observed in "microscopic" interactions such as in photon-atoms collisions (i.e., stimulated-emission, black-body spectrum etc.) or in hadron-hadron scattering [14].

3) According to our theory, the (B-E or F-D) "quantum" partition laws are no longer formally derived by making use of the indistinguishability hypothesis, as generally done[4], but rather by assuming "aleatory" (viz. stochastic) weight-factors W_i in the Bernoulli partition law. This implies an "aleatory" character of the corresponding interactions which is provided, quite naturally, by the uncertainties affecting the "microscopic" scattering events according to the Heisenberg Principle. This is in agreement with the above proposition regarding the physical condition in which quantum statistical effects manifest themselves. Note that in the case of scattering events taking place in a "Measurement-Apparatus" the aleatory character disappears, W_i are to be taken as constants and the M-B "classical" partition law is obtained, as shown in Sect.1[15]. As an example, let us derive here the (B-E) statistics for a beam of particles split over two scattering channels as, for instance, in a two-way IF. By the method of Sec.1 we obtain the beam partition probability P_{nl} by integrating the corresponding Bernoulli expression over the range of existence of the aleatory weight-variables W_i with the condition: $W_1+W_2=1$. The partition law is given by:

$$P_{nl}=(n!/(l!(n-l)!)^* \int_0^1 W_1{}^l \, dW_1 \int_0^1 W_2{}^{(n-l)} \, dW_2 \cdot \delta(1-W_1-W_2).$$

The double integral is then reduced to the following: $I= \int_0^1 W_1{}^l \, (1-W_1)^{(n-l)} \, dW_1 =$

$\Gamma(l+1)\cdot\Gamma(n-l+1)/\Gamma(n+2) = l!(n-l)!/(n+1)!$, through the definition of the Gamma-Function, $\Gamma(x)$[16]. The final result is $P_{nl}=1/(n+1)$, the two-way B-E partition law. It can be shown that the method is general and accounts for any number of scattering channels. Furthermore, if the formal conditions expressing the Pauli Principle are introduced,the correct F-D partition law is obtained[17].

4) SUMMARY AND CONCLUSIONS

Our theory is fully self-consistent. Its overall consequences are the following: There *are not* "quantum" and "classical" particles in physics. As far as distinguishability is concerned, all particles are the same. Their quantum-statistical behaviour is determined *uniquely* by the relevance taken in their motion by the dynamical implications of the

Heisenberg Principle (and of Pauli Principle). In other words, these two Principles are the *only* basic theoretical assumptions of quantum statistics. It can also be shown that in the context of QST the formal implications of these Principles consist precisely of the wavefunction-symmetrization (or anti-symmetrization) method. The formal laws of quantum theory and their range of application are left unchanged by our solution. As a conclusion, we believe that in a theoretical framework cleared from old, unnecessary and misleading assumptions of vague epistemological significance, quantum theory can reveal itself in a more clear fashion as a most powerful and complete theory of all natural phenomena. We acknowledge stimulating and enlightening discussions with E.D. Commins, P.W.Milonni and C.H.Townes.

5) REFERENCES

1) R.P Feynman, R.B.Leighton and M.Sands, *The Feynman Lectures on Physics*, Vol.3, Ch.4, Addison Wesley, Reading, 1965; S.Prasad, M.O.Scully and W.Martienssen, Opt. Comm. 62, 139, 1987; Z.Y.Ou, C.K.Hong and L.Mandel, Opt.Comm. 63, 118, 1987.
2) R.Loudon, *The Quantum Theory of Light*, Clarendon Press, Oxford, 1983, Ch.6.
3) D.F.Walls, Am.J.Phys. 45, 952, 1977.
4) R.C.Tolman, *The Principles of Statistical Mechanics*, Oxford U.Press, 1967, Ch.X; K.Huang, *Statistical Mechanics*, Wiley, N.Y.1963, Ch.9.
5) M.Born and E.Wolf, *Principles of Optics*, Macmillan, N.Y.1964, Ch.7, 8; L.Mandel, Opt.Soc.Am.51, 797, 1961, reports light-degeneracy parameters for various types of incoherent sources. Typically, $\bar{n}$=0.005 for a strong high-pressure Hg-arc.
6) W.Martienssen and E.Spiller, Am.J.Phys, 32, 919, 1964.
7) F.De Martini, Di Fonzo, P.Mataloni and K.H.Strobl, submitted for publication.
8) The two statements are found, respectively in: A.Einstein, Sitzungsberichte, Preussische Akademie der Wissenschaften: 1924, pag.261; and 1925, pag.18. Also in: A.Pais, *"Subtle is the Lord.."The Science and Life of Albert Einstein*, Oxford University Press, 1982, pag.430; and in Revs.Mod.Phys. 51, 4, 1979, pag.896. P.A.M.Dirac, Proc.R.Soc. A 112, 661, 1926, first related (B-E) statistics to symmetrization of particle microstates (A.Pais, Rev.Mod.Phys. op.cit. pag.894). Note that the formulation of the (B-E) partition law which is generally used has been introduced by A.Einstein, Sitzungsber.1925, op.cit. pag.3.
9) T.S.Kuhn, *The Structure of Scientific Revolutions*. The Chicago U. Press, 1962.
10) The origin of the "intrinsic indistinguishability" concept belongs to the pre-quantal era as it may be traced down to the work by L.Boltzmann, "Weiter studien uber die Warmegleichgewicht unter Gasmolekulen", Wiener Ber., II, 66, 1872. The first formulation of a partition law that agrees with Planck's law is actually due to M.Planck: "Uber eine Verbesserung der Wien'schen Spektralgleichung" in Verh. d.D. Phys.Ges, 2, 202, 1900; and Ann.d. Phys.4, 553, 1901. According to T.S.Kuhn (op.cit. below), the origin of the first formal derivation of the partition law could be identified as a combinatorial theory ("Combinationslehre") that Planck could have found in: E.Netto, "Combinatorik" in: *"Encyclopadie der Mathematischen Wissenschaften*, Vol.1, Part 1, Leipzig, 1898 or in: L.Boltzmann, "Uber die Beziehung zwischen dem Zweiten Hauptsatze der mechanischen Warmetheorie und der Warhrscheinlichkeitsrechnung respektive den Satzen uber das Warmegleichgewicht", Wiener Ber., II, 76, 373, 1877. A full historical account of the origin of Planck's theory is found in: T.S.Kuhn, *Black-Body Theory and Quantum*

Discontinuity, Oxford, 1978, (Ch.4, text and note 20); and in works by M.J.Klein and M.J.Kangro, quoted therein.

11) Consider the *first* part of the following aphorism by Ludwig Wittgenstein: "We cannot think what we cannot think; so what we cannot think we cannot say either". (L.Wittgenstein, *Tractatus Logicus Philosophicus*, 5.61. The Humanities Press, New Jersey, 1974). In our case, if the hypothesis of intrinsic particle indistinguishability is adopted, this hypotesis *cannot* be violated in *any* step of the development of a valid formal theory. It appears evident to us that, if a *correct* theory (like the wavefunction-symmetrization theory) cannot be formulated without making recourse to a side assumption that violates an hypothesis, that hypothesis itself is false. Then, in our case the "indistinguishability hypothesis" should be rejected, on logical grounds. Although this way of reasoning may appear somewhat artificial, it is nevertheless similar to the one used by N.Bohr in his debate with Einstein on the EPR argument. Cfr. N.Bohr, in: P.A.Schlipp ed., *Albert Einstein Scientist Philosopher*, Cambridge University Press, 1982.

12) N.Bohr: "Can quantum-mechanical description of physical reality be considered complete?", Nature, 136, 65, 1935 and Phys.Rev. 48, 696, 1935, reprinted in: J.A. Wheeler and W.Zurek (eds.), *Quantum Theory and Measurement*, Princeton University Press, 1983. In these papers, which open up the field of "Quantum Measurement Theory" and reply to the argument of Einstein, Podolsky and Rosen (EPR), Bohr anticipates the detection of a "Visibility loss" if a light and mobile screen is adopted in an IF. This effect, that according to Heisenberg Principle counterbalances the gain of information on the paths taken by photons within IF, is precisely the one shown by our B-E curve $V(\bar{n})$, in Fig.2.

13) Highly-degenerate beams of particles may be obtained, in practice, only with photons and phonons. In general, other Bose-particles, like π-mesons, are not created at a sufficient rate in the actual high-energy experiments.

14) The validity of this proposition is being tested experimentally in our Laboratory by reproducing "microscopic" photon-momentum transfer effects in a macroscopic IF, under controlled conditions. Preliminary results validate our proposition. (B-E) Hanbury Brown-Twiss effects in hadron-hadron scattering have been investigated by: G.Goldhaber, W.B.Fowler, S.Goldhaber, T.F.Hoang, T.E. Kalogropoulos and W.R.Powell, Phys.Rev.Lett. 3, 181, 1959; G.Goldhaber, W.Lee and A.Pais, Phys.Rev. 120, 300, 1960; A.Giovannini, G.C.Mantovani and S.P.Ratti: "Old and New Variables, Old and New Optical Concepts in High-Energy Hadron-Hadron Scattering", Rivista del Nuovo Cimento, 2, 1, 1979.

15) Note that our present theory *is not* an "Hidden-Variable" theory, nor implies the exotic assumptions of "Stochastic-Electrodynamics" or the ones of the Bohm's "Quantum-Potential Theory". (Cfr. T.Marshall and E.Santos in: *New Techniques and Ideas in Quantum Measurement Theory*, ed.by D.Greenberger, The N.Y. Academy of Sciences, N.Y. 1986; D.Bohm, Phys.Rev.85, 166, 1952).

16) H.B.Dwight, *Tables of Integrals*, MacMillan, N.Y., 1961, pg.209.

17) J.Tersoff and D.Bayer, Phys.Rev.Lett. 50, 553, 1983.

GIANT QUANTUM OSCILLATORS FROM RYDBERG ATOMS: ATOMIC COHERENT STATES AND THEIR SQUEEZING FROM RYDBERG ATOMS

R.K. Bullough

Department of Mathematics
UMIST, P.O. Box 88
Manchester M60 1QD, U.K.

and

G.S. Agarwal[1], B.M. Garraway[2], S.S. Hassan[3]
G.P. Hildred[2], S.V. Lawande[4], N. Nayak[1], R.R. Puri[5]
B.V. Thompson[2], J. Timonen[6] and M.R.B. Wahiddin[7]

[1] UMIST and University of Hyderabad, [2] UMIST
[3] UMIST and Ain Shams University, Egypt
[4] BARC Bombay [5] BARC, Bombay and UMIST
[6] UMIST and University of Jyväskylä, Finland
[7] UMIST and Malaya University

1. INTRODUCTION

This paper summarises work since about 1979 by all the authors indicated: RKB is given prominence *only* because he bears the responsibility for the present paper. All the work has proved relevant to Rydberg atoms. Here we lay particular stress on recent results for squeezing by Rydberg atoms.

A typical Rydberg atom consists of a Hydrogen-like atom (with a single "optical" electron) excited to a level with a high principal quantum number n. Examples are Na in the $30S_{1/2}$ state[1] and ^{85}Rb in the $63P_{1/2}$ state.[2] These atoms can make microwave transitions to adjacent Rydberg states e.g. $30S_{1/2} \rightarrow 30P_{1/2}$ (λ = 2·22 mm.) for Na or $63P_{3/2} \rightarrow 61D_{3/2}$ (λ = 13.8 mm.) for ^{85}Rb. Systems of one or many Rydberg atoms, typically in microwave cavities, give us opportunity for creating and studying novel microwave fields. Examples for N = 1 ^{85}Rb atoms were described by Walther[3] and Meystre.[4] The cavity temperatures were $\lesssim 2^{o}K$ [3] or zero [4], the Q's were large ($\gtrsim 10^{9}$) and maser action took place. It is then important to know the effects of a few black-body photons (2 or 3). These small photon numbers mean that despite the long wave-lengths involved the quantum character of the fields can still be evident. (In the experiments[3] the maser field may contain about 50 photons and this field may be sub-Poissonian[4]). In practice it is then difficult to measure directly such weak field or scattered field intensities. In this connection the paper by Yurke[5] on the use of Josephson parametric amplifiers suggests we shall now be able to detect individual microwave photons contributing to the fluorescence spectrum from single ^{85}Rb atoms. We have already calculated these spectra for *arbitrary* $Q \lesssim 5 \times 10^6$ [6,7] and we are now able to reach the higher Q values $Q \lesssim 10^{10}$. We show a few results in this connection later in this paper.

Of equal interest are the collective behaviours of $N > 1$ Rydberg atoms in low-Q cavities.[1,8–12] Typically the atoms enter the cavity in a beam and everything is inferred (including the state of the field, compare Refs. 1-4) from the numbers N_+ of atoms leaving the cavity in the upper Rydberg state and the number N_- leaving in the lower state. Implicitly resonance on two-state atoms is involved (or there is a simple degeneracy[1]) and $N_+ + N_- = N$, the total number of atoms typically inside the cavity. Because of the long wavelengths compared to atomic sizes (despite the large radii (~ 3600 a_0 for ^{85}Rb [3]) of Rydberg atoms) the Dicke model of N 2-level atoms on the same physical site proves to be a good theoretical model (cf. e.g. Refs. 8,9 and references). This model can be driven by the vacuum[13], by coherent, chaotic or mixed coherent-chaotic light[14–23], and indeed by squeezed light.[24] For an unsqueezed vacuum the model is essentially an exactly solvable model — which means that everything, spectrum, photon statistics, squeezing, for example, can be calculated *and compared with experiment.* Experiments on one, several, and many atoms have been successfully carried out for the N atoms in single mode black-body fields in low-Q cavities.[1,10–12] We shall see in §2 how the Dicke model reduces to a giant quantum oscillator in this context: it has apparently been seen.[1] We shall then give new results for giant quantum oscillators made from N atoms making 2-photon transitions in low-Q cavities and oscillators making k-photon transitions. These results are all open to future experiments and the technology is apparently available.[25,26]

For the present meeting the squeezing in coherent fields is probably of greatest interest. We note that the Dicke model of N 2-level atoms on the same site, driven by an applied *coherent* harmonic field is an N+1 dimensional representation of the SU(2) group. Its Hamiltonian H is a generator of an SU(2) (= atomic) coherent state and for $N < \infty$ this coherent state is squeezed.[27,53] Unfortunately, physics requires we couple the Dicke model in a cavity to a heat bath, while in the (unsqueezed) vacuum the vacuum modes play a comparable role. A major theme of this paper is to investigate the generation of SU(2) coherent states by driving N atoms coherently and to investigate the squeezing associated with this. The situation is complicated by a phase transition[7,14,15,22], and there is much still to understand about the squeezing off-resonance.

By focussing on the SU(2) group rather than the usual SU(1,1) group[28] we gain some new insights into squeezing. We refer to squeezing in the SU(1,1) group in §§5 and 6.

2. GIANT QUANTUM OSCILLATORS OF THE INCOHERENTLY DRIVEN DICKE MODEL: BOSON REPRESENTATIONS OF THE DICKE MODELS AND THEIR EXPERIMENTAL OBSERVATION

The fundamental master equation for the total density operator ρ for N 2-level atoms, each with resonant frequency ω_0, coupled to a single mode of the radiation field of frequency ω, itself coupled to a heat bath at temperature $\beta^{-1} > 0$, is (for example Refs. 6-9,23).

$$\frac{d\rho}{dt} = -i[H,\rho] - (\bar{n}+1)\kappa(a^\dagger a\rho - 2a\rho a^\dagger + \rho a^\dagger a) - \bar{n}\kappa(aa^\dagger\rho - 2a^\dagger\rho a + \rho aa^\dagger) \quad (1)$$

and

$$H = \omega a^{\dagger}a + \omega_0 S^z + g(a^{\dagger}S^- + aS^+) \tag{2}$$

in rotating wave approximation. The operators $S^z, S^{\pm}$ are collective Dicke operators $S^z = \sum_{i=1}^{N} S^z_i$, $S^{\pm} = \sum_{i=1}^{N} S^{\pm}_i$ satisfying angular momentum commutation relations

$$[S^{\pm}, S^z] = \mp S^{\pm} \ , \ [S^+, S^-] = 2S^z \ ; \tag{3}$$

$a, a^{\dagger}$ are the usual single mode field operators, $[a, a^{\dagger}] = 1$. Temperature enters through $\bar{n} \equiv (e^{\beta\omega} - 1)^{-1}$, and Q through $\kappa = \tfrac{1}{2}\omega Q^{-1}$.

To get a representation of SU(2) we would like to eliminate the field operators $a, a^{\dagger}$. This can be done under the low-Q condition $Ng^2\kappa^{-2} \ll 1$.[7-9,23] The density operator ρ in (1) is replaced by the *reduced* operator ρ_a for the atoms alone and for exact resonance $\Delta\omega \equiv \omega_0 - \omega = 0$ the master equation becomes[7-9,23]

$$\begin{aligned} \frac{d\rho_a}{dt} = -i[H_0, \rho_a] &- (\bar{n} + 1)g^2\kappa^{-1}(S^+S^-\rho_a - 2S^-\rho_a S^+ + \rho_a S^+S^-) \\ &- \bar{n}g^2\kappa^{-1}(S^-S^+\rho_a - 2S^+\rho_a S^- + \rho_a S^-S^+) \ . \end{aligned} \tag{4}$$

For the purposes of the present section the important point is that the Hamiltonian H_0 in (4) is simply $H_0 \equiv \omega_0 S^z$. This has interesting quantum consequences.

To avoid any possibility of confusion we note that there is no external coherent driving field: this will be introduced in §3. The master equation (1) shows that the atoms are driven *incoherently* (by the black-body radiation in the single mode).

The Dicke states $|S,m\rangle$, $S = \tfrac{1}{2}N$, $-S \leqslant m \leqslant S$, are simultaneous eigenstates of total spin S and the Hamiltonian H_0 : H_0 has eigenvalues $m\omega_0$. It is well known (e.g. Ref. 27) that there is a boson description of the Dicke operators in which (Primakoff-Holstein transformation)

$$S^- = (2S - \hat{n})^{\frac{1}{2}} b \ ; \ S^+ = b^{\dagger}(2S - \hat{n})^{\frac{1}{2}} = (2S + 1 - \hat{n})^{\frac{1}{2}} b^{\dagger} \ . \tag{5}$$

Then $[S^+, S^-] = 2S^z$ with $S^z = \hat{n} - S$: $\hat{n} \equiv b^{\dagger}b$ and $[b, b^{\dagger}] = 1$. The functions of $\hat{n}$ are read in an eigenvalue or power series sense. Likewise $[S^{\pm}, S^z] = \mp S^z$. If we define $\bar{S}^{\pm}, \bar{S}^z$ by $(2S)^{-\frac{1}{2}} S^{\pm}, (2S)^{-1} S^z$ then

$$[\bar{S}^+, \bar{S}^-] = 2\bar{S}^z = (\hat{n}S^{-1} - 1) \rightarrow -1 \tag{6}$$

as S (i.e. $\tfrac{1}{2}N$) $\rightarrow \infty$. Thus the $\bar{S}^{\pm}$ are bose operators for $S = \tfrac{1}{2}N \rightarrow \infty$. With no external coherent driving field in a low-Q cavity and $N < \infty$, $H = H_0 \equiv \omega_0 S^z$: the eigenstates are the Dicke States $|S,m\rangle$, $-S \leqslant m \leqslant S$. In boson representation these are the states

$$|S,m\rangle = ((m+S)!)^{-\frac{1}{2}} (b^{\dagger})^{m+S} |0\rangle \ , \ -S \leqslant m \leqslant S \ . \tag{7}$$

These span a basis for a $(2S+1) = (N+1)$ dimensional representation of SU(2).

The Hamiltonian H_0 is still coupled to black-body radiation. In equilibrium it is intuitive that the situation is that described in Fig. 1 (taken from Ref. 9). For $N \rightarrow \infty$ the atomic system is a gigantic quantum oscillator in equilibrium with the radiation and the number of atoms N_+ in the upper state will have asymptote $\bar{n}$ as $N \rightarrow \infty$. But for small N, $N_+ \sim N\bar{n}/$

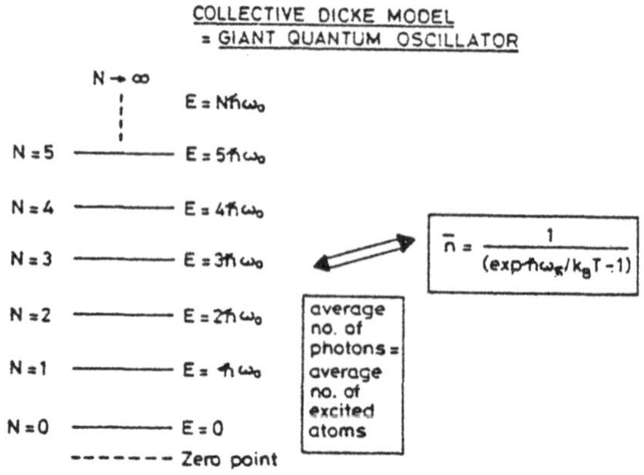

Fig. 1. Illustrating how, as $N \to \infty$, N 2-level atoms can form a macroscopic quantum oscillator of frequency ω_0, the atomic frequency.

$(1+2\bar{n})$, the Einstein result, and the curve will then rise monotonically with N to $\bar{n}$. The Fig. 2 shows experimental results for Na atoms making $30S_{1/2} \to 30P_{1/2}$ transitions. The transition has a degeneracy of two[1] so the expected asymptote is $N_+ = 2\bar{n}$: $\beta^{-1} = T = 300°K$ and $900°K$ so $\bar{n}$ is large (47 and 137). In practice a Dicke factor N on the rate constant is needed to ensure that the atoms (which enter the cavity in $30S_{1/2}$) reach equilibrium within the cavity transit time of 2.5 μ sec. Consequently the small values of N_+ observed for small N describe failure to reach equilibrium rather than the quantum mechanics implied in Fig. 1. However, we have shown[8,9] that there are realisable choices of cavity Q and transit time which permit examination of the equilibrium curve of N_+ against N for $N < \infty$ and $N_+ < \bar{n}$.

To find an analytical expression for this curve one can either solve (4) in equilibrium ($d\rho_a/dt = 0$) by going to a Fokker-Planck equation in atomic coherent state description[21], or one can note that a particular function of S^z solves this equation and evaluate the partition function $Z = \mathrm{Tr}\exp(-\beta H_0)$. One finds either way[21,29] that

$$N_+ = \sum_{j=1}^{N} jX^j \Big/ \sum_{j=0}^{N} X^j = \left[\frac{1-X}{1-X^{N+1}}\right] \sum_{j=1}^{N} jX^j \tag{8}$$

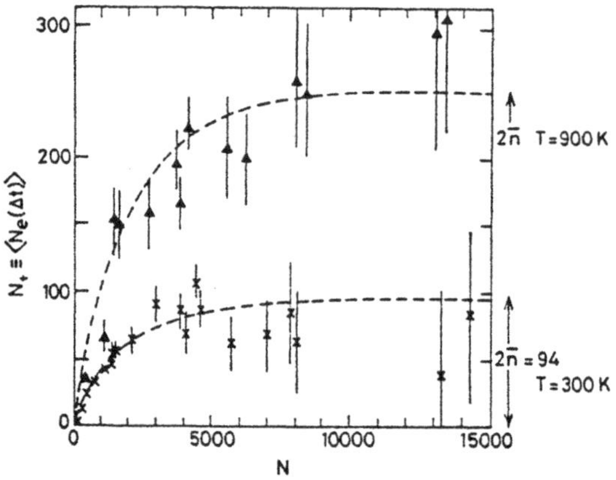

Fig. 2. Plot of the number of atoms N_+ leaving the cavity after 2.5 μ sec. as observed in Ref. 1.

where

$$X \equiv \bar{n}/(1+\bar{n}) = e^{-\beta\omega_0} \quad . \tag{9}$$

One then has that[8,9,21,29]

$$N_+ = X[1 - (N+1)X^N + NX^{N+1}]/[1 - X - X^{N+1} + X^{N+2}] \quad . \tag{10}$$

Thus, for $N = 1$, $N_+ = X/(1+X) = \bar{n}\,[1+2\bar{n}]^{-1}$, the Einstein result, and for $N \to \infty$, $N_+ = X/(1-X) = \bar{n}$. This is a real quantum result which describes a macroscopic quantum oscillator (bosons) for $N \to \infty$ and a single fermion-like atom for $N = 1$. For $1 < N < \infty$ there is mixed fermi-bose behaviour.

The agreement with experiment which Fig. 2 indicates shows that the Dicke model is a good model for N Rydberg atoms coupled to a single mode of black-body radiation in low-Q cavities. Further support for the model comes from solving (4) in Dicke basis for the dynamical evolution of microwave super-radiance into the cavity. In the experiments[10] the Na atoms enter the cavity in the $29S_{1/2}$ state and make $29S_{1/2} \to 28P_{1/2}$ transitions (λ = 1·85 mm.). Evidently $\langle S,m|\rho_a(t)|S,m\rangle \equiv P_m(t)$ is the probability of finding $N_+ = \frac{1}{2}N+m$ atoms in the upper state at time t (or $N_- = \frac{1}{2}N-m$ in the lower state). From (4) (which has $\Delta\omega = 0$) transferred to a frame rotating at ω_0 this probability satisfies[8,9]

$$\frac{dP_m(\bar{t})}{d\bar{t}} = (1+\bar{n})\nu_{m+1}P_{m+1}(\bar{t}) + \bar{n}\nu_m P_{m-1}(\bar{t}) - [(1+\bar{n})\nu_m + \bar{n}\nu_{m+1}]P_m(\bar{t}) \tag{11}$$

where $\nu_m \equiv (\frac{1}{2}N+m)(\frac{1}{2}N-m+1)$; $\bar{t} \equiv g^2\kappa^{-1}t$. We show only the Fig. 3 in this paper: the close agreement with the experiments[10] is discussed in Refs. 8,9.

We now give wholly new results by extending the argument to Dicke-like atomic systems making 2-photon, and then k-photon, transitions. For

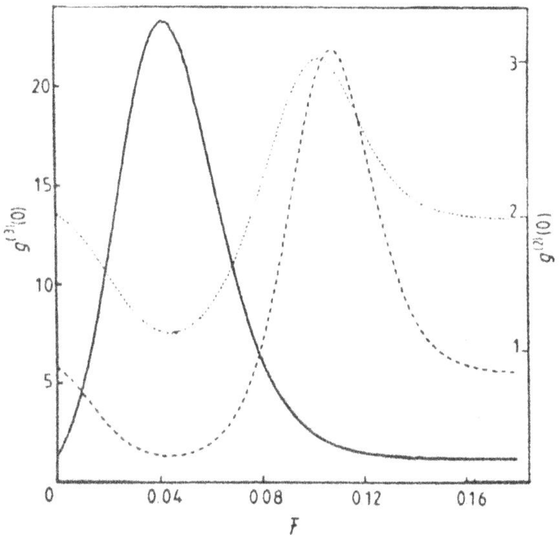

Fig. 3. Plot (from Ref. 8) of the coherence functions $g^{(2)}(0)$ and $g^{(3)}(0)$ (two dotted curves) as well as the transient intensity $G^{(1)}(0)$ as functions of $\bar{t}$. The dotted curve for $g^{(2)}(0)$ has the asymptote 2.

$k = 2$ (2-photon transitions) one can show that for three level atoms with ground, intermediate, and excited states $|g\rangle, |i\rangle, |e\rangle$ spaced by frequencies $\omega_0 - \Delta(|g\rangle \rightarrow |i\rangle)$ and $\omega_0 + \Delta(|i\rangle \rightarrow |e\rangle)$ with coupling constants g_1 and g_2 respectively, then, for $|\Delta|$ much greater than the 1-photon Rabi frequencies, there is an effective 2-photon Hamiltonian

$$H_{ef} = 2\omega_0 S^z + \{\omega_0 + NG_1\}a^\dagger a + G_2 a^\dagger a S^z + g\{a^{\dagger 2}S^- + S^+ a^2\} \qquad (12)$$

where $S^z = \sum_{i=1}^{N} S_i^z$, $S^\pm = \sum_{i=1}^{N} S_i^\pm$, $S_i^z = \frac{1}{2}(|e\rangle\langle e| - |g\rangle\langle g|)$, $S_i^+ = |e\rangle\langle g|$, $S_i^- = |g\rangle\langle e|$: $G_1 \equiv (g_1^2 + g_2^2)/2\Delta$, $G_2 \equiv (g_1^2 + g_2^2)/\Delta$ and $g \equiv g_1 g_2/\Delta$.

The terms in G_1 and G_2 are 'Stark shift' terms. In the 40S – 39P, 39P → 39S transitions of ^{85}Rb (for example) g_1 and g_2 differ by 2% [25,26]. Thus $g_2 \approx g_1$ so $G_2 \approx 0$. In (12) the cavity mode is "resonant" at ω_0 and is shifted by $NG_1 \sim N \times 10^3$ Hz. compared with $\omega_0 \sim 2\pi \times (3{\cdot}4 \times 10^{10}$ Hz.). If we assume the situation general for selected k-photon transitions the effective interaction Hamiltonian which neglects Stark shift terms is

$$g\{(a^\dagger)^k S^- + S^+(a)^k\} \quad . \qquad (13)$$

Under the low-Q condition $Ng^2\kappa^{-2} << 1$ one can then show that the master equation for the reduced atomic density operator ρ_a in the laboratory frame is

$$\frac{d\rho_a}{dt} = -i[H_0,\rho_a] - g^2\kappa^{-1}(\bar{n}+1)^k(k-1)!(S^+S^-\rho_a - 2S^-\rho_a S^+ + \rho_a S^+S^-) - g^2\kappa^{-1}(\bar{n})^k(k-1)!(S^-S^+\rho_a - 2S^+\rho_a S^- + \rho_a S^-S^+) \qquad (14)$$

where $H_0 = k\omega_0 S^z$. It is now plain that in the steady state the argument for N_+ in the 1-photon case goes over: the formula (10) still applies and the only change is $X = [\bar{n}/(1+\bar{n})]^k$. Thus for $N = 1$

$$N_+ = X/(1+X) = (\bar{n})^k/[(1+\bar{n})^k + \bar{n}^k] \sim \tfrac{1}{2} \qquad (15a)$$

for large enough $\bar{n}$, the k-photon generalisation of the Einstein result. But for $N \rightarrow \infty$

$$N_+ \sim X/(1-X) \sim \bar{n}/k \quad . \qquad (15b)$$

The curve of N_+ at equilibrium plotted against N is therefore a monotonically rising curve with about the same initial slope as in the 1-photon case but with the asymptote at large N reduced by a factor k^{-1}. An experiment to observe this quantum result, which assumes $g_1 \approx g_2$, seems well worth doing.

We should really show that the Primakoff-Holstein transformation exists in the k-photon case. However, if we set $S^z = [\![\hat{n}/k]\!] - S$ where $[\![x]\!]$ is interpreted as the largest integer less than x then[27]

$$S^+ = \{(2S+1) - [\![\hat{n}/k]\!]\}f_k(\hat{n})(b^\dagger)^k$$

$$f_k(\hat{n}) = \{[\![\hat{n}/k]\!](\hat{n}-k)!/(\hat{n})!\}^{\frac{1}{2}}$$

$$S^- = \{2S - [\![\hat{n}/k]\!]\}^{\frac{1}{2}}(b)^k f_k(\hat{n}) \quad . \qquad (16)$$

There are bose operators

$$B^\dagger = f_k(\hat{n})(b^\dagger)^k \quad , \quad B = (b)^k f_k(\hat{n}) \qquad (17a)$$

such that

$$[B,B^{\dagger}] = 1 \quad . \tag{17b}$$

As before $(\sqrt{2})^{-1} S^{+} \to B^{\dagger}$, $(\sqrt{2}S)^{-1} S^{-} \to B$ as $S \to \infty$. The k-photon system is therefore a gigantic k-photon quantum oscillator as $N \to \infty$. The eigenstates $|Sk,mk\rangle$ of H_0, with eigenvalues $mk\omega_0$, $-S \leqslant m \leqslant S$, form a basis for an N+1 dimensional representation of the group SU(2).

It is also plain from (14) that there is an analogue of (11) in which $(1+\bar{n}) \to (1+\bar{n})^k(k-1)!$ and $\bar{n} \to (\bar{n})^k(k-1)!$ We shall given an analysis of the statistics of this k-photon super-radiance and results comparing with Fig. 3 elsewhere.

3. THE COHERENTLY DRIVEN DICKE MODEL: EXPERIMENTAL GENERATION OF SU(2) COHERENT STATES AND THEIR SQUEEZING

Since the 1-photon Dicke model with $S = \tfrac{1}{2}N$ provides an N+1 dimensional representation of the group SU(2) it could be used as a generator of SU(2) coherent states: these are squeezed for $N < \infty$.[27,53] Unfortunately this ideal situation assumes no heat bath. The point of this section is to discover how far SU(2) coherent states are still generated in the presence of the real physics of the heat bath.

In principle we extend (4) so that

$$H_0 = \omega_0 S^z + \Omega\, S^{+}\, e^{-i\omega t} + \Omega\, S^{-}\, e^{i\omega t} \quad ; \tag{18}$$

we have added an external harmonic driving field of frequency ω and amplitude expressed by a Rabi frequency Ω. If, first of all, we ignore the heat bath (the damping terms in $g^2\kappa^{-1}$ in (4)) we can work with the Schrödinger equation with Hamiltonian (18). It is convenient to make a canonical transformation to a frame rotating at ω : operators $\hat{O} \to e^{i\omega S^z t}\, \hat{O}\, e^{-i\omega S^z t}$. We are then interested in

$$i \frac{\partial}{\partial t} \left[e^{i\omega S^z t}|\Psi(t)\rangle\right] = H_R\left[e^{i\omega S^z t}|\Psi(t)\rangle\right] \tag{19a}$$

$$H_R = \Delta\omega S^z + \Omega(S^{+} + S^{-}) \;;\; \Delta\omega \equiv \omega_0 - \omega \quad . \tag{19b}$$

The Dicke ground state is the state $|S, -S\rangle$. We are therefore interested in

$$|\mu(t),S\rangle = e^{-iH_R t}|S, -S\rangle \tag{20}$$

where $\mu(t)$ is to be found. Evidently the SU(2) group elements $e^{-iH_R t}$, acting on the Dicke ground state, generate time dependent SU(2) coherent states.

Application of non-canonical parametrisation theory for SU(2)[30] (Baker-Campbell-Hausdorf theory) shows that

$$\begin{aligned} |\mu(t),S\rangle &= \exp(\mu(t)S^{+})\, \exp(\beta(t)S^{z})\, \exp(-\mu^{*}(t)S^{-})|S,-S\rangle \\ &= \mathcal{N}^{-1} \exp(\mu(t)S^{+})\; |S, -S\rangle \quad . \end{aligned} \tag{21}$$

This shows $|\mu(t),S\rangle$ is an SU(2) coherent state (cf. eg. Ref. 31) while we find

$$\mu(t) = \frac{-i\tilde{\omega}\Omega \tan\tilde{\omega}t - \tfrac{1}{2}\Omega\Delta\omega \tan^2\tilde{\omega}t}{\tilde{\omega}^2 + \tfrac{1}{4}\Delta\omega^2 \tan^2\tilde{\omega}t} \;;\; \tilde{\omega} = \sqrt{\tfrac{1}{4}(\Delta\omega)^2 + \Omega^2} \quad . \tag{22}$$

The normalisation is

$$|\mathcal{N}| = (1 + |\mu|^2)^S$$

$$\mathcal{N} = |\mathcal{N}| e^{i\Phi} \quad ; \quad \Phi = S \tan^{-1} \left[\frac{\Delta\omega\bar{\Omega} \tan \bar{\Omega} t}{\bar{\Omega}^2 - \tfrac{1}{4} \omega^2 \tan^2(\bar{\Omega} t)} \right] \quad . \tag{23}$$

The state is therefore *oscillating* at twice the Rabi frequency $\bar{\Omega}$. At the singularities of the tan function off-resonance (i.e. where $\Delta\omega \neq 0$) μ goes through the point $2\Omega/\Delta\omega$ each half period. The SU(2) coherent state is then the state $|2\Omega/\Delta\omega, S\rangle$. On resonance ($\Delta\omega = 0$) $\mu(t) = -i \tan \Omega t$ and $\Phi = 0$. The state is therefore annihilated at $t = \pi/2\Omega$ and every period π/Ω subsequently. This crucial difference between resonance $\Delta\omega = 0$, and off-resonace *might* just have some bearing on the 'real' situation where the heat bath is coupled in. The role of the heat bath is ultimately to eliminate the oscillation and drive the system to a steady state. Nevertheless, we show that on resonance what is essentially an SU(2) coherent state, with its squeezing, remains in the steady state below a certain phase transition of second order type. Off resonance the squeezing is significantly different from that of an SU(2) coherent state.

Before demonstrating this we note that Katriel et al.[27] use (7) to give the interesting boson form

$$|\mu, S\rangle = \mathcal{N}^{-1} \sum_{\ell=0}^{2S} \left[\frac{\mu^\ell}{\ell!}\right] \left[\prod_{p=0}^{\ell-1} \{2S+1 - (\hat{n}-p)\}^{1/2} \right] (b^\dagger)^\ell |0\rangle \tag{24}$$

for the coherent state $\mathcal{N}^{-1} e^{\mu S^+} |S,-S\rangle$. They compute the squeezing as squeezing in the pseudo-fields $x = (\sqrt{2})^{-1}(b^\dagger + b)$, $p = i(\sqrt{2})^{-1}(b^\dagger - b)$:

$$(\Delta x)^2 = \tfrac{1}{2} + \langle b^+ b\rangle - \langle b^+\rangle\langle b\rangle + \mathrm{Re}\,[\langle (b^+)^2\rangle - \langle b^+\rangle^2]$$

$$(\Delta p)^2 = \tfrac{1}{2} + \langle b^+ b\rangle - \langle b^+\rangle\langle b\rangle - \mathrm{Re}\,[\langle b^+)^2\rangle - \langle b^+\rangle^2] \tag{25}$$

from which[27]

$$(\Delta x)^2 \sim \tfrac{1}{2} - \rho^2/4S \quad , \quad (\Delta p)^2 \sim \tfrac{1}{2} + \rho^2/4S \tag{26}$$

where, since $\langle \mu, S| b^\dagger b |\mu, S\rangle = 2S|\mu|^2/(1+|\mu|^2)$, $\rho^2 = 2S|\mu|^2$ and is held fixed as $S \to \infty$. Evidently large ρ means substantial squeezing for $S < \infty$. Other points are that the state $|\mu, S\rangle$ is a minimum uncertainty state upto $O(S^{-1})$: there is no squeezing for $S \to \infty$ (where the state is an α-(Glauber)-coherent state): and for finite S (i.e. finite N) the *squeezing decreases with N.*

So far we have attempted to use the atomic system to generate its own SU(2) (= atomic) coherent states. If for the moment we leave aside the experimental problem of placing N atoms in an atomic coherent state with finite μ at $t = 0$ this state $|\mu, S\rangle$ will evolve according to the master equation (4) (which is in the laboratory frame and has the cavity mode chosen resonant with the atoms). The sources of the fluorescent field are $S^x = \tfrac{1}{2}(S^+ + S^-)$. Since $\langle S^z\rangle \nless 0$ the condition for squeezing in the fluorescent field is therefore

$$F_{x,y} \equiv (\Delta S^{x,y})^2 + \tfrac{1}{2}\langle S^z\rangle < 0 \tag{27a}$$

or $$F^{x,y} \equiv 2(\Delta S^{x,y})^2/|\langle S^z\rangle| < 1 \quad . \tag{27b}$$

As usual all of the squeezing in $(\Delta S^y)^2$ is due to terms $\langle S^+S^+\rangle$, $\langle S^-S^-\rangle$ and $\langle S^\pm\rangle^2$ in

$$(\Delta S^y)^2 = \tfrac{1}{4} (-\langle S^+S^+\rangle - \langle S^-S^-\rangle + \langle S^+S^-\rangle + \langle S^-S^+\rangle$$
$$+ \langle S^+\rangle^2 + \langle S^-\rangle^2 - 2 \langle S^+\rangle\langle S^-\rangle) \quad . \tag{28}$$

The signs of these quantities change in $(\Delta S^x)^2$ which may also be squeezed. We have solved (11), moved to the laboratory frame, numerically with the initial atomic condition $|\mu,S\rangle$.[32] The Fig. 4a shows evolution of $(\Delta S^y)^2$ oscillating at $2\omega_0$ in the laboratory frame about the evolution of $\frac{1}{2}|\langle S^z\rangle|$: N = 17 atoms and $\bar{n} = 10^{-7}$. Perhaps the most striking, and most depressing, result is that shown in Fig. 4b. Squeezing is almost eliminated except for very short times even though $\bar{n}$ is still only $\bar{n} = 0{\cdot}1$.

We return to the main theme of this section and drive the Dicke model with a coherent driving field. In a frame rotating at ω the Hamiltonian is H_R, eqn.(19b). For a strong coherent driving field it may be adequate to ignore black-body photons. Anyway we shall take $\bar{n} = 0$ ($\beta^{-1} = T = 0$) and we shall use γ_0, half the A-coefficient, instead of $g^2\kappa^{-1}$. In practice this means either we are in the low-Q cavity at zero temperature with the cavity mode at frequency ω, not ω_0,[9] or we are in free space i.e. there is no cavity. It is well known that, without coherent driving field, the vacuum modes alone are equivalent to a single resonant mode in a low-Q cavity with γ_0 replacing $g^2\kappa^{-1} \equiv \eta_{cav}\ \gamma_0$: η_{cav} is the Purcell factor.[8,9,23,33]

The master equation in the frame rotating at ω is now

$$\frac{d\rho_a}{dt} = -i[H_R,\rho_a] - \gamma_0[S^+S^-\rho_a - 2S^-\rho_a S^+ + \rho_a S^+S^-] \quad . \tag{29}$$

For exact resonance, $\Delta\omega = 0$, Puri and Lawande[14,15] solved this equation exactly in the steady state by using the $|\mu,S\rangle$ as basis. They found the Fokker-Planck equation equivalent to (29) and then found the steady state result

$$\rho_{ss} = N_f \sum_{m,n=0}^{2S} (\tilde{g}^*)^{2S-m}(\tilde{g})^{2S-n}(S^-)^m(S^+)^n \tag{30a}$$

where $\tilde{g} = i\Omega\gamma_0^{-1}$ and

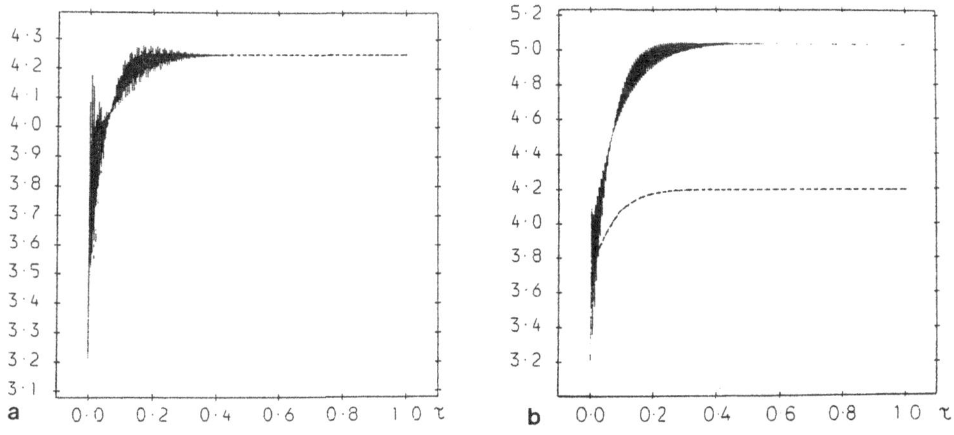

Fig.4(a) Plot from (Ref.32) of the fluctuation $(\Delta S^y)^2$ and inversion $\frac{1}{2}|\langle S^z\rangle|$ for N = 17 atoms as a function of $\tau \equiv 2g^2\kappa^{-1}(1+\bar{n})t$. The atoms are in the state $|\mu,S\rangle$ at t = 0 ($\mu = \frac{1}{2}e^{i\phi}$ and $\theta = \pi/6$; $\phi = \pi/2$) and super-radiate into a cavity containing black-body photons with $\bar{n} = 10^{-7}$. The inversion is the dotted line and $(\Delta S^y)^2$ fluctuates about it.

Fig.4(b) As (a) but now $\bar{n} = 0.1$.

$$N_f^{-1} = \sum_{p=0}^{2S} \sum_{k=0}^{p} \frac{(2S-p+k)!}{(p-k)!(2S-p)!} |\tilde{g}|^{2(2S-k)} \quad . \tag{30b}$$

For weak driving fields $|\tilde{g}| \to 0$

$$\rho_{ss} \sim (S^-)^{2S} (S^+)^{2S}/(2S!)^2 \quad ; \tag{31a}$$

for $|\tilde{g}| \to \infty$

$$\rho_{ss} \sim \hat{I}\ (2S+1)^{-1} \tag{31b}$$

($\hat{I}$ is the unit operator). It is then possible to calculate the nth order intensity-intensity correlation function[16]

$$G^{(n)}(0) \equiv \langle (S^+)^n (S^-)^n \rangle = \mathrm{Tr}\ ((S^+)^n (S^-)^n \rho_{ss})$$

$$= N_f \sum_{p=n}^{2S} \sum_{k=0}^{p-n} \frac{(2S-p+k+n)!p!}{(p-k-n)!(2S-p)!} |\tilde{g}|^{2(2S-k)} \quad . \tag{32}$$

In particular, for $N = 1$ (one atom resonance fluorescence)

$$G^{(1)}(0) = |\tilde{g}|^2/(1+2|\tilde{g}|^2)$$

$$G^{(2)}(0) = 0 \quad , \quad g^{(2)}(0) \equiv G^{(2)}(0)/|G^{(1)}(0)|^2 = 0 \tag{33a}$$

and for $N = 2$

$$G^{(1)}(0) = 4|\tilde{g}|^2(1+|\tilde{g}|^2)/(3|\tilde{g}|^4 + 4|\tilde{g}|^2 + 4)$$

$$G^{(2)}(0) = 4|\tilde{g}|^4/(3|\tilde{g}|^4 + 4|\tilde{g}|^2 + 4)$$

$$g^{(2)}(0) = (3|\tilde{g}|^4 + 4|\tilde{g}|^2 + 4)/(4(1+|\tilde{g}|^2)^2) \quad . \tag{33b}$$

A most interesting result, well worth experimental investigation as now seems feasible, is that there is a phase transition of second order type at a critical value of $\Omega\ N^{-1}$. Define $\theta = 2|\tilde{g}|N^{-1} = 2\Omega/\gamma_0 N$. Then the normalised fluorescence *intensity* $G^{(1)}(0)/(\tfrac{1}{2}N)^2$ plotted against θ develops a cusp at $\theta = 1$ as $N,\Omega \to \infty$ and this is actually well pronounced for the realisable values $N \gtrsim 100$ (say). The Fig. 5 (taken from Ref. 16) shows how $g^{(2)}(0)$ (defined in (33a)) varies with θ : for $N \to \infty$, $g^{(2)}(0) = 1$ for $\theta < 1$ and the system is in a coherent α-state. (We prove $g^{(n)}(0) = 1$ for $\theta < 1$ for all *fixed* n). This is the limiting case of an SU(2) atomic coherent state but ($N \to \infty$) there is no squeezing. However we show below there *is* squeezing for $N < \infty$ for suitably small θ and that this is essentially the squeezing in an SU(2) coherent state: for $\theta > 1$ and $N < \infty$ this squeezing completely disappears (see the Fig. 7 below where F^y is plotted against θ). Returning to the Fig. 5, at $\theta = 1$ (and $\Omega,N \to \infty$) $g^{(2)}(0)$ rises rapidly from unity to $g^{(2)}(0) = 1{\cdot}2$. Indeed one can show[16] that

$$\lim_{|\tilde{g}|\to\infty} g^{(2)}(0) = \frac{6}{5}\,\frac{(N-1)(N+3)}{N(N+2)} \longrightarrow 1{\cdot}2 \text{ as } N \to \infty \quad . \tag{34}$$

This compares with the substantial *anti*-bunching, $g^{(2)}(0) < 1$, for $1 \leqslant N \lesssim 10$ and the value $g^{(2)}(0) = 2$ for normal bunching: the value 6/5 is therefore *partial coherence of the fluorescent field* and the phase transition at $\theta = 1$ is characterised by this — as well as by the other features described below.

By defining[34]

$$F(N,\theta) = (2/N\theta)^{2N}\, D \ ; \ \theta = 2|g|N^{-1} = 2\Omega/\gamma_0 N$$

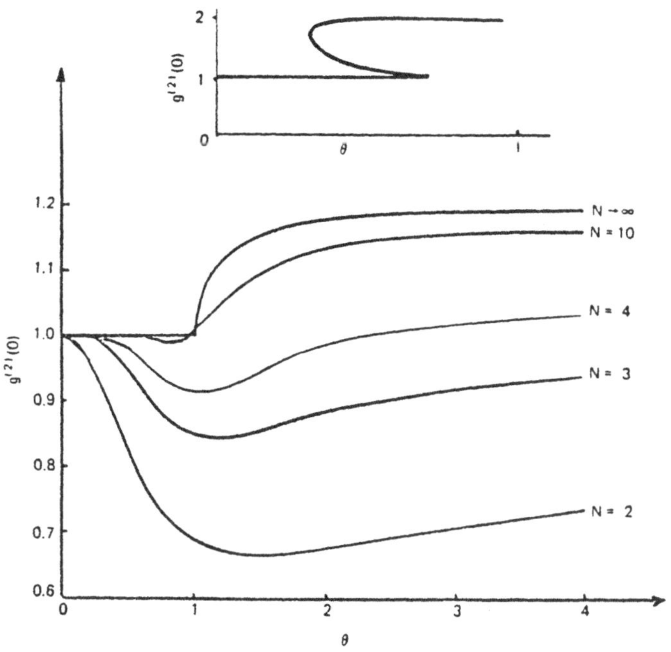

Fig. 5. Plots of $g^{(2)}(0)$ (from Ref. 16) as a function of $\theta = 2\Omega\gamma_0^{-1} N^{-1}$ for different N. Note the antibunching ($g^{(2)}(0) < 1$) for $N \lesssim 10$ and the phase transition at $\theta = 1$. Inset is $g^{(2)}(0)$ in a semiclassical approximation.

$$D = \sum_{n=0}^{N} |g|^{2(N-n)} \frac{(N+n+1)!(n!)^2}{(N-n)!(2n+1)!} \tag{35}$$

(in which $|g| \equiv \frac{1}{2}N\theta$) it is possible to show[34] that in the "thermodynamic limit" Ω, $N \to \infty$

$$\langle S^z\rangle N^{-1} = (\theta/4N)(\partial(\ell n\, F(N,\theta)/\partial\theta)$$

$$= -\tfrac{1}{2}\sqrt{1-\theta^2}\ ,\quad \theta < 1\ ;\quad = 0,\ \theta > 1 \tag{36a}$$

$$\langle S^y\rangle N^{-1} = \tfrac{1}{2}\theta\ ,\ \theta < 1\ ;\ = \tfrac{1}{2}\theta[1 - \sqrt{\theta^2-1}/(\theta^2 \sin^{-1}(\theta^{-1}))]\ ,\ \theta > 1 \tag{36b}$$

while $\langle S^x\rangle N^{-1} = 0$, for all θ. Thus both $\langle S^z\rangle N^{-1}$ and $\langle S^y\rangle N^{-1}$ develop cusps at $\theta = 1$, and it is plain that the phase transition there is of second order type. The Fig. 6 plots $\langle S^y\rangle N^{-1}$ against θ for different N and $N \to \infty$. We conclude that by placing the atoms in a low-Q cavity and monitoring (by ionisation detectors)[1-4] the numbers $N_\pm$ of atoms in their upper (lower) states in the atomic beam leaving the cavity it should be possible to observe this phase transition in $\langle S^z\rangle N^{-1}$ with $N \sim 100$ Rydberg atoms as the incident field amplitude Ω is increased (input of coherent power ~ μ W cm.$^{-2}$ [9]). This conclusion assumes finite temperatures do not change it: we have few results so far at finite β^{-1}.

One could also expect to check the fluctuations in S^z: in thermodynamic limit[34]

$$N^{-2}(\Delta S^z)^2 = 0\ ,\ \theta < 1\ ;\ = \tfrac{1}{4} - \frac{\theta^2}{4}\ (1 - \sqrt{\theta^2-1}/(\theta^2 \sin^{-1}(\theta^{-1}))],\ \theta > 1. \tag{37}$$

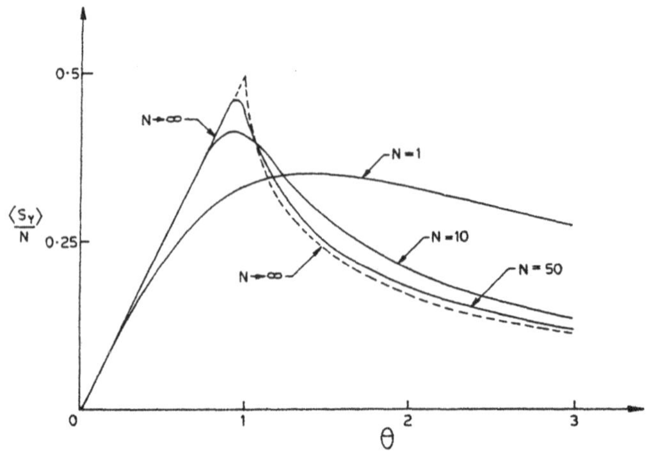

Fig. 6. Plot of $\langle S^y\rangle N^{-1}$ against θ for different N.

Thus the quantum fluctuations $N^{-2}(\Delta S^z)^2$ disappear for $\theta < 1$ consistent with the view that the system is in a coherent α-state: the quantum fluctuations $N^{-2}(\Delta S^{x,y})$ also both disappear for $\theta < 1$ in thermodynamic limit. The phase transition at $\theta = 1$ is thus also characterised by the onset of quantum fluctuations $(\Delta S^{x,y,z})^2 N^{-2}$.

There is no squeezing for $\theta \lessgtr 1$ and for $\theta > 1$ in thermodynamic limit at exact resonance. There is squeezing for finite $N < \infty$ for $\theta \lesssim 1$ but not for $\theta \gtrsim 1$. Moreover, for $\theta \lesssim 1$ *the squeezing decreases with N* as the Fig. 7 shows. Indeed in the region $\theta \lesssim 1$ with $N < \infty$ the system is essentially in an SU(2) coherent state as the following analysis of the squeezing indicates. For finite N and Ω one finds[34] that

$$(\Delta S^x)^2 N^{-2} = -\tfrac{1}{2}\langle S^z\rangle N^{-2} + (N+1)(N+2)/12NF(N,\theta)$$

$$(\Delta S^y) N^{-2} = -\tfrac{1}{2}\langle S^z\rangle N^{-2} - (N+1)(N+2)/12NF(N,\theta) + (\tfrac{1}{2}\theta\langle S^y\rangle/N - \langle S^y\rangle^2 N^{-2}) \qquad (38)$$

Since $\langle S^y\rangle > 0$ (Fig. 6) $(\Delta S^y)^2$ is squeezed only for sufficiently small θ:$(\Delta S^x)^2$ is never squeezed. The squeezing reduces with N as for the SU(2) coherent state (eqn.(26)). However, the state is not quite minimum uncertainty to $O(N^{-1})$ (compare (26)) because of the extra term in $(\Delta S^y)^2$. But the Fig. 6 shows that for small enough θ for each finite N this term is negligible. The state is therefore very close to minimum uncertainty and close to an SU(2) coherent state. We cannot easily quote the actual error.

The main message is that on resonance there is squeezing in $(\Delta S^y)^2$ for $\theta \lesssim 1$. The largest squeezing is for $N = 1$ where $F^y \lesssim 0{\cdot}85$ (Fig.7). For $\theta \gtrsim 1$ there is no squeezing (Fig.7). Moreoever when $\theta \to \infty$ (with $N < \infty$)

$$(\Delta S^x)^2 \to \tfrac{1}{2}\langle S^+S^-\rangle - \tfrac{1}{2}\langle S^z\rangle \to \frac{1}{12} N(N+2)$$

$$(\Delta S^y)^2 \to \tfrac{1}{2}\langle S^+S^-\rangle - \tfrac{1}{2}\langle S^z\rangle - \langle S^y\rangle^2 \to \frac{1}{12} N(N+2) \qquad (39)$$

while $\tfrac{1}{2}|\langle S^z\rangle| \to 0$. This is not minimum uncertainty and there is no squeezing in either component. We also find $F^x \sim \tfrac{1}{2} N^2\theta^2$, $F^y \sim \tfrac{1}{2}N^2\theta^2$, both greater than one.

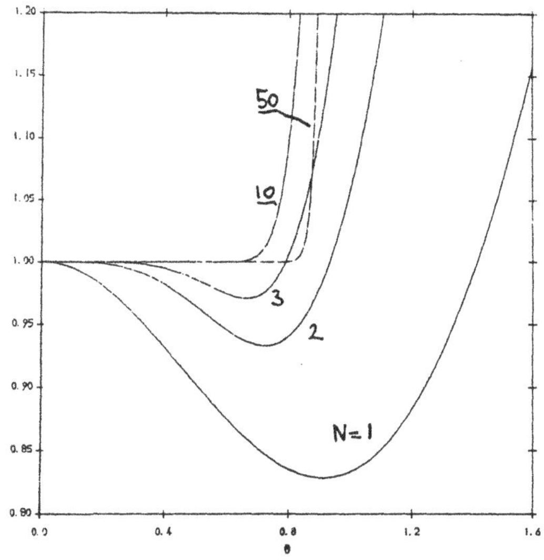

Fig. 7. Squeezing in on-resonant resonance fluorescence, $F^y = 2(\Delta S^y)^2/|\langle S^z \rangle|$ against θ.

4. THE COHERENTLY DRIVEN DICKE MODEL: SQUEEZING IN OFF-RESONANT RESONANCE FLUORESCENCE

Although we noted some difference on and off resonance of the coherent state $|\mu(t),S\rangle$, eqn.(21), there is a remarkable change in the squeezing off-resonance from the coherently driven Dicke model. The Fig. 8 shows that squeezing in the off-resonant resonance fluorescence *increases* with N reaching $F^y = 0{\cdot}75$ for N = 40 and the chosen detuning. The interesting region is apparently $\theta = 2\Omega/\gamma_0 N \sim 1$, but the actual phase

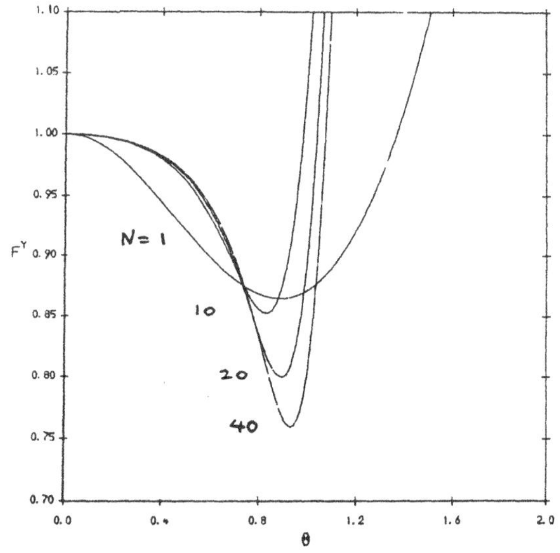

Fig. 8. Squeezing in $F^y(F^y < 1)$ in off-resonant resonance fluorescence: $\Delta\omega = ½(0{\cdot}524 \times N\gamma_0)$. Squeezing increases with N.

transition at $\theta = 1$ when $\Omega, N \to \infty$, on resonance, disappears off resonance[18-20]. We can say no more about the $\theta = 1$ region at this stage.

To investigate the squeezing off-resonance in its entirety we should bear in mind that the phase transition at $\theta = 1$ on resonance is a quantum form of a bistable phase transition[8,9,35]: this is clear from the inset to Fig. 5, while we find there is a natural 'transmitted' field which switches on only for $\theta \gtrsim 1$. Moreover[8,9,35] a Dicke model of $N \sim 50$ Rydberg atoms, frequencies ω_a, in a low-Q cavity at zero temperature, coupled to a single cavity mode, frequency ω_c, and driven by a coherent external field of frequency ω_f displays at least one *first order* phase transition for suitable detunings $\delta_a \equiv \omega_a - \omega_f$, $\delta_c \equiv \omega_c - \omega_f$: for $\delta_a = \delta_c = 0$ the model is equivalent to (29) on resonance, so it has the second order transition at $\theta = 1$ there (on resonance θ takes the form $2\kappa E_0/gN = 2\Omega/gN$ for the cavity model[9,35] instead of $\theta = 2\Omega/\gamma_0 N$ for the vacuum used so far). It follows that off-resonance the squeezing will exhibit all of the complications described by Lugiato[36] in his paper on squeezing in optical bistability.

An analysis of off-resonant squeezing in resonance fluorescence in the region $\theta > 1$, which proves to be squeezing in $(\Delta S^x)^2$ (rather than in $(\Delta S^y)^2$ for $\theta < 1$ shown in Fig. 8) can be developed in the following way. With Δ written for $\Delta\omega = \omega_0 - \omega$ and $\Delta \neq 0$ we use the solution of (29) in a frame rotating at ω [18,19]

$$\rho_{ss} = D^{-1} \sum_{m,n=0}^{N} a_{mn} \, (-i\Omega/\gamma_0)^{-m} (i\Omega/\gamma_0)^{-n} (S^-)^m (S^+)^n \tag{40a}$$

in which (compare eqns. (30))

$$a_{mn} = \frac{\Gamma(m - i\Delta\gamma_0^{-1}+1)\Gamma(n+i\Delta\gamma_0^{-1}+1)}{m!n!\Gamma(1+i\Delta\gamma_0^{-1})\Gamma(1-i\Delta\gamma_0^{-1})}$$

$$D = \sum_{m=0}^{N} |\Omega\gamma_0^{-1}|^{-2m} \frac{(N+m+1)!\Gamma(m-i\Delta\gamma_0^{-1}+1)\Gamma(m+i\Delta\gamma_0^{-1}+1).}{(N-m)!(2m+1)!\Gamma(1+i\Delta\gamma_0^{-1})\Gamma(1-i\Delta\gamma_0^{-1})} \tag{40b}$$

The expectation values $\langle (S^+)^p (S^z)^r (S^-)^q \rangle$ are given by[18,19]

$$\begin{aligned}\langle (S^+)^p (S^z)^r (S^-)^q \rangle &= \mathrm{Tr}[\rho_{ss} (S^+)^p (S^z)^r (S^-)^q] \\ &= D^{-1} \sum_{n=\max(p,q)}^{N} (-i\Omega\gamma_0^{-1})^{-n+q} (i\Omega\gamma_0^{-1})^{-n+p} \\ &\quad \times a_{n-p,n-q} \sum_{m=0}^{N-n} \frac{(m+n)!(N-m)!(\tfrac{1}{2}N-n-m)^r}{(N-m-n)!m!} .\end{aligned} \tag{40c}$$

From this both $(\Delta S^x)^2$ and $(\Delta S^y)^2$ can be calculated and some results for squeezing, now in $(\Delta S^x)^2$, found from the numerical evaluations of eqns.(40) are given in Fig. 9: F_x is given by (27a) and squeezing has $F_x < 0$: the quantity $\delta = \omega - \omega_0 = -\Delta$, and from (29) the results are in the frame rotating at ω — the relevant frame for observation of squeezing by homodyning with a local oscillator.[9,32] In contrast with the resonance results this squeezing in $(\Delta S^x)^2$ increases with N and there is no squeezing in $(\Delta S^y)^2$.

For further understanding it is helpful to develop analytic (if approximate) methods: H_R is diagonalised by the unitary transformation

$$\hat{O} \to \tilde{\hat{O}} \equiv \exp(i\Phi S^y)\hat{O} \exp(-i\Phi S^y) \tag{41}$$

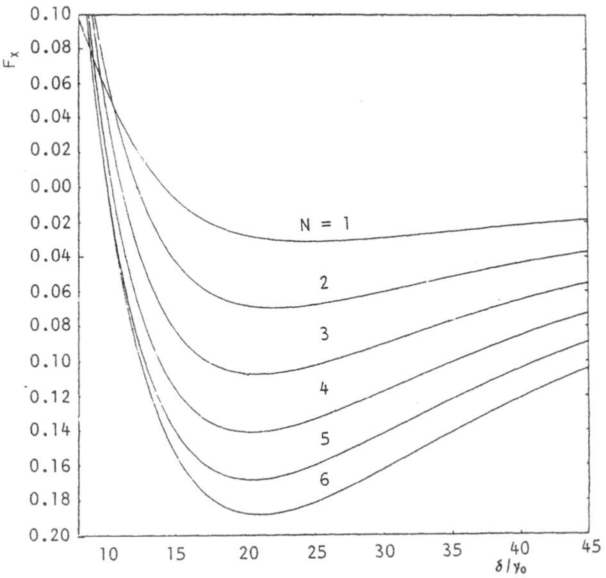

Fig. 9. The function F_x based on an exact solution of the master equation (29) for off resonant resonance fluorescence. Squeezing has $F_x < 0$. The normalised Rabi frequency $2\Omega/\gamma_0 = 20$ and $\theta \gtrsim 3$: $\delta = \omega - \omega_0 = -\Delta$.

for any operator $\hat{0}$ provided that

$$\tan\Phi = 2\Omega\Delta^{-1} \tag{42}$$

One finds[32] $\tilde{S} = S^z\cos\Phi - S^x\sin\Phi$, $\tilde{S}^x = S^x\cos\Phi + S^z\sin\Phi$, $\tilde{S}^y = S^y$ and $[\tilde{H}_R,\tilde{\rho}] = 2\,\tilde{\Omega}[S^z,\tilde{\rho}]$ where $\tilde{\Omega}^2 = \frac{1}{4}\Delta^2 + \Omega^2$ (as in §2). The master equation for $\tilde{\rho}$ involves diagonal terms like $S^z\,\tilde{\rho}\,S^z$ and off-diagonal terms like $S^z\,\tilde{\rho}\,S^+$. A secular approximation is made which drops these off-diagonal terms and the master equation is then

$$\frac{d\tilde{\rho}}{dt} = -\,2i\tilde{\Omega}[S^z,\tilde{\rho}] + \gamma_0\,\{2\sin^2\Phi S^z\tilde{\rho}S^z - S(S+1)(1+\cos^2\Phi)\tilde{\rho} +$$
$$+ \tfrac{1}{2}(\cos\Phi-1)^2\,S^+\tilde{\rho}S^- + \tfrac{1}{2}(\cos\Phi+1)^2\,S^-\tilde{\rho}S^+$$
$$- [\cos\Phi\,S^z + \tfrac{1}{2}(1-3\cos^2\Phi)(S^z)^2,\tilde{\rho}]_+\} \quad . \tag{43}$$

($[\ ,\ ]_+$ is anti-commutator). The secular approximation is good if $\tilde{\Omega}$ (ie. Ω) is large enough: the error is $O(\gamma_0 S/\tilde{\Omega})$.

The steady-state solution $\tilde{\rho}_{ss}$ of (43) proves to be $\tilde{\rho}_{ss} = N_0^{-1}C^{S^z}$ with $N_0 = \sum_{j=-S}^{S} C^j = \sinh((S+\frac{1}{2})\ln C)/\sinh(\frac{1}{2}\ln C)$ so that $\mathrm{Tr}\tilde{\rho}_{ss} = 1$. The constant C is

$$C = [(1-\cos\Phi)/(1+\cos\Phi)]^2 \tag{44}$$

with Φ given by (42). We find[32]

$$F_x = \sin^2\Phi\{S(S+1) - [f_n + \tfrac{1}{8}\cos\Phi]^2 + [(2+3\cos^2\Phi)/8\cos\Phi]^2\}$$
$$F_y = [\sin^2\Phi/4\cos\Phi][f_n - (1+\cos^2\Phi)/(4\cos\Phi)] \quad . \tag{45}$$

Squeezing requires one of the $F_i < 0$ $(i = x,y)$. The number f_n is computed from

$$f_j = \tfrac{1}{2}j \frac{[(1+\cos\Phi)^{2j} + (1-\cos\Phi)^{2j}]}{[(1+\cos\Phi)^{2j} - (1-\cos\Phi)^{2j}]} \tag{46}$$

for $j = n = 2S+1 = N+1$. We find $\langle S^z\rangle = -\cos\Phi\,(f_n - f_1)$ and in approaching resonance $\cos\Phi \to 0$ and $f_n \to 1/(4\cos\Phi) + O(\cos\Phi)$. Thus $\langle S^z\rangle \to 0$ and Ω is such that we are firmly in the region $\theta > 1$.

Since F_y is never negative there is no squeezing in $(\Delta S^y)^2$ while F_x can be negative and there is squeezing in $(\Delta S^x)^2$ in agreement with Fig. 9: this contrasts with Fig. 8 where the squeezing is in $(\Delta S^y)^2$ off-resonance and $\theta < 1$. It also contrasts with the whole situation on-resonance (Fig. 7). For large enough N this approximate analysis shows that the value of F_x at its minimum (i.e. at maximum squeezing) is proportional to N and the position of this minimum moves towards resonance. However, at resonance itself $F_x = F_y = {}^1/_{12}\,N(N+2)$, > 0 in both cases and in agreement with (39) for large θ. Thus F_x becomes very sensitive to detuning as N becomes larger. Curves based on the master equation (43) displaying these features are given in Fig. 10 where F^x (not F_x) is plotted as a function of $\delta = -\Delta$ for $N = 1, 2, \cdots, 8$. Squeezing requires $F^x < 1$.

We have checked these approximated results in several different ways[32] — especially both F_i were computed numerically from eqns. (40) and agreement is complete. Some other results based on (40) are shown in the Fig. 9. One can also calculate the bunching (anti-bunching): one finds from the steady solution of (43) that

$$G_1(0) = \sin^2\Phi\{S(S+1) + [(3\cos^2\Phi+1)/4\cos\Phi][(1+\cos^2\Phi/4\cos\Phi) - f_n]\}$$

$$G_2(0) = \sin^4\Phi\Big\{[S(S+1)]^2 + \frac{5}{16}\,\frac{S(S+1)}{\cos^2\Phi}\,[9\cos^4\Phi + 2\cos^2\Phi + 1]$$

$$+ \frac{3}{256}\cdot\frac{1}{\cos^4\Phi}\,[1+\cos^2\Phi][35\cos^6\Phi + 5\cos^4\Phi + 5\cos^2\Phi + 3]$$

$$- fn\left\{\frac{S(S+1)}{2\cos\Phi}\,[5\cos^2\Phi+1] + \frac{3}{64}\cdot\frac{1}{\cos^3\Phi}\,[35\cos^6\Phi+5\cos^4\Phi+5\cos^2\Phi+3]\right\}\Big\}. \tag{47}$$

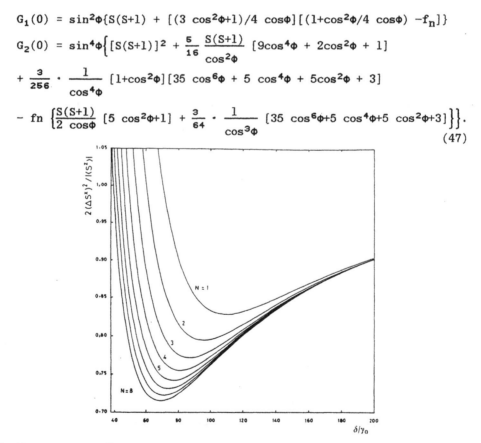

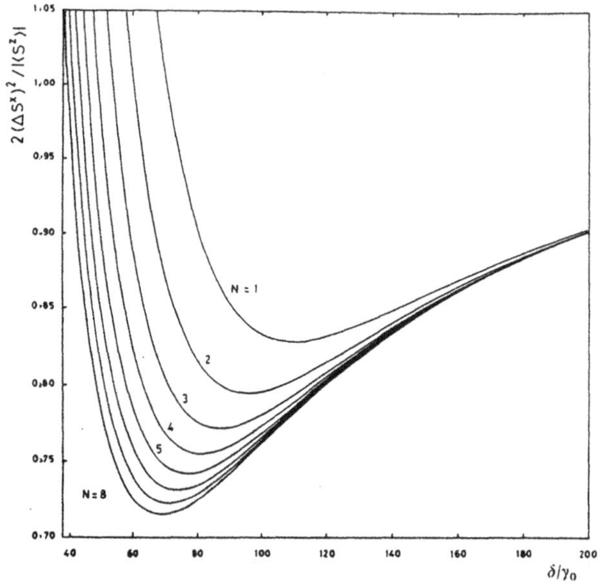

Fig. 10. Plot of F^x for $2\Omega/\gamma_0 = 100$. The secular solution for the density matrix is used. Squeezing has $0 \leqslant F^x < 1$; $\theta \gtrsim 12{\cdot}5$.

For $\Delta \to 0$ expansion of f_n in powers of $\cos^2\Phi$ yields $g^{(2)}(0) \to \frac{6}{5}[(N+1)(N+3)/N(N+2)]$ in agreement with (34), confirming again that we are in the partially coherent regime $\theta > 1$.

In this regime $\theta > 1$, and off resonance, the squeezing in $(\Delta S^x)^2$ can apparently be made as large as one likes simply by increasing N. However we find F^x (the *ratio* (27b)) has value 0·715 at minimum for N = 8, and as both numerator and denominator increase linearly with N for large enough N the minimum of $F^x \to$ finite as $N \to \infty$. This finite value is one half. We conclude that by working *off*-resonance in the regime where $\theta > 1$ an experiment on 50 - 100 Rydberg Na atoms will show squeezing in microwave resonance fluorescence comparable with the largest values so far observed (in parametric down conversion[37,38]). As noted the experiment may be perfomed in a low-Q microwave cavity[9] and, for example, the fluorescence emerging at right angles to the off-resonant cavity mode can apparently be detected by the set-up adopted for the Josephson parametric amplifier as described by Yurke[5]: thus a homodyne detection of the squeezing now seems entirely possible.

Summary of §§3 and 4

To sum up these two sections §§3 and 4 on the coherently driven Dicke model: despite the phase transition of second order type at $\theta \equiv 2\Omega/\gamma_0 N = 1$ on resonance when $\Omega, N \to \infty$, it is possible to find regions of $\theta < 1$ for $N < \infty$ where the steady state is in all essentials an SU(2) (i.e. atomic) coherent state: the squeezing is small ($F^y \sim 0{\cdot}83$ for N = 1 and $\theta = 0{\cdot}91$) and decreases with N as for the SU(2) coherent state. There is no squeezing for $\theta \gtrsim 1$ on resonance.

In contrast off resonance the squeezing increases with N apparently in both the $\theta \lessdot 1$ regime (Fig. 8) and the $\theta \gtrsim 1$ regimes (Figs. 9,10) and there is therefore no SU(2) coherent state for $\theta > 1$. For $\theta \lesssim 1$ it is still $(\Delta S^y)^2$ which is squeezed however ($F^y \sim 0{\cdot}76$ for N = 40 in Fig. 8) though any bound on this squeezing is still to be investigated. On the other hand for $\theta \gtrsim 1$ it is $(\Delta S^x)^2$ which is squeezed: this squeezing can be large though $F^x \gtrdot 0{\cdot}5$.

We have also investigated the squeezing in resonance fluorescence near the first order phase transition which occurs[9,35] at $\Delta_a \equiv 2\delta_a/N\Gamma = 1$, $\Delta_c = -\delta_c\kappa^{-1} = -5$ and $\theta \approx 3$ ($\Gamma \equiv \kappa g^2/(\kappa^2+\delta_c^2)$ [9,35]). There is no squeezing in $(\Delta S^y)^2$, but there is small squeezing in $(\Delta S^x)^2$ for $\theta \lesssim 3$: here F^x falls to about 0·96 for $\theta \sim 2{\cdot}63$ and N = 50. It is surprising (perhaps) that the squeezing is so small near this phase transition since the quantity $g^{(2)}(0)$ *for the cavity field* reaches quite large values[9,35]. However, we find the squeezing in the cavity field is also small ($F^y \gtrsim 0{\cdot}96$, no squeezing in F^x).

Evidently there is still much to learn about the squeezing displayed by the coherently driven Dicke model — particularly to understand *why* it takes the particular forms it does in the different regimes.

We note again that there is a rich source of rewarding experiments to be found in systems of N Rydberg atoms in low-Q microwave cavities driven coherently. Atomic coherent states can (apparently) be constructed on resonance; the squeezing in resonance fluorescence is complicated, but not necessarily small; and the phase transitions at $\theta = 1$ (on resonance) and $\theta \approx 3$ (off resonance) are very well worth investigating.

At this stage we do not know what other phase transitions or squeezings may occur.

Finally, we note that everything we have said for 1-photon coherent states $|\mu,S\rangle$ carries over to the k-photon coherent states of Ref. 27 by driving the k-photon master equation (14) by adding to H_0 the external field contribution $\Omega^{(k)}(e^{-ik\omega t}S^+ + e^{ik\omega t}S^-)$ where $\Omega^{(k)} = \Omega^k/D^{(k-1)}$ and $D^{(k-1)}$ has the dimensions of frequency to the power (k-1). In particular in a frame rotating at $k\omega$ the Hamiltonian is

$$H_R^{(k)} = k\Delta\omega\, S^z + \Omega^{(k)}S^+ + \Omega^{(k)}S^- \tag{48}$$

in which S^z, $S^\pm$ are the k-photon operators of §2. This Hamiltonian is a generator of the k-photon coherent states of Ref. 27, while reference to the k-photon master equation (14) shows that at zero temperature, where $\bar{n} = 0$, the theory of the coherently driven Dicke model developed for 1-photon transitions in §3 and this §4 goes over *en bloc* (with $\Omega \to \Omega^{(k)}$ and $\gamma_0 \to \gamma_0^{(k)}$, one half the k-photon A-coefficient). In particular the phase transition at $\theta = 1$, with $\theta = 2\Omega^{(k)}/\gamma_0^{(k)}N$ survives on resonance, while the regions $\theta < 1$ are essentially in k-photon SU_2 coherent states[27] $|\mu;k,S\rangle = \mathcal{N}^{-1}\sum_{\ell=0}^{2S}\mu^\ell\left\{\begin{bmatrix}2S\\ \ell\end{bmatrix}\right\}^{1/2}|kS,k(\ell-S)\rangle$. However, except that any experiment would be a k-photon experiment, there seems to be no essential difference btween the results in the k-photon case and the results in the 1-photon case. In particular both on and off resonance the squeezing (as computed in terms of the operator fluctuations $(\Delta S^x)^2$ and $(\Delta S^y)^2$ in §3 and this §4 for the 1-photon case) is unchanged. Katriel et al. compute k-photon coherent state squeezing in terms of the pseudo-field quantities x and p as is done in eqn. (25) in the 1-photon case. It does not seem to be possible to detect the k-photon structure — essentially one can observe the operators B, B^+ eqns. (17) but not the operators b, b^+ this way.

5. RYDBERG ATOMS IN A HEAT BATH WITH SQUEEZED VACUUM

This section will be very short and is designed only to point out that everything we have done so far can be done all over again for the incoherently driven or coherently driven Dick model in a *squeezed* vacuum. We have used our methods of reaction field theory[39-42] to derive the following master equation for the reduced density operator in a frame rotating at ω

$$\begin{aligned}\frac{d\rho_a}{dt} &= -i\Omega[S^+ + S^-,\rho_a] - i\Delta[S^z,\rho_a]\\ &-\gamma_0(1+\tilde{N}+\bar{n})(S^+S^-\rho_a - 2S^-\rho_aS^+ + \rho_aS^+S^-)\\ &-\gamma_0(\tilde{N}+\bar{n})(S^-S^+\rho_a - 2S^+\rho_aS^- + \rho_aS^-S^+)\\ &-\gamma_0|\tilde{M}|e^{-i\Phi}(2S^+\rho_aS^+ - S^+S^+\rho_a - \rho_aS^+S^+)\\ &-\gamma_0|\tilde{M}|e^{i\Phi}(2S^-\rho_aS^- - S^-S^-\rho_a - \rho_aS^-S^-)\end{aligned} \tag{49}$$

in which $S^\pm, S^z$ are the collective Dicke operators introduced by (3) and $\Delta \equiv \omega_0-\omega$. The parameters $\tilde{N}$, $\tilde{M} = |\tilde{M}|e^{i\Phi}$ are the parameters of the squeezing of the vacuum: each mode $(\underline{k},\lambda)$ of the vacuum is supposed in the state $|0,\varepsilon_{\underline{k},\lambda}\rangle$ obtained by the action of the SU(1,1) group operator $\exp(\tfrac{1}{2}\varepsilon^*_{\underline{k},\lambda}a^2_{\underline{k},\lambda} - \tfrac{1}{2}\varepsilon_{\underline{k},\lambda}a^{+2}_{\underline{k},\lambda})$ with $\varepsilon_{\underline{k},\lambda} = r_{\underline{k},\lambda}e^{i\Phi_{\underline{k},\lambda}}$ $(r_{\underline{k},\lambda}\in[0,\infty))$. Then the correlations are

$$\langle a^\dagger_{\underline{k},\lambda}a_{\underline{k}',\lambda'}\rangle = \delta_{\underline{kk}'}\delta_{\lambda\lambda'}\sinh^2 r_{\underline{k},\lambda} \equiv N_{\underline{k},\lambda}\delta_{\underline{kk}'}\delta_{\lambda\lambda'}$$

$$\langle a_{\underline{k},\lambda} a_{\underline{k}',\lambda'} \rangle = -\delta_{\underline{k}\underline{k}'}\delta_{\lambda\lambda'} \sinh r_{\underline{k},\lambda} \cosh r_{\underline{k},\lambda} e^{i\Phi_{\underline{k},\lambda}}$$

$$\equiv M_{\underline{k},\lambda}\delta_{\underline{k}\underline{k}'}\delta_{\lambda\lambda'} \quad . \tag{50}$$

Evidently $|M_{\underline{k},\lambda}|^2 = N_{\underline{k},\lambda}(N_{\underline{k},\lambda} + 1)$(and $\tilde{M},\tilde{N}$ are the resonant mode values). Thus every squeezed vacuum state is a minimum uncertainty state. Simple additivity of $\bar{n}$ and $\tilde{N}$ is achieved so far by imposing the black body radiation as a classical stochastic field in the fashion of Appendix 1 of Ref. 55. A less general form of the master equation (49) for N = 1 was independently reported before this by Walls and colleagues[44,45] (unfortunately for us!)

We have relatively few detailed results so far. However we can note immediately that for $\Omega = 0$ (driving by a *partly* incoherent squeezed vacuum only i.e. by the squeezed vacuum and the black-body field) the function of S^z is no longer a steady state solution of (49) so that (compare §2) there is no longer a simple giant quantum oscillator in this case. The squeezing parameter $\tilde{N}$ increases the effective mean number $\bar{n}$ of black-body photons in the resonant mode, but the effect of the phase information in the squeezing parameter $\tilde{M}$ on what otherwise would be a giant quantum oscillator with asymptote $\bar{n} + \tilde{N}$ for N_+ is still to be worked out.

For N = 1 we have carried out an analysis of Mandel's Q-parameter[46] and find interesting variations on the results already reported[47] for the unsqueezed vacuum at finite temperatures. These results are not suitable for a simple survey here.

It is easy to solve the set of Bloch equations equivalent to (49) for the polarisation and inversion in Heisenberg representation in the case N = 1. For the polarisation we find in the steady state ($t = \infty$) (and compare Refs. 44,45).

$$\langle S^x(t = \infty)\rangle = \frac{-\Omega(\Delta+2\gamma_0|\tilde{M}|\sin\Phi)}{(1+2N+2\bar{n})[\gamma_0^2(1+2N+2\bar{n})^2-4\gamma_0^2|\tilde{M}|^2+\Delta^2+\Omega_1^2} \tag{51a}$$

in which

$$\Omega_1^2 \equiv 4\Omega^2(\tfrac{1}{2} + (|\tilde{M}| \cos\Phi)/(1+2\tilde{N} + 2\bar{n})) \quad . \tag{51b}$$

Plainly this polarisation is $O(\bar{n}^{-3})$ for large enough $\bar{n}$ so the black-body radiation destroys the coherent polarisation in this limit.

We have calculated the spectrum on and off resonance for N = 1. On resonance we find the line narrowing reported in Refs. 44,45. We have also calculated the level shifts by reaction field theory. Results agree with Ref. 48. Details of both calculations will be given elsewhere.

It is apparently evident that there is a single mode low-Q cavity model with master equation (49) in which γ_0 is replaced by $g^2\kappa^{-1}$. Thus it will be interesting to see the effect of the squeezing term in $\tilde{M}$ on all of the results reported in §§2-4 of this paper. For such a single mode problem the squeezing lies in the single cavity mode and this situation would seem to be experimentally realisable more easily than that required for a vacuum squeezed over a sufficiently broad band (for example at the much shorter values $\lambda = 1{\cdot}06\mu$ and large squeezing of Kimbles's experiments[37,38] the band width ~ 10MHz. only).

We must report the results of these different investigations elsewhere.

6. OTHER ASPECTS OF RYDBERG ATOMS IN CAVITIES

We complete this paper with a few additional figures illustrating other properties of Rydberg atoms in cavities.

The Fig. 11 (taken from Ref. 6) shows the saturation of the spectral width of fluorescence from a single Rydberg (i.e. 2-level) atom coupled to a single resonant mode in a cavity of moderate Q ($g\kappa^{-1} = 0{\cdot}5$) at increasing temperature (increasing $\bar{n}$): $\Delta = \omega_0 - \omega = 0$. The straight line is the low-Q result $\gamma = \bar{\gamma} \equiv g^2\kappa^{-1}(1+2\bar{n})$. The lower curve with asymptote near unity is the actual value of γ. The phenomenon describes for incoherent driving fields the features reported by Lewenstein[49] for coherent driving fields.

The Fig. 12 (taken from Ref. 9) is the spectrum at $Q = \infty$ computed from the Jaynes-Cummings model[50] (single atom coupled to a single mode) on resonance ($\Delta = 0$). A small detector width ($\gamma = 0{\cdot}1$) is imposed. The field is in an α-state initially ($\bar{n} = |\alpha|^2 = 1{\cdot}0$). All frequencies are in units of g. The numbers refer to the Jaynes-Cummings eigenvalues. In general there are *four* transitions $\omega_{n+1,\pm} - \omega_{n,\pm} = \omega_0 \pm \sqrt{n+1}\,g \pm \sqrt{n}\,g$ on resonance.[9]

The Fig. 13 (taken from Ref. 9) is the fluorescence spectrum for realisable $Q < \infty$: $g\kappa^{-1} = 5{\cdot}0$, the black body field has $\bar{n} = 1{\cdot}0$ and successive detunings $\Delta\kappa^{-1} = 0{\cdot}5$, $1{\cdot}0$, $2{\cdot}0$ and $5{\cdot}0$ clearly show the J-C "1-photon" peak at $(\nu-\omega_0)\kappa^{-1} = 1{\cdot}9$. (This occurs at $\pm \sqrt{2g^2+\frac{1}{4}\Delta^2} \pm \sqrt{g^2+\frac{1}{4}\Delta^2} = \pm 3\sqrt{25} \pm \frac{5}{2}\sqrt{5} = \pm 1{\cdot}9$ or $\pm 13{\cdot}09$. The peaks at $\pm 13{\cdot}09$ and $-1{\cdot}9$ cannot be seen).

The Fig. 14 (taken from Ref. 51) plots $\langle S^z \rangle$ against gt for the single atom in a cavity at finite Q and temperature. The atom starts in its upper state at $t = 0$. Values are detuning $\Delta g^{-1} = 2$, $\kappa g^{-1} = 1$ (Q's ~ 10^6), $\bar{n}$ (black-body photons) = 5. There is also an initial coherent field with $|\alpha|^2 = 5$, but there is little evidence of revivals at this Q. The asymptote is $-\frac{1}{2}(1+2\bar{n})^{-1}$, so this is a non-trivial evolution from an initial non-equilibrium state at $t = 0$ to actual equilibrium (almost).

The Fig. 15 (taken from Ref. 51) increases the Q of Fig. 14 (decreases κg^{-1} to $0{\cdot}001$): $\bar{n} = 2$, $\Delta g^{-1} = 1$, $|\alpha|^2 = 2$. Revivals are now well pronounced and at close to the times expected for $Q = \infty$.

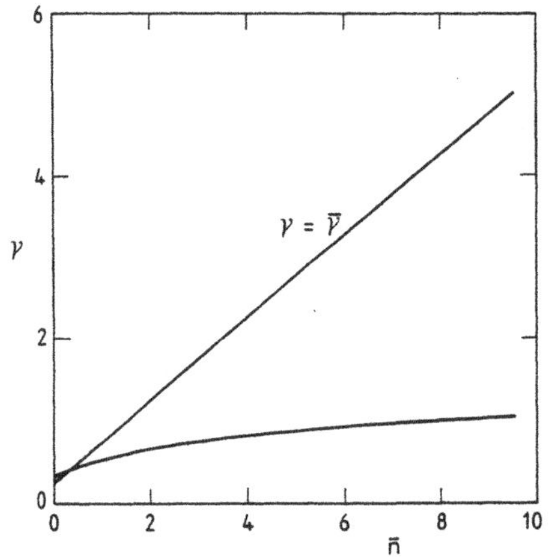

Fig. 11. Spectral width γ of resonance fluorescence for a single atom in a cavity at moderate Q ($g\kappa^{-1} = 0{\cdot}5$) with $\bar{n}$ varying. The straight line represents $\gamma = \bar{\gamma} \equiv g^2\kappa^{-1}(1+2\bar{n})$, the low Q result.

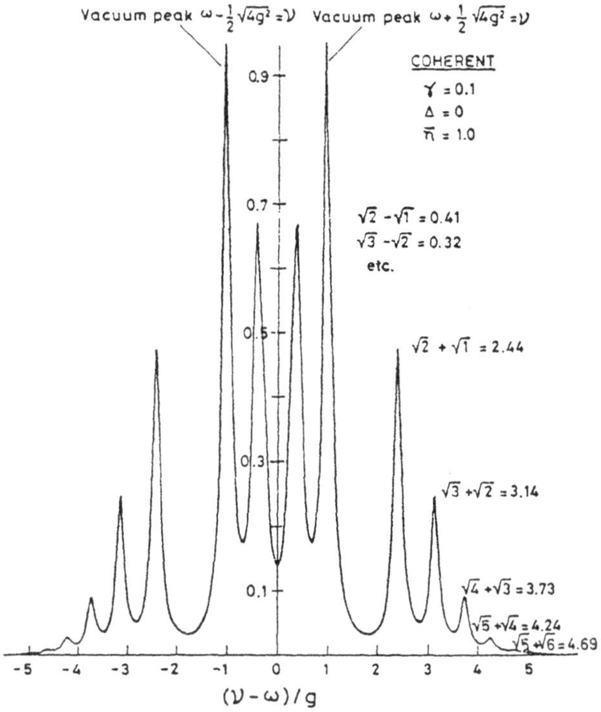

Fig. 12. The fluorescence spectrum of a single Rydberg atom coupled to a resonant mode in an ideal cavity.

The Fig. 16 (taken from Ref. 51) plots $\langle S^z \rangle + 0{\cdot}4$ as a function of gt with (below) the results for $Q = \infty$ with $\Delta g^{-1} = 5$ but otherwise as for Fig. 15. Amplitudes do not agree at gt $\gtrsim$ 10 but times do. However revival

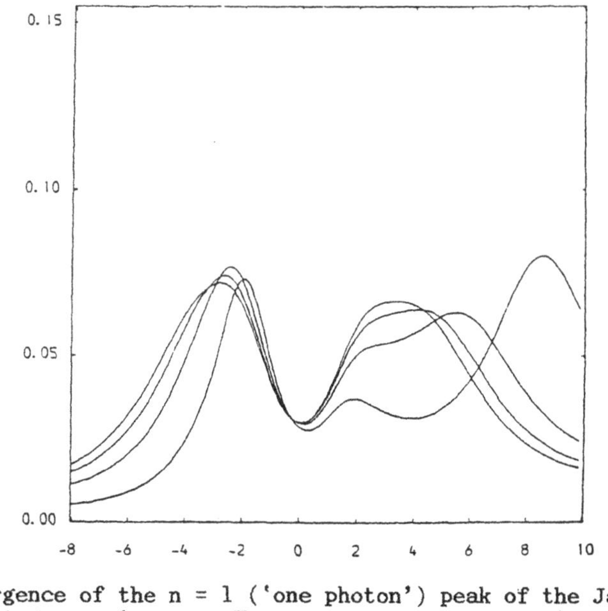

Fig. 13. Emergence of the n = 1 ('one photon') peak of the Jaynes–Cummings model for $g\kappa^{-1} = 5{\cdot}0$, $\bar{n} = 1{\cdot}0$ and detunings $\Delta g^{-1} = 0{\cdot}5$, $1{\cdot}0$, $2{\cdot}0$ and $5{\cdot}0$.

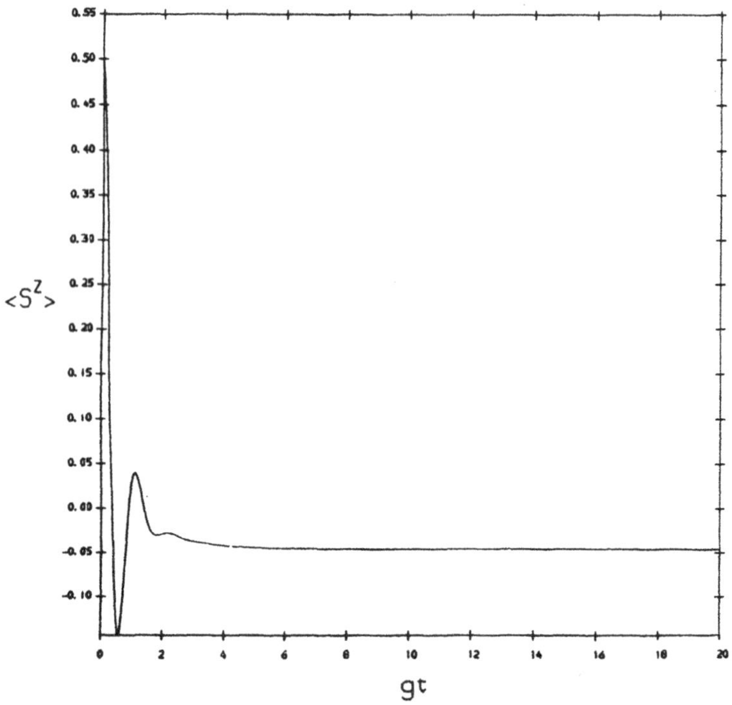

Fig. 14. $\langle S^z \rangle$ plotted against gt for a single atom in a cavity starting in its upper state at t = 0.

times also disagree for gt $\gtrsim$ 45. These results which take black-body photons into account otherwise seem to confirm the basis of the zero temperature maser theory given by Meystre[4]: they also indicate (by

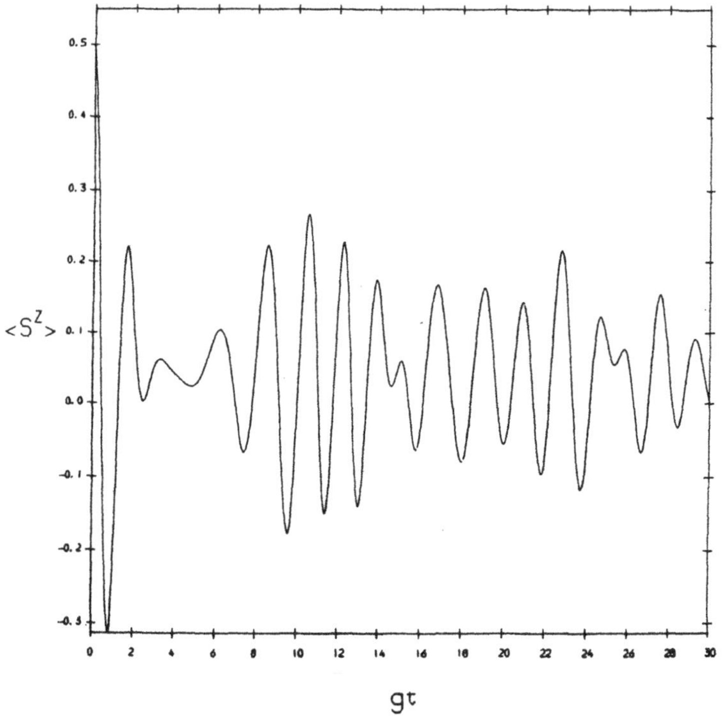

Fig. 15. Evolution of $\langle S^z \rangle$ for Q increased to about Q ~ 10^9 (κg^{-1} = 0·001).

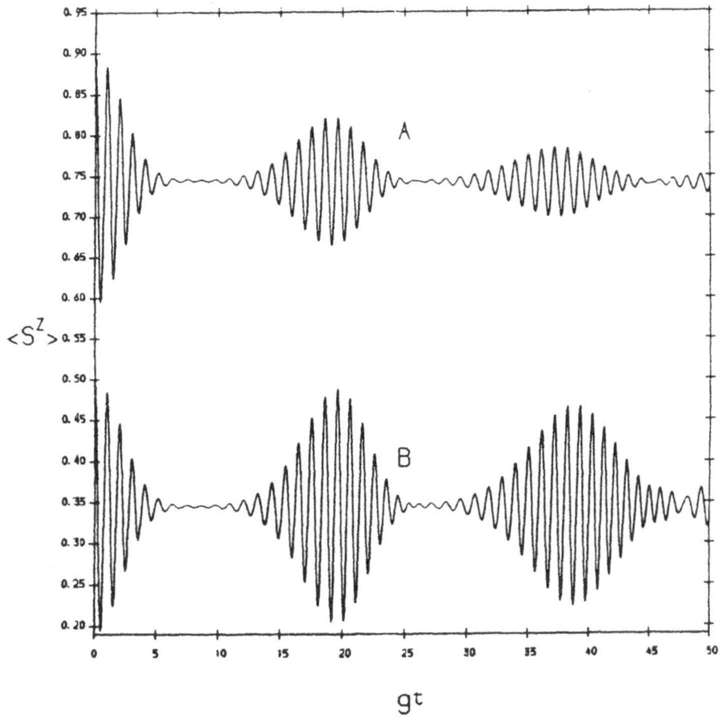

Fig. 16. Comparison of the evolution of $\langle S^z \rangle$ plotted as $\langle S^z \rangle + 0{\cdot}4$ for $\kappa g^{-1} = 0{\cdot}001$ (A) with the evolution for $\kappa g^{-1} = 0$ ($Q = \infty$): $\bar{n} = 2$, $\Delta g^{-1} = 5$, $|\alpha|^2 = 2$ (B) .

comparison with Fig. 14) the judicious choice of Q and temperature made by Walther[2,4,52] in his experiments. However, the effect of the few black body photons present *on the photon statistics* at these high Q is still to be investigated.

The Fig. 17 shows revivals in $g^{(2)}(0)$ for a single atom making 2-photon transitions with $Q = \infty$. The Hamiltonian is given by (12) for N = 1 and Stark shifts are dropped ($G_1 = G_2 = 0$). The atom starts in the upper state $|e\rangle$ and the initial field is a coherent state field with $|\alpha|^2 = 50$. These revivals are rather compact and periodic and this is because although the relevant Rabi frequencies are still irrational they are of the form $\Omega_n = g\sqrt{(n+1)(n+2)}$ (in the absence of Stark shifts and for an atom initially in its excited state). By choosing $|\alpha|^2 = 50$ (i.e. large) the relevant photon numbers n are also large and the frequencies on resonance essentially ng, i.e. rational.

The final two figures Figs. 18a,18b plot squeezing in $S_1 = \langle :(a+a^+)^2: \rangle - (\langle a+a^+ \rangle)^2$ and squeezing in $S_2 = \langle :i(a-a^+)^2: \rangle - \langle i(a-a^+) \rangle^2$ respectively as functions of gt. Although there is squeezing for N = 1 the two figures show that these squeezings have disappeared for N = 10. The Hamiltonian is (12) with $G_1 = G_2 = 0$, namely

$$H = 2\omega_0 S^z + \omega_0 a^+ a + g\{a^{+2} S^- + S^+ a^2\} \tag{52}$$

On the other hand the Hamiltonian H_{SCF}

$$H_{SCF} \equiv \omega_0 a^+ a + g\{a^{+2} \langle S^- \rangle + \langle S^+ \rangle a^2\} \tag{53}$$

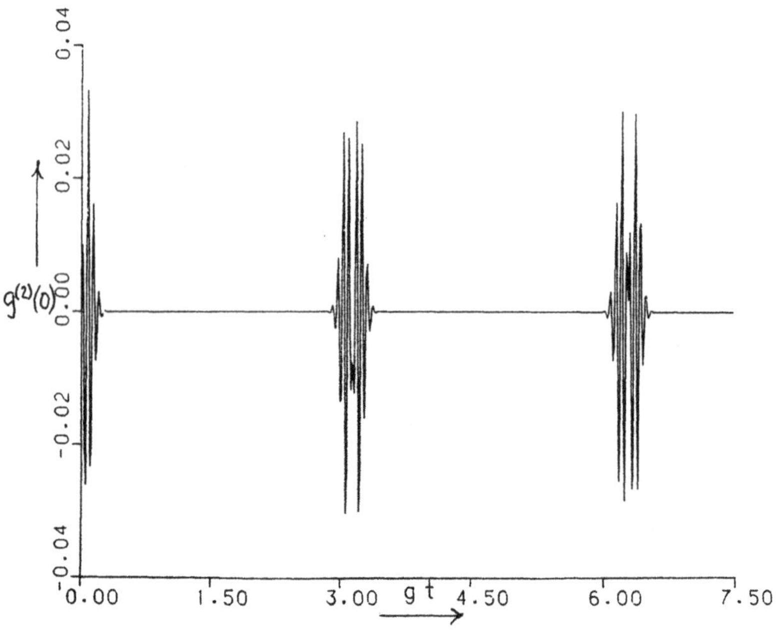

Fig. 17. Plot of $g^{(2)}(0) = (\langle (a^+)^2 a^2 \rangle / \langle a^+ a \rangle^2) - 1$ for a single atom making 2-photon transitions in an ideal cavity.

is a generator of the SU(1,1) group (a^+a, a^{+2}, a^2 form an SU(1,1) Lie algebra). Representations are labelled by a number S but are infinite dimensional. Just as in §2 for the SU(2) group, $e^{-iH_{SCF}t}$ acting now on the Dicke state $|S,S\rangle$ generates an SU(1,1) coherent state (which is squeezed[27,53]). For comparison with the Figs. 18 $e^{-iH_{SCF}t}$ must act on the initial α-state but the result (of course[54]) is still a squeezed state. It is therefore clear that the quantum fluctuations in the coupling to the

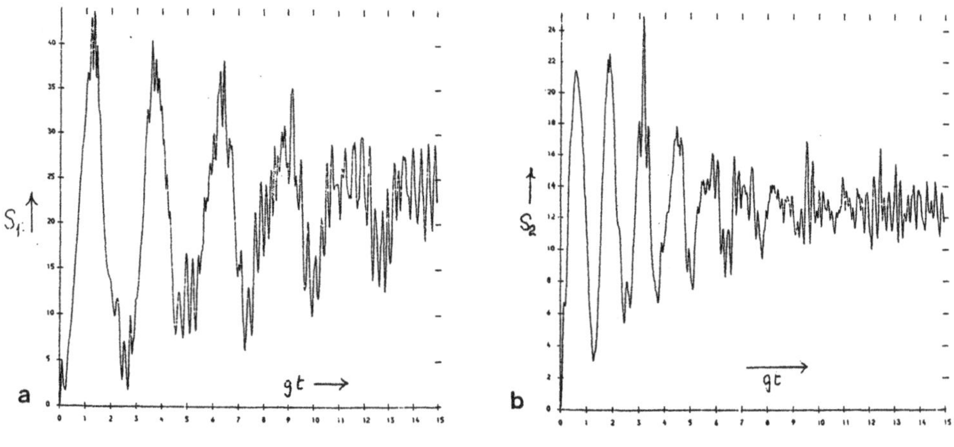

Figs. 18. Plots of squeezing S_1 (a) and S_2 (b) in 2-photon transitions as functions of gt: N = 10, $|\alpha|^2 = 5$ and there are no Stark shift terms ($G_1 = G_2 = 0$).

matter operators of H destroy the squeezing. Comparison with §§2-4 shows that to reduce H to H_{SCF} we must adiabatically eliminate the *atoms* (as e.g. in the 2-photon maser). It may then be possible to generate SU(1,1) coherent states in the laboratory in some analogy with the generation of SU(2) coherent states described in §3 of this paper.

REFERENCES

1. J. M. Raimond, P. Goy, M. Gross, C. Fabre, and S. Haroche, Phys. Rev. Lett. 49:117 (1982).
2. D. Meschede, H. Walther, and G. Müller, Phys. Rev. Lett. 54:551 (1985).
3. H. Walther, "Non-classical radiation from one-atom oscillators". This meeting.
4. P. Meystre, "Generation and detection of sub-Poissonian fields in micromasers". This meeting.
5. B. Yurke, "Squeezing thermal microwave radiation". This meeting. Also B. Yurke, P. G. Kaminsky, R. E. Miller, E. A. Whittaker, A. D. Smith, A. H. Silver and R.W. Simon,"Observation of 4·2K equilibrium noise squeezing via a Josephson-parametric amplifier". To be published.
6. G. S. Agarwal, R. K. Bullough, and G. P. Hildred, Optics Comm. 59:23 (1986).
7. G. S. Agarwal, R. K. Bullough, and G. P. Hildred, "Spectral and statistical properties of radiation from an atom in cavities of arbitrary Q and temperature". To be published.
8. G. P. Hildred, R. R. Puri, S. S. Hassan and R. K. Bullough, J. Phys B: At. Mol. Phys. 17: L538 (1984).
9. R. K. Bullough, "Photon, quantum and collective effects from Rydberg atoms in cavities", Hyperfine Interactions, J. C. Baltzer, AG, Basel, 37:71 (1987).
10. J. M. Raimond, P. Goy, M. Gross, C. Fabre and S. Haroche, Phys. Rev. Lett. 49:1924 (1982).
11. P. Goy, J. M. Raimond, M. Gross, and S. Haroche, Phys. Rev. Lett. 50: 1903 (1983).
12. Y. Kaluzny, P. Goy, M. Gross, J. M. Raimond, and S. Haroche, Phys. Rev. Lett. 51:1175 (1983).
13. R. H. Dicke, Phys. Rev. 93:99 (1954).
14. R. R. Puri and S. V. Lawande, Phys. Lett. 72A: 200 (1979).
15. R. R. Puri and S. V. Lawande, Physica A: 101:599 (1980).
16. S. S. Hassan, R. K. Bullough, R. R. Puri, and S. V. Lawande, Physica A 103:213 (1980).
17. S. S. Hassan and R. K. Bullough in: "Optical Bistability", eds. C. M. Bowden, M. Ciftan and H. R. Robl, Plenum, New York (1981), p.367.
18. R. R. Puri, S. V. Lawande, and S. S. Hassan, Optics Comm. 35:179 (1980).
19. S. V. Lawande, R. R. Puri, and S. S. Hassan, J. Phys. B: At. Mol. Phys. 14:4171 (1981).
20. S. S. Hassan, G. P. Hildred, R. R. Puri, and S. V. Lawande, J. Phys. B: At. Mol. Phys. 15:1029 (1982).
21. S. S. Hassan, G. P. Hildred, R. R. Puri, and R. K. Bullough, J. Phys. B: At. Mol. Phys. 15: 2635 (1982).
22. S. S. Hassan, G. P. Hildred, R. R. Puri, and R. K. Bullough, in: "Coherence and Quantum Optics 5", eds. L. Mandel and E. Wolf, Plenum, New York (1984), p.491.
23. R. R. Puri, G. P. Hildred, S. S. Hassan, and R. K. Bullough: "Coherence and Quantum Optics 5", eds. L. Mandel and E. Wolf, Plenum, New York (1984), p.527.
24. S. S. Hassan, R. R. Puri, and R. K. Bullough. To be published.
25. M. Brune, J. M. Raimond, and S. Haroche, Phys. Rev. A 35:154 (1987).
26. M. Brune, J. M. Raimond, P. Goy, L. Davidovich, and S. Haroche, Phys. Rev. Lett. 59:1899 (1987).

27. J. Katriel, A. I. Solomon, G. D'Ariano, and M. Rasetti, Phys. Rev. D 34:2332 (1986). Also M. Rasetti "Multiphoton squeezed states". This meeting.
28. M. Caves, Phys. Rev. D 23:1693 (1981).
29. R. K. Bullough, "Some remarks on the organisation of living matter and its thermal disorganisation" in: "Seminar on the living state III" ed. R. K. Mishra, World Scientific, Singapore (1988).
30. R. Gilmore, "Lie Groups, Lie algebras and some of their applications", John Wiley and Sons, New York (1972), p.149.
31. J. M. Radcliffe, J. Phys. A: Gen. Phys. 4:313 (1971)
32. M. R. B. Wahiddin, B. M. Garraway and R. K. Bullough, J. Mod. Optics 34:1007 (1987).
33. E. M. Purcell, Phys. Rev. 69:681 (1946).
34. R. R. Puri, "Stochastic behaviour of driven atomic systems. Some exact results", Ph.D. thesis, University of Bombay (1981).
35. R. K. Bullough, S. S. Hassan, G. P. Hildred, and R. R. Puri, in: "Optical Bistability III" eds. H. M. Gibbs, P. Mandel, N. Peyghambarian, and S. D. Smith, Springer-Verlag, Heidelberg, (1986) p.235.
36. L. A. Lugiato, "General analysis of the spectrum of squeezing in optical bistability". This meeting.
37. Ling-An Wu, H. J. Kimble, J. L. Hall, and Haifa Wu, Phys. Rev. Lett. 57:2520 (1986).
38. H. J. Kimble, "Quantum fluctuations in optical measurements". This meeting.
39. R. K. Bullough, in: "Coherence and quantum optics", eds. L. Mandel and E. Wolf, Plenum, New York (1973); p.121.
40. R. K. Bullough, R. Saunders, and F. Ahmad, J. Phys. A: Gen. Phys. 68:759 (1975).
41. R. Saunders "Coherence in the spontaneous emission of radiation; super-radiance" Ph.D. thesis University of Manchester (1973).
42. S. S. Hassan, "Reaction field theory, coherence and cooperation in quantum optics." Ph.D. thesis, University of Manchester (1975).
43. S. S. Hassan and R. K. Bullough, J. Phys. B: At. Mol. Phys. 8:L147 (1975).
44. C. W. Gardiner, Phys. Rev. Lett. 56:1917 (1986); H. J. Carmichael, A. S. Lane, and D. F. Walls, Phys. Rev. Lett. 51:2539 (1987).
45. H. J. Carmichael, A. S. Lane, and D. F. Walls, J. Modern Optics 34:821 (1987).
46. L. Mandel, Optics Letts. 4:205 (1979); R. Short and L. Mandel, Phys: Rev. Lett. 51:384 (1982).
47. S. S. Hassan, R. K. Bullough, G. P. Hildred and M. R. B. Wahiddin, "Thermal field effects of photon statistics in resonance fluorescence" J. Phys. B: At. Mol. Phys. In press 1988.
48. G. J. Milburn, Phys. Rev. A 34:4882 (1986).
49. M. Lewenstein, "Quantum statistical properties of atoms coupled to frequency dependent reservoirs". This meeting.
50. E. T. Jaynes and F. W. Cummings, Proc. IEEE 51:89 (1963).
51. N. Nayak, R. K. Bullough, B. V. Thompson, and G. S. Agarwal, "Quantum collapse and revival of Rydberg atoms in cavities of arbitrary Q and temperature" IEEE J. Quantum Electronics: Special issue on "Nonlinear optics of single atoms, ions and electrons". In the press for May 1988.
52. G. Rempe, H. Walther, and N. Klein, Phys. Rev. Lett. 58:353 (1987).
53. K. Wódkiewicz and J. H. Eberly, J. Opt. Soc. Am. B 2:458 (1985). This paper treats squeezing in both SU(2) and SU(1,1) coherent states and applications of the SU(2) and SU(1,1) groups in quantum optics.
54. H. P. Yuen, Phys. Rev. A 13:2226 (1976)
55. G. P. Hildred, S. S. Hassan, R. R. Puri and R. K. Bullough, J. Phys. B: At. Mol. Phys. 16:1703 (1983)

QUANTUM STATISTICAL PROPERTIES OF STRONGLY DRIVEN ATOMS COUPLED TO FREQUENCY-DEPENDENT RESERVOIRS

Maciej Lewenstein
Institute for Theoretical Physics, Polish Academy of Sciences
Aleja Lotników 32/46, Warsaw 02-668 Poland

Thomas W. Mossberg
Department of Physics, University of Oregon
Eugene, Oregon 97403 USA

INTRODUCTION

In this paper, we analyze the spectral and statistical properties of atoms driven by a strong, single-mode, light field and coupled to a reservoirs of electromagnetic field modes whose spectral density displays a strong frequency dependence. One realisation of this system consists of a driven atom confined within an optical cavity. We shall discuss in detail some new aspects of this problem, recently discovered by us[1,2]:

a) dynamical modification of spontaneous emission;
b) squeezing and dressed-state-polarization effects.

Before we turn to discussion of these effects, let us mention recent work in the area of cavity quantum electrodynamics in order to place the present results in a proper context with it.

The effect of modification of spontaneous emission has been discovered by Purcell in 1946[3]. He predicted that the spontaneous emission rate of an atom located in a cavity tuned to the atomic transition frequency is substancially larger that in free space. The enhancement results from a cavity induced increase in photon mode density at the atomic transition frequency. Employing the same idea au rebours, Kleppner[4] predicted the supression of spontaneous emission, that occurs if a cavity or a waveguide is used to reduce the density of photon modes in the spectral region of the atomic transition. Kleppner's paper stimulated a series of experimental works on this subject[5] in both the microwave and optical regimes. In most of experiments, the dimension of the cavity was comparable to the wavelength. Quite recently, however, Heinzen et al.[6] showed that analogous effects can be observed in confocal cavities of large dimensions. These authors have been also able to measure cavity induced radiative shifts of the atomic transition frequency[7].

We have shown in Ref. 1 and 2 that modifications of spontaneous emission may be effected not only with essentially passive means described above, but also through a dynamical means, i.e. by imposing a strong driving field on the atoms. In order for such dynamical effects to occur, the atoms must reside in a region of space in which the density of photon modes varies appreciably on a frequency scale set by the Rabi frequency of the driving field. Cavities provide a natural setting for finding such frequency-dependent mode densities, but they may also arise in waveguides, bulks of solid state etc.

Another area of cavity quantum electrodynamics which is of interest for our present purposes deals with quantum statistical aspects of collective and single-atom behavior. Studying the role of quantum fluctuations in optical bistability Casagrande and Lugiato[8] predicted small photon antibunching effects. Later Lugiato and Strini[9] predicted small squeezing effects in the same process. As has been shown by Carmichael[10] all these effects are closely related to the photon antibunching[11] and squeezing[12] discovered in the resonance fluorescence of a single two-level atom in a free space. Quite recently, Raizen et al. have shown that squeezing of the light transmitted through an atom-containing cavity can in fact become quite substantial, provided cavity width Γ becomes comparable or smaller than the atomic spontaneous emission rate. Such large squeezing effects occured in a regime of weak driving fields[13]

As we shall discuss below, large atomic squeezing can be found also under the conditions of strong or moderate driving fields and appropriate tunings of the cavity, atomic and laser frequencies. Under the same conditions novel effects such as polarization of the atom-field dressed-states populations occurs. Although the squeezing of the scattered light is not as large as atomic squeezing, the optical squeezing arises in a regime quite unexpected on the basis of free space results.

DYNAMICAL MODIFICATIONS OF SPONTANEOUS EMISSION

Although the effects we are going to discuss are of a fundamentally quantum nature, we may understand them intuitively in the framework of semi-classical Bloch equations. In the resonant case and in the absence of damping these equations take a form

$$\frac{d\vec{\sigma}}{dt} = \vec{\Omega}x\vec{\sigma}, \quad (1)$$

where $\vec{\sigma} = (\sigma_1, \sigma_2, \sigma_3)$ is a Bloch vector while $\vec{\Omega} = (\Omega, 0, 0)$ is a pseudo-field vector and Ω is a Rabi frequency. If we prepare the system in such a state that $\vec{\sigma}$ is initially parallel to $\vec{\Omega}$, the Bloch vector will, in the absence of damping, stay in this position forever. This phenomenon is usually referred to as spin-locking. In our present case the relevant damping mechanism is spontaneous decay which is triggered by quantum fluctuations of the vacuum field. We may try to model such fluctuations semi-classically introducing small fluctuating part of the vector $\vec{\Omega}$

$$\Omega(t) = \Omega + \delta\Omega(t). \quad (2)$$

Linear stability analysis of the Bloch vector precession descibed by Eq. (1) with respect to field fluctuations $\delta\Omega(t)$ leads to the following result:

a) Fluctuations induced variation of the in-phase component of the polarization σ_1 is significant if and only if the field fluctuations $\delta\Omega(t)$ contain Fourier components at the frequencies $\pm\ \Omega$.
b) The σ_2 and σ_3 components of the Bloch vector are sensitive to field fluctuations at the frequencies $0, \pm\ \Omega$.

As we see Rabi frequency provides means of controlling which reservoir spectral components contribute to Bloch vector damping. Suppose that a driven atom is contained in a cavity of the width Γ. Assume that the atomic frequency ω_a, laser frequency ω_L and cavity frequency ω_c are all equal. If we prepare an atom in one of its dressed states (so that initially $\vec{\sigma}$ is parallel to $\vec{\Omega}$) two scenarios are possible. If Rabi frequency is smaller than cavity width Γ, the Bloch vector will be driven away from the locked position. On the other hand if Rabi frequency is much larger than cavity width, there will be practically no vacuum fluctuations in the cavity to trigger the decay. The σ_1 component of the Bloch vector will remain constant for a very long time.

Similar analysis can be made concerning decay rates of the other components of the Bloch vector σ_2 and σ_3. Although their decay will always be triggered by the vacuum fluctuations at the zero frequency (cavity resonance), the damping of σ_2 and σ_3 may be significantly reduced, depending on the strength of the driving field Ω.

Analogous results are obtained in the nonresonant case. Depending on the values of the laser-atom detuning $\Delta_1 = \omega_L - \omega_a$ and the laser-cavity detuning $\Delta_2 = \omega_L - \omega_c$, the effects of dynamically induced inhibition or enhancement of decay rates are observed.

The above discussed effects exhibit themselves on the level of a spectrum of resonance fluorescence. In order to describe it quantitatively we introduce the hamiltonian of our system

$$H = \frac{\omega_a}{2}\sigma_3 + \Omega(e^{i\omega_L t}\sigma + e^{-i\omega_L t}\sigma^+) + \int|k|c_k^+c_k dk + \int|k|b_k^+b_k dk$$
$$\int g_c(k)(c_k^+\sigma + \sigma^+c_k)dk + \int g_b(k)(b_k^+\sigma + \sigma^+b_k)dk, \tag{3}$$

where the $\sigma_\pm$'s are the usual Pauli matrices describing a two-level atom. The operators $c_k^\pm$, $b_k^\pm$ correspond respectively to modes associated and unassociated with the cavity resonance. The coupling constants $g_b(k)$, $g_c(k)$ are connected to the appropriate photon mode densities. Since $|g_b(k)|^2$ is needed only in the neighborhood of ω_a, ω_L, and ω_c it may be treated as a constant. The coupling constant $|g_c(k)|^2$ is taken to be a single Lorentzian of half-width Γ.

The first step in solving the problem of time evolution of the system, described by hamiltonian (3), requires the derivation of atomic Bloch equations, i.e. equations describing the evolution of the quantum averages of σ, σ^+, and σ_3. In derivation it is convenient to introduce the reservoir response functions

$$\int|g_c(k)|^2\exp\big(i(|k|-\omega_c)t\big)dk \cong \gamma_c\Gamma e^{-\Gamma t}, \tag{4a}$$

$$\int|g_b(k)|^2\exp\big(i(|k|-\omega_c)t\big)dk \cong \gamma_b\delta(t). \tag{4b}$$

Note that the equation (4a) expresses the fact that photon modes associated with the cavity resonance have a finite response time Γ^{-1}. Clearly the interaction of the atom with the cavity (background) modes contributes an amount γ_c (γ_b) to the overall spontaneous emission rate at the center of the resonance ($\omega_a = \omega_c$).

The parameters of the model should fulfill the following conditions:

a) The total density of photon modes in the cavity has a well developed maximum at ω_c (i.e. $\gamma_c \gg \gamma_b$).
b) The width of the cavity, Γ, is larger than atomic spontaneous emission rates γ_b and γ_c.
c) The Rabi frequency Ω is of the order or larger than cavity width Γ.

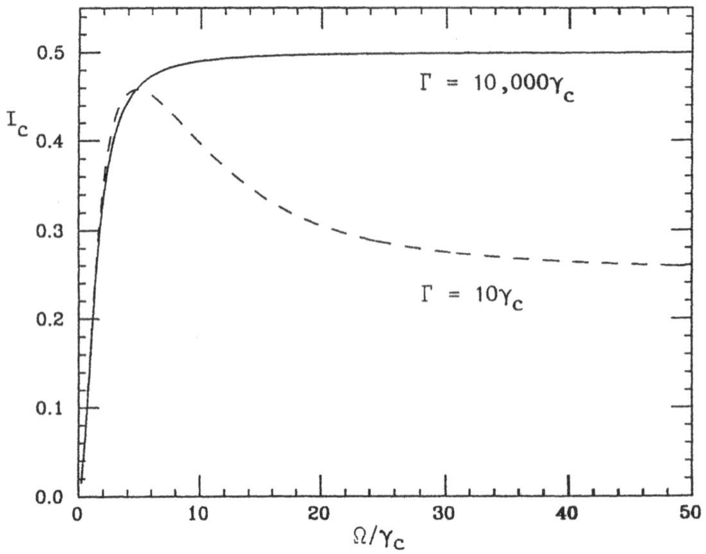

Fig. 1. Fluorescence intensity I_c of the cavity field as the function of Rabi frequency. Solid line represents free space result. Both detunings are set equal to zero and $\gamma_b/\gamma_c = 0.1$.

d) The laser-atom detuning Δ_1 may be, but is not neccessarily comparable with Rabi frequency.

e) The laser-cavity detuning Δ_2 may also be comparable with Ω. Note that this condition is meaningful only for cavity widths of the order or less than Ω.

Under such conditions the Bloch equations may be derived by eliminating the photon operators through the first order expansion in γ_b and γ_c(Born approximation). Note, however, that in doing so one cannot employ a Markov approximation for the cavity mode, due to the conditions c) and e). A similar approach was used to calculate single-time atom-field correlation functions. We have calculated few quantities characterizing scattered (or transmitted) radiation. In particular we evaluated total intensities of the radia-

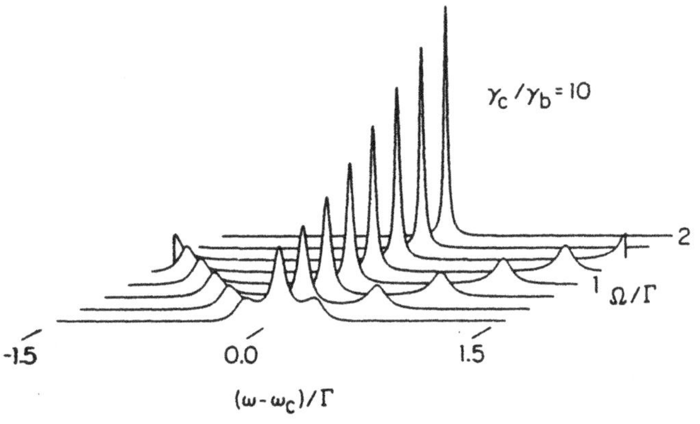

Fig. 2. Resonance fluorescence spectra of the background field. The driving field, atom and cavity are exactly resonant. Horizontal: observation frequency; vertical: relative fluorescence intensity. Successive traces correspond to increasing driving field strength. $\gamma_b+\gamma_c=\Gamma/20$.

tion scattered into cavity or background modes. Corrrespondingly two different power spectra have been calculated.

The Fig. 1 shows the intensity of the radiation scattered into cavity modes (dashed line) as a function of driving field Rabi frequency Ω. Cavity width is $\Gamma = 10\gamma_c$ in this case. For the reference free space results are plotted (solid line for practically infinite cavity width). As we see intensity of fluorescence grows as in free space with increasing Ω for $\Omega \ll \Gamma$. I_c has a maximum at $\Omega \cong \Gamma$ and decreases to its saturation value, twice smaller than in free space, for $\Omega \to \infty$. The reasons for the decrease of I_c are following: i) for $\Omega \gg \Gamma$ no photons can be emitted at the sidebands of the Mollow spectrum[14] (since the density of the cavity modes practically vanishes at these frequencies); ii) for $\Omega \gg \Gamma$ the emission at the central frequency is also slightly supressed.

The Fig. 2 shows the resonance fluorescence spectra in the resonant case. The supression of the decay rate of the in-phase component of the Bloch vector σ_1 leads to a substantial narrowing of the central peak of the Mollow triplet. The width and positions of the sidebands are also modified in comparison to free space results.

More results can be found in Ref. 1. There we discuss non-resonant case as well as resonance fluorescence of atoms located within a waveguide.

SQUEEZING AND DRESSED-STATE-POLARIZATION EFFECTS

Let us now concentrate on stationary squeezing effects. Atomic squeezing is characterized by variances of different components of the Bloch vector. The relative squeezing parameter of the σ_1 component is given by

$$r_1 = \lim_{t\to\infty} \frac{1 - \langle\sigma_1(t)\rangle^2}{\langle\sigma_3(t)\rangle} \tag{5}$$

It is this parameter r_1 which shows squeezing for Δ_1 comparable to Ω in free space and is expected to do so in the cavity as well.

We shall now consider the behavior of the steady-state values of σ_1 and σ_3 in free space and in a cavity in order to obtain some insight into resulting atomic squeezing. In free space and for a fixed, but large Ω, the atomic inversion, σ_3, vanishes approximately for $\Delta_1 = 0$ and approaches -1 for $\Delta_1 \gg \Omega$. On the other hand, the atomic coherence component, σ_1, vanishes for both $\Delta_1 = 0$ and $\Delta_1 \gg \Omega$, but attains a non-zero maximum somewhere in between. The ratio r_1 achieves its minimum value (i.e. squeezing is maximized) in the region $\Delta_1 \cong \Omega$ where σ_1 is maximized. The extend of squeezing in free space is determined by the quantitative behavior of σ_1 and σ_3 as a function of Δ_1. As we shall see below, the Ω- and Δ_1-dependences of these quantities may be substabtially modified in a cavity. Thereby a cavity may be employed to modify the extent and regime of atomic squeezing.

The result illustrating the above analysis is presented in Fig. 3 and 4. Let us now consider the case when the atom is coupled to a cavity field. Assume also that the background contribution is negligible. In this case, for

$$\sqrt{\Omega^2 + \Delta_1^2} \gg \Gamma$$

it is possible to tune the cavity close to particular dressed-state transition frequencies (i.e. ω_L, $\omega_L \pm \sqrt{\Omega^2+\Delta_1^2}$). By tuning close to the sideband frequencies ($\omega_L \pm \sqrt{\Omega^2+\Delta_1^2}$ one enhances one of the transitions between the dressed states. In this way one is able to modify the steady-state dressed-state inversion which is proportional to

$$w_{dr} = \frac{\Omega}{\Omega'}\langle\sigma_1\rangle + \frac{\Delta_1}{\Omega'}\langle\sigma_3\rangle, \tag{6}$$

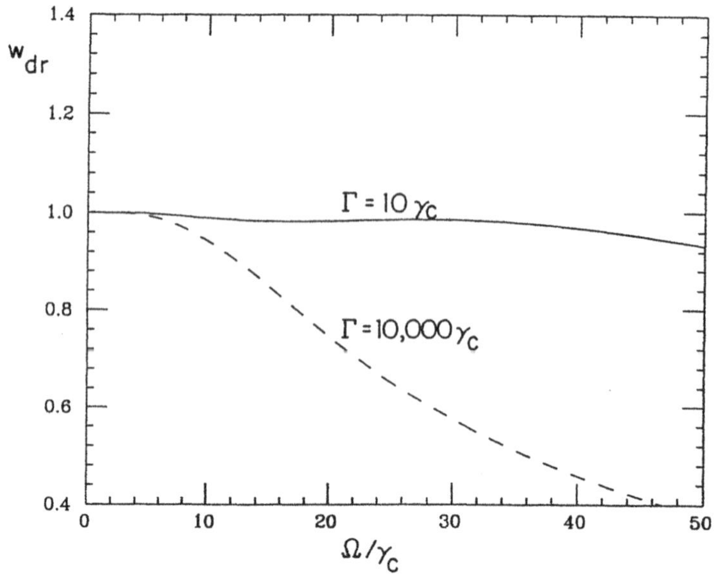

Fig. 3. Dressed-state inversion w_{dr} versus Ω/γ_c for $\Delta_1 = 10\gamma_c$. In free space w_{dr} is 1 for small Ω (because of large Δ_1) and decays to zero as Ω grows. In the cavity ($\Delta_2 = 30\gamma_c$, $\Gamma = 10\gamma_c$) w_{dr} stays close to 1 for much larger Ω due to a broad maximum at $\Omega' \cong \Delta_2$.

where $\Omega' = \sqrt{\Omega^2+\Delta_1^2}$. In effect, one can polarize the dressed-state populations. Note that this polarization effect (an assymetry in populations within dressed-state doublets) is induced by a difference between the density of photon modes at $\omega = \omega_L + \Omega'$ and $\omega = \omega_L - \Omega'$. For this reason the cavity does not have to be tuned exactly at $\omega = \omega_L \pm \Omega'$ in order to dressed-state polarize the atoms.

The effects of dressed-state polarization are shown in Fig. 3. For $\Delta_1 \neq 0$, the dressed-state inversion is no more equal to σ_1, but it becomes ever more nearly so as Ω grows larger than Δ_1. Thus in the $\Omega > \Delta_1$ region at least, we may see from Fig. 3 that the cavity enhances the value of σ_1. It turns out that the behavior of $\sigma_3$as a function of Ω is only weakly affected by the cavity.

Results plotted in Fig. 4 indicate that the cavity increases both the extent of squeezing and the range of values of Ω over which it occurs. As mentioned before, free-space squeezing is optimized when $\Omega \cong \Delta_1$ and attains a maximum value of about 16%. In the cavity, optimal squeezing occurs for $\Omega' \cong \Delta_2$ and attains a relative value of 60%.

Unfortunately, the large atomic squeezing effects described above do not carry over to the light scattered by atoms. If one considers the spectrum of squeezing associated with the cavity field, one finds that squeezing occurs mainly for sideband frequencies, and that maximum squeezing observed is roughly 15%. Intersetingly, however, peak squeezing occurs in a region of parameters different from that which maximizes squeezing in free space.

This work has been supported by the Office of the Naval Research under the contract (N00014-85-K-0724) and by NSF grant PHY85-04620.

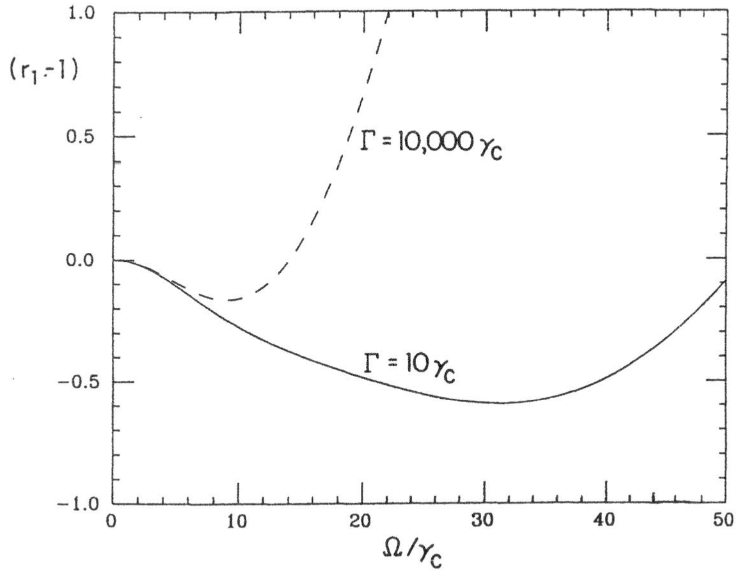

Fig. 4. Relative variance $(r_1 - 1)$ versus Ω/γ_c for $\Delta_1 = 10\gamma_c$. In free space the variance has a minimum at $\Omega \cong \Delta_1$ amounting to 16% of squeezing. In the cavity ($\Delta_2 = 30\gamma_c$, $\Gamma = 10\gamma_c$) the minimum appears at $\Omega' \cong \Delta_2$ amounting to 60% of squeezing.

REFERENCES

1. M. Lewenstein, T. W. Mossberg, R. J. Glauber, Phys. Rev. Lett. 59, 775 (1987); M. Lewenstein, T. W. Mossberg, Phys. Rev. A, in press 1988.
2. M. Lewenstein, T. W. Mossberg, Phys. Rev. A, submitted 1987.
3. E. M. Purcell, Phys. Rev. 69, 681 (1946).
4. D. Kleppner, Phys. Rev. Lett. 47, 233 (1981).
5. P. Goy, J. M. Raimond, M. Gross, S. Haroche, Phys. Rev. Lett. 50, 1903 (1983); R. G. Hulet, E. S. Hilfer, D. Kleppner, Phys. Rev. Lett. 55, 2137 (1985); W. Jhe, A. Anderson, E. A. Hinds, D. Meschede, L. Moi, S. Haroche, Phys. Rev. Lett. 58, 666 (1987); F. deMartini, G. Innocenti, in *Quantum Optics IV*, ed. J. D. Harvey, D. F. Walls, Springer, New York 1986.
6. D. J. Heinzen, J. J. Childs, J. E. Thomas. M. S. Feld, Phys. Rev. Lett. 58, 1320 (1987).
7. D. J. Heinzen, M. S. Feld, Phys. Rev. Lett., in press 1987.
8. F. Casagrande, L. A. Lugiato, Nuovo Cimento B55, 173 (1980); see also P. D. Drummond, D. F. Walls, Phys. Rev. A23, 2563 (1981).
9. L. A. Lugiato, G. Strini, Opt. Comm. 41, 67 (1982).
10. H. J. Carmichael, Phys. Rev. Lett. 55, 2790 (1985).
11. H. J. Carmichael, D. F. Walls, J. Phys. B9, L43 and 1199 (1976).
12. D. F. Walls, P. Zoller, Phys. Rev. Lett. 47, 709 (1981).
13. M. G. Raizen, L. A. Orozco, M. Xiao, T. L. Boyd, H. J. Kimble, Phys. Rev. Lett. 59, 198 (1987).
14. B. R. Mollow, Phys. Rev. 188, 1969 (1969).

GENERATION AND DETECTION

OF SUBPOISSONIAN FIELDS IN MICROMASERS

P. Meystre

Optical Sciences Center
University of Arizona
Tucson, AZ85721

I. INTRODUCTION

Subpoissonian fields, and in particular number states of the electromagnetic field, exhibit intensity fluctuations below the classical limit. The last few years have witnessed considerable interest in the generation of such states. To our knowledge, the first observation of subpoissonian fields was performed by Short and Mandel[1] in single-atom resonance fluorescence, following a prediction of Carmichael and Walls.[2] More recently, Saleh and Teich[3] and Walker and Jakeman[4] have produced subpoissonian fields by using antibunched electron sources and detection-event-triggered deadtimes in light beams, respectively. An important technological breakthrough was achieved by Machida et al,[5] who demonstrated subpoissonian (or intensity squeezed) fields in a pump-noise-suppressed semiconductor laser. This method is closely related to the generation of subpoissonian light in a micromaser[6] as well as to the recent proposal of a squeezed-pump laser by Marte and Walls.[7] High number states of the electromagnetic field were recently generated by Walther[8] following a prediction by Filipowicz et al.[9]

This paper reviews the generation and detection of subpoissonian fields in micromasers. Section II considers a lossless resonator, a situation under which the generation of number states of the electromagnetic field is predicted. Because of the exceeedingly high Q-factors presently available in superconducting microwave cavities, this is a good approximation of micromasers for times as long as hundreds of milliseconds. Section III discusses the case of a lossy cavity and reviews the generation of sub-poissonian and squeezed states in both one-photon[9–11] and two-photon[12–14] micromasers.

Because of the lack of efficient photon counting techniques in the microwave regime, the dynamics of micromasers is typically monitored by measuring the state of the successive atoms as they exit the resonator. The concomitant projection of the correlated atom-field density matrix on the relevant atomic state allows to extract information on the reduced density matrix for the field alone. A succession of such measurement on a single quantum system is discussed in Section IV for the case of a lossless cavity.[15] We show how successive measurements lead to the evolution of the cavity field towards a number state. Section V discusses simple examples of measurements-induced dynamics in the case of a finite-Q cavity, and shows in particular measurements-induced quantum diffusion. Finally, Section VI is a summary and conclusion.

II. LOSSLESS CAVITY

In a micromaser[4] a (monoenergetic) beam of Rydberg two-level atoms is injected inside a single-mode high-Q cavity at such a low rate that at most one atom at a time is present inside the resonator. Because of the single-mode nature of the atom-field interaction, the atoms are not subjected to irreversible spontaneous emission, which relies on the availability of a continuum of modes of the electromagnetic field. Intuitively, one expects therefore the rate of change of the mean photon number $\langle n\rangle$ in the cavity to be given by[6]

$$\langle \dot{n}\rangle = R \sin^2(\kappa\sqrt{\langle n\rangle+1}\, t_{\mathrm{int}}/2) - \gamma\langle n\rangle \ , \tag{1}$$

where R is the injection rate of atoms inside the cavity of damping rate γ, t_{int} is the time of flight of the atom through the cavity and κ the atom-field coupling constant. The first term in Eq. (1) is the gain due to the change in atomic inversion as deduced from the Rabi oscillations formula, the "+1" accounting for spontaneous emission into the resonator mode, while the second term describes cavity losses. The possible mean photon numbers $\langle n\rangle$ are approximately given by the stable stationary solutions of Eq. (1), and multistability is clearly possible. The maser threshold occurs when the linearized (stimulated) gain for $\langle n\rangle \cong 0$ compensates the cavity losses:

$$R \frac{d}{d\langle n\rangle} \sin^2(\kappa\sqrt{\langle n\rangle}\, t_{\mathrm{int}}/2\Big|_{\langle n\rangle=0} \cong R\,(\kappa t_{\mathrm{int}})^2/4 = \gamma \ . \tag{2}$$

This is precisely the result obtained in the more precise analysis of Ref. 6.

Eq. (1) is quite different in spirit from the rate equations usually encountered in maser and laser theories, which are extensions of Einstein's discussion of radiative interactions. Typically, the amplification process is described by a rate equation of the general form

$$\langle \dot{n}\rangle = \alpha\,(N_2-N_1)\langle n\rangle - \gamma\langle n\rangle \ , \tag{3}$$

where α is some generalized cross-section proportional to the square of the dipole moment and (N_2-N_1) is the population inversion. More sophisticated master equation versions of Eq. (3) exist, but they are still parametrized by the same α. This indicates that in conventional lasers and masers, the dipole coupling, which is proportional to both dipole moment and field strength, is replaced by a cross-section which is quadratic in both.

An indication of why this is the case is given by Lamb[17] in his original semi-classical theory of the maser, and is also quite clear from the quantum theory of the laser presented in Ref. 18. This theory does start by including quantum Rabi-oscillations in the atom-field interaction, but rapidly proceeds to integrate the resulting equations over the exponential atomic level decay, leading to an result of type (3). This step, which is certainly justified in normal laser operation, has a fundamental impact: it averages out the quantum Rabi phases governing the atom-field interaction, introduces incoherent saturation and is ultimately at the origin of the poissonian output of the laser far above threshold.[19]

In contrast, the micromaser uses a single mode cavity and no irreversible spontaneous emission takes place. It is characterized by a continuous, *reversible* exchange of excitation between the atom and the cavity mode. In practice, fluctuations associated with weak cavity losses, atomic velocity spread and arrival times are unavoidable and produce at least some washing out of the quantum mechanical phases. This is discussed in Section III, which makes contact with conventional lasers and masers.

The most favorable situation to study the impact of atom-field coherence on micromaser operation is obviously the lossless case. This situation was originally discussed by Jaynes and Cummings.[20] At time t_i, an atom in a state given by the density matrix ρ_A

enters the cavity containing the field in state ρ_F, so that the density operator of the combined system is $\rho(t_i) = \rho_A \rho_F$. After the interaction, the state of the system is $\rho(t_i+t_{int}) = U(t_{int})\rho(t_i)U^\dagger(t_{int})$. If the density matrices of both the field and the atom are initially diagonal they remain so for all times. When the atom leaves the resonator, the field is left in the state

$$\rho_F(t_i+t_{int}) = Tr_A[U(t_{int})\rho(t_i)U^\dagger(t_{int})] , \tag{4}$$

where Tr_A denotes trace over the atomic states. At resonance, the matrix elements $p_n = \langle n|\rho_F|n\rangle$ are explicitly given by[20]

$$p_n(t_i+t_{int}) = \rho_{aa}(t_i) [(\alpha_{n+1} + \alpha_n e^x)p_n(t_i) + p_{n+1}e^x p_{n+1}(t_i) + \beta_n p_{n-1}(t_i)]. \tag{5}$$

Here

$$\beta_n = 1 - \alpha_n = (\kappa^2 n/\Omega_n^2) \sin^2(\kappa\sqrt{n}t_{int}/2) , \tag{6}$$

and

$$\frac{\rho_{aa}(t_i)}{\rho_{bb}(t_i)} = e^{-x} . \tag{7}$$

Jaynes and Cummings showed that if the injected atoms have a positive temperature $x > 0$ ($\rho_{bb} > \rho_{aa}$), a steady state can always be found. It has the photon statistics of a thermal field. It is more surprising that under appropriate conditions, a steady state can also be reached if *inverted* atoms are injected inside the cavity. Ref. 9 shows that for a given interaction time t_{int}, the lossless micromaser presents a number of "trapping levels" $|n_q-1>$ satisfying the resonance condition

$$\sin^2(\kappa\sqrt{n_q}\, t_{int}/2) = 0 \tag{8a}$$

or

$$\kappa\sqrt{n_q}\, t_{int} = 2q\pi, \quad q \text{ integer} \tag{8b}$$

which separates the phase-space of the field into disconnected blocks. This permits to generate number states of the electromagnetic field, as well as coherent or incoherent mixtures of such states. Indeed, states of the form

$$\Psi_{field} = \sum_q \alpha_q |n_q> \tag{9}$$

are readily seen to be stable states of the cavity mode. Physically, the stable states $|n_q>$ are such that the successive atoms experience a $2q\pi$-pulse as they fly through the cavity in the interaction time t_{int}. In general, $2q\pi$ pulses are not possible for quantum fields because of their inherent intensity fluctuations. Number states are an exception to this rule.

Eq. (9) shows that the lossless micromaser offers in principle a way to generate macroscopic quantum states of the electromagnetic field — optical Schrödinger cats. But how this can be achieved in practice remains unclear. Ref. 21 proposes an operational way to generate *mixtures* of the type

$$\rho_{\text{field}} = \sum_q p_q \, |n_q\rangle\langle n_q| \tag{10}$$

and indeed, we can numerically produce such density matrices routinely, see Section IV.

Coupling to the environment causes quantum coherences to be destroyed on a very fast time scale for macroscopic objects. The influence of dissipation on quantum superpositions and the consequent destruction of quantum mechanical interference phenomena have recently been discussed.[21] In the micromaser the suppression of quantum coherences is caused both by fluctuations in the atomic injection time[22] and by cavity damping. The decay of quantum coherences seems to scale at the rate[21,22] $\gamma_{coh} \cong \mathcal{D}\gamma$, where $\mathcal{D}$ is a measure of the "distance" between the states who are in quantum superposition. Although it is not completely evident that this result is generic, it holds for all examples considered so far. For coherent states $|\alpha\rangle$ and $|\beta\rangle$, one has roughly[21] $\mathcal{D} \cong |\langle\alpha|\beta\rangle|^2$ while for number states[22] $|n\rangle$ and $|m\rangle$ with a quadratic coupling to the bath, $\mathcal{D} \cong |n-m|^2$.

The lossless micromaser offers, at least in principle, a possibility of generating coherent superpositions of trapped number states $a|n\rangle + b|m\rangle$, with m and n arbitrary large. Thus it represents an interesting potential system to verify these results by measuring the decay of quantum superpositions for varying $\mathcal{D}$. We are presently investigating to which extent such tests can be carried on in practice, the most difficult question being the actual experimental preparation of such states in a controled way.

III. CAVITY DAMPING

With the inclusion of weak cavity damping, the role of the atom-field coherences becomes less pronounced. Still, the micromaser remains governed by coherent dynamics of the type (1). Detailed theories of the one- and two-photon micromasers are given in Ref. 6 and 14. These papers show that to a good approximation the photon number distribution is governed by the Fokker-Planck equation

$$\frac{\partial}{\partial t} p(n,t) = -\frac{\partial}{\partial n}[Q(n)\, p(n,t)] + \frac{1}{2}\frac{\partial^2}{\partial n^2}[G(n)\, p(n,t)] \ , \tag{11}$$

where

$$Q(n) = R \sin^2(\kappa\sqrt{n}t_{\text{int}}/2) - \gamma(n - n_b) \ , \tag{12}$$

and

$$G(n) = R \sin^2(\kappa\sqrt{n}t_{\text{int}}/2) + \gamma(n + n_b + 2nn_b) \tag{13}$$

for the one-atom micromaser. Here, the discrete variable n is approximated by a continuous function and we have included the possibility that the cavity is at a finite temperature with corresponding mean thermal photon number n_b. A similar equation holds for the two-photon case, the major change being that $\sqrt{n}$ is replaced by n.

The Fokker-Planck equation (10) and the corresponding effective potential

$$V(n) = -\int dn \, \frac{Q(n)}{G(n)} \tag{14}$$

provide an intuitively appealing understanding of many of the characteristics of the micromaser. Quantum and thermal fluctuations drive the maser towards steady state. If one minimum of $V(n)$ dominates, the steady state photon statistics $p(n)$ will be essentially single-peaked and, as it turns out, subpoissonian.

The zeros of $V(n)$ are solutions of

$$n_0/N_{ex} = \sin^2\left[\sqrt{n_0/N_{ex}}\ \Theta\right], \tag{15}$$

where we have introduced the parameters $N_{ex} = R/\gamma$ and $\Theta = \sqrt{N_{ex}}\ \kappa t_{int}/2$. As discussed in Ref. 6, Θ plays the role of a pump parameter for the micromaser. For $\Theta < 1$ the (unique) minimum of $V(n)$ is at $n = 0$. At $\Theta = 1$ the minimum $n_0 = 0$ turns into a local maximum and the mean photon number in the cavity starts to grow. This is the maser threshold. As Θ is further increased, the effective potential $V(n)$ acquires an increasing number of minima. This is illustrated in Fig. 1, where $V(n)$ is drawn for the three pump parameters $\Theta = 4$, 2.116π and 8. The intermediate value of Θ is precisely such that the global minimum at $n = 0.167N_{ex}$ is replaced by a global minimum at $n = 0.647N_{ex}$. The micromaser exhibits a sharp jump in the intracavity mean photon number each time a minimum of $V(n)$ loses its global character and is replaced in this role by the next one. In the vicinity of these transitions, the photon statistics are clearly double-peaked and super-poissonian, as illustrated in Fig. 2.

Davidovich *et al* find similar results in the two-photon micromaser.[13] Although the corresponding potential is quite different, and characteristic of a system exhibiting a first-order rather than a second-order phase transition, they obtain both super-poissonian photon statistics when two minima of the potential have comparable depths and subpoissonian light when one minimum is dominant. They also predict a small amount of transient squeezing, which might become more important for increasing N_{ex}.

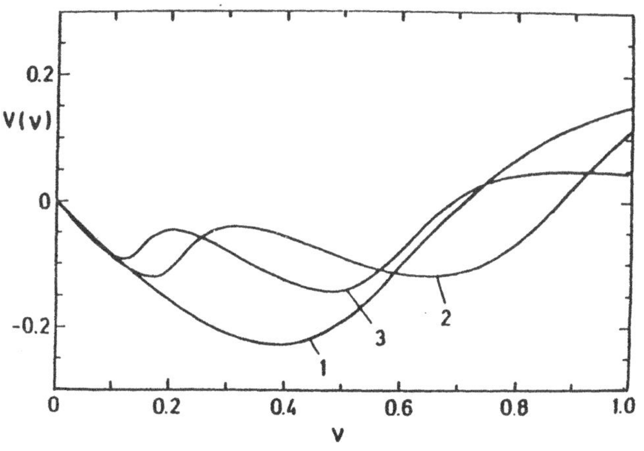

Figure 1: Potential $V(\nu)$, where $\nu = n/N_{ex}$ for 3 values of the pump parameters (1) Θ = 4., (2) Θ = 2.116π, (3) Θ = 8.(From Ref. 6.)

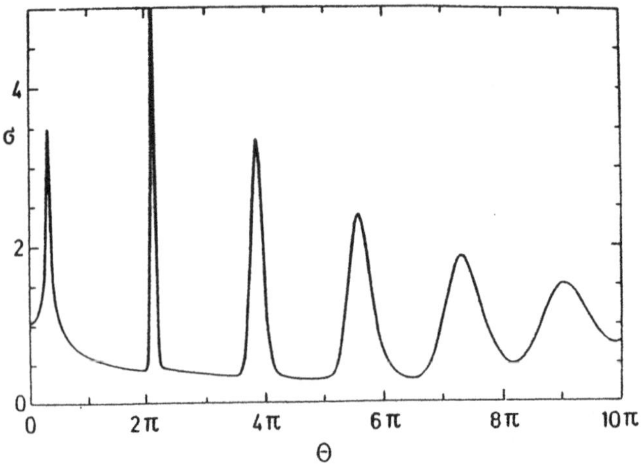

Figure 2: Normalized standard deviation σ for the photon distribution for $N_{ex} = 200$ and $n_b = 0.1$. Poissonian photon statistics correspond to $\sigma = 1$. (From Ref. 6.)

Connection with conventional masers and lasers

The behaviour of the micromaser is governed by the *coherent* light-matter interaction, as evidenced by the appearance of Rabi-type terms in all equations of the preceding Section. This is further reflected in the steady-state photon statistics[6]

$$p_n = C \left[\frac{n_b}{1 + n_b}\right]^n \prod_{\ell=1}^{n} \left[1 + \frac{R}{\gamma n_b \ell} \sin^2(\kappa\sqrt{\ell}\, t_{int}/2)\right] , \tag{16}$$

where C is a normalization constant. This should be contrasted to the results for a single-mode laser,[18] whose dynamics is dominated by incoherent saturation an for which

$$p_n = C' \frac{(A^2/B\gamma)^{n+A/B}}{(n + A/B)!} . \tag{17}$$

Here $A = \frac{R}{2}(\kappa/\Gamma)^2$, $B = (\kappa/\Gamma)^2 A$ are the linear gain and saturation coefficient, respectively and C'' is another normalization constant. We have taken for simplicity $\gamma_a = \gamma_b = \Gamma$, where γ_i is the decay rate of level i. Far above threshold ($A >> B$), the photon statistics (17) become

$$p_n \cong \exp(-\langle n\rangle) \frac{\langle n\rangle^n}{n!} , \tag{18}$$

that is, approach those of a coherent state. Note however that p_n is *always* super-Poissonian. In this sense, conventional lasers are classical devices, and the origin of this behaviour can be seen in incoherent saturation. In contrast, micromasers do not really saturate. They are much more like free-electron lasers, where the precise value of the interaction time between successive atoms and the cavity mode determines if energy is transfered to the field or not. This Rabi-flopping dominated dynamics is seen to be at the origin of subpoissonian light generation in the micromaser.

The micromaser can readily be modified to operate in a way undistinguishable from regular lasers: Ref. 18 shows that in the case $\gamma_a = \gamma_b \equiv \Gamma$ the effect of spontaneous emission is equivalent to introducing a stochastic interaction t_{int} with distribution

$$P(t_{int}) = \Gamma \exp(-\Gamma t_{int}) \quad . \tag{19}$$

In this context, spontaneous emission acts very much like a classical source of noise. A similar averaging procedure must be introduced in the micromaser if the atomic beam has a (maxwellian) velocity distribution, instead of being monoenergetic as assumed until now. In the experimental realization of the micromaser a Fizeau velocity selector is used to narrow the atomic velocity distribution at will.[4] By varying the velocity spread $\Delta v/v$, it is thus be possible to investigate the transition between the regime discussed in this paper ($\Delta v/v \to 0$) and the regime of operation of conventional lasers ($\Delta v/v \cong 1$).[19]

IV. FIELD MEASUREMENTS — LOSSLESS CASE

The preceeding Section shows that micromasers operate as generators of subpoissonian fields. The question remains to detect such fields. There are no efficient photon detectors in the microwave regime, and so far, the only measurements experimentalists have at their disposal are on the state of the successive atoms as they exit the resonator, e.g. by the technique of filed ionization. Note that the atoms play therefore a dual role in this system, both as a pump and as detectors. What information can be gained by such measurements is the primary question addressed in the following Sections.

Ensemble averages are unable to predict the behaviour of the atoms exiting the resonatorin a single realization of the experiment. To determine a the outcome of a typical experiment, it is necessary instead to explicitly take into account the measurements performed on the successive atoms and their back-action on the state of the cavity mode. We illustrate how this works first in the case of a lossless cavity.[15] We assume for simplicity that the atoms are injected inside the cavity in their excited state and that the field density matrix is initially diagonal. If no measurement is performed on the atoms other than checking that they indeed left the cavity, the density matrix of the field just after atom i exits the cavity is given by Eq. (5) with $x = -\infty$ as

$$p_n(t_i+t_{int}) = p_{n-1}(t_i)\sin^2\Omega_n t_{int}/2 + p_n(t_i)\cos^2\Omega_{n+1} t_{int}/2 \quad . \tag{20}$$

Things are quite different if a measurement of the state of the atom is performed. If it is found to be in the state $s = a$ or b the field density matrix after measurement becomes[23]

$$\rho_F(t_i+t_{int}) = Tr_A[|s\rangle\langle s|\; U(t_{int})\rho(t_i)U^\dagger(t_{int})] \;, \tag{21}$$

with corresponding photon statistics

$$p_n^a\,(t_i+t_{int}) = \mathcal{N}_a\, p_n(t_i)\cos^2\Omega_{n+1} t_{int}/2 \;, \tag{22a}$$

or

$$p_n^b\,(t_i+t_{int}) = \mathcal{N}_b\, p_{n-1}(t_i)\sin^2\Omega_n t_{int}/2 \;. \tag{22b}$$

Here $\mathcal{N}_a$ and $\mathcal{N}_b$ are constants which must be introduced to guarantee that the field density matrix remains normalized after the measurement.

The probability for the i-th atom to exit the cavity in its upper state is

$$p_a\,(t_i+t_{int}) = \sum_{n=0}^{\infty} p_n(t_i)\,\cos^2\Omega_{n+1}t_{int}/2\,, \tag{23}$$

and the outcome of a given measurement will yield (22a) or (22b) with probabilities p_a and $1-p_a$, respectively. Comparison of Eqs. (22) and (23) shows that $\mathcal{N}_a = 1-\mathcal{N}_b = p_a$.

It is important to realize that measuring the atom in its upper state at the exit of the resonator *does not* imply that the field remains unchanged. Rather, its photon statistics are "reshuffled" according to Eq. (21). In general, this also implies that the expectation value $\langle n\rangle$ of the photon number in the cavity mode is changed, since

$$\sum_n n p_n(t_i) \neq \mathcal{N}_a \sum_n n p_n(t_i)\,\cos^2\Omega_{n+1}t_{int}/2\,. \tag{24}$$

This result is actually not surprising: before the measurement, $\langle n\rangle$ is known only to within its standard deviation σ^2, and not conserving $\langle n\rangle$ does not violate any law of physics. It is only if the field happens to be in a number state $|m\rangle$, with $p_n = \delta(n-m)$ at the time of injection of the i-th atom, that an exact conservation of $\langle n\rangle$ is guaranteed if the atom exits the cavity in the upper state.

Similar considerations apply if the atom exits the cavity in the lower state. In this case, $\langle n\rangle$ does not necessarily increase by one, but may either increase *or decrease* within the limits permitted with the energy spread.

Numerically, we proceed by chosing an initial field density matrix, typically a thermal field with average photon number n_b, so that $p_n(0) = 1/(1+n_b)\,(n_b/(1+n_b))^n$. This allows to compute the probability p_a (Eq.(23)) for the first atom to exit the cavity in the upper state. A random number generator returns a uniform random deviate r between 0 and 1. We say that the atom was measured to be in the lower state if $r < p_a$ and in the upper state if $r > p_a$. The field density matrix after measurement is then given by either (22a) or (22b), depending on the outcome of the measurement, and the procedure is repeated.

Table 1, which is reprinted from Ref. 15, gives selected results from typical runs with $n_b = 20$ and $\kappa t_{int} = \pi/2$. These are interesting parameters in that they lead to the existence of the low trapping states $n_1 = 15$ and $n_2 = 63$, (as well as $n_q = 16q^2-1$, q integer), so that the numerics are easily manageable. Taking n_b large gives a relatively large initial probability for the field to be above the first trapping state, and thus to investigate the possibility that the intracavity field evolves towards the mixture $\alpha|n_1\rangle\langle n_1| + \beta|n_2\rangle\langle n_2|$, $\alpha + \beta = 1$, which, if reached, should be stable as readily seen from Eqs.(5) and (6). Although the examples of the Table indicate the convergence of the system towards a trapping number state $|n_q\rangle$, we have found numerous instances where mixtures of such states were reached. Early indications that this was not the case[15] had to be revised after improved accuracy was introduced in the numerics.

The two examples listed in the Table give in each case the measurement (atom) number, the *a priori* probability that it gives "atom up" as an answer, the actual outcome of the measurement, and the corresponding inferred values of $\langle n\rangle$ and of the standard deviation $\sigma^2 \equiv (\langle n^2\rangle - \langle n\rangle^2)/\langle n\rangle$, with $\langle n\rangle = 20$ and $\sigma^2 = 21$ initially.

The Table illustrates a number of features of the successive measurement processes: First, as already mentioned $\langle n\rangle$ needs not be conserved if the atom is measured to be in the upper state, or increased by one if the atom is found in the lower state. This is illustrated particularly strongly in the measurements of atoms 9 and 11 of the first sequence. Second, we find that a measurement typically tends to reduce the

Table 1: Extract from the records of two typical numerical experiments. Atom indicates the number of the atom measured, *P(a)* the probability that the measurement yields an atom in upper state as an outpu, Result the actual output, $\langle n \rangle$ the inferred intacavity expectation value of the photon number and σ^2 its variance. (From Ref. 15.)

Atom	$P(a)$	Result	$\langle n \rangle$	σ^2	Atom	$P(a)$	Result	$\langle n \rangle$	σ^2
1	0.467	down	19.5	21.9	1	0.467	up	21.7	18.7
2	0.240	down	20.1	21.6	2	0.728	up	22.4	17.7
3	0.185	down	21.2	20.9	3	0.820	up	22.8	17.2
4	0.179	down	23.0	20.0	4	0.867	up	23.1	16.9
5	0.193	down	25.5	18.7	5	0.896	up	23.2	16.7
6	0.211	down	29.0	16.9	6	0.914	up	23.3	16.6
7	0.223	down	33.4	14.5	7	0.927	up	23.3	16.6
8	0.218	down	38.0	12.0	8	0.937	down	24.0	16.4
9	0.196	up	25.4	16.9	9	0.832	up	24.0	16.4
10	0.450	down	32.6	14.1	10	0.858	up	23.8	16.5
11	0.423	down	40.2	10.2	11	0.875	up	23.6	16.5
12	0.367	up	32.3	13.4	12	0.887	down	26.1	15.4
13	0.526	down	41.6	8.58	13	0.799	down	28.6	12.6
14	0.453	down	47.5	5.57	14	0.653	up	28.8	13.9
15	0.411	down	50.1	5.15	15	0.703	up	28.7	15.2
		⋮					⋮		
51	0.884	up	56.8	0.707	51	0.974	up	19.5	13.0
52	0.886	up	56.8	0.634	52	0.975	up	19.3	12.8
53	0.888	up	56.9	0.571	53	0.975	up	19.2	12.6
54	0.890	up	56.9	0.516	54	0.976	up	19.1	12.4
55	0.892	up	56.9	0.467	55	0.976	up	18.9	12.2
		⋮					⋮		
106	0.977	up	60.3	0.328	106	0.989	up	16.3	6.22
107	0.977	up	60.3	0.313	107	0.989	up	16.3	6.16
108	0.977	up	60.3	0.299	108	0.989	up	16.3	6.10
109	0.978	down	61.3	0.868	109	0.989	up	16.2	6.04
110	0.984	up	61.3	0.830	110	0.989	up	16.2	5.98
		⋮					⋮		
1107	1.000	up	63.0	0.0586	1107	1.000	up	15.1	0.414
1108	1.000	up	63.0	0.0585	1108	1.000	up	15.1	0.413
1109	1.000	up	63.0	0.0584	1109	1.000	up	15.1	0.413
1110	1.000	up	63.0	0.0583	1110	1.000	up	15.1	0.412

spread σ^2, although this needs not always be the case. As σ^2 is reduced, one reaches a regime of much better "conservation of the mean energy", as should be expected.

Although at the end of the printed sequences the intracavity photon statistics is exceedingly narrow and hence the field close to a number state, there is still no guarantee that the system will eventually be trapped to this point. Indeed, a finite σ^2 indicates that the probability *P(a)* is not exactly equal to unity (the printed value is rounded). Under conditions leading to low $|n\rangle$ trapping as shown here, further iterations up to 3000 atoms only lead to a further reduction of σ^2, but in other cases where no resonant condition is met for small $|n\rangle$, such as e.g. $\kappa t_{int} = 2$, we found some cases where the system eventually passed the near-trap $|n\rangle = 9$, despite the fact that the variance σ^2 was on the order of $3 \cdot 10^{-6}$ for several hundred measurements! This is because in such marginally stable[9] situations quantum fluctuations can never be neglected, however small they may be.

V. MEASUREMENTS-INDUCED DYNAMICS

Quantum diffusion

We conclude this paper by further illustrating how repeated measurements yield information on the dynamics of a single realization of the system in the case of finite cavity losses. Consider the situation of case (2) in Figure 1. where the ensemble average photon statistics of the micromaser exhibits a double-peaked distribution. These predictions bear little relevance to a single realization of the micromaser. Rather, they apply either to the average characteristics of a large number of identical micromasers or to the outcome of a large number of experiments performed on a single system starting from identical initial conditions. In particular, one would intuitively expect that as far as a

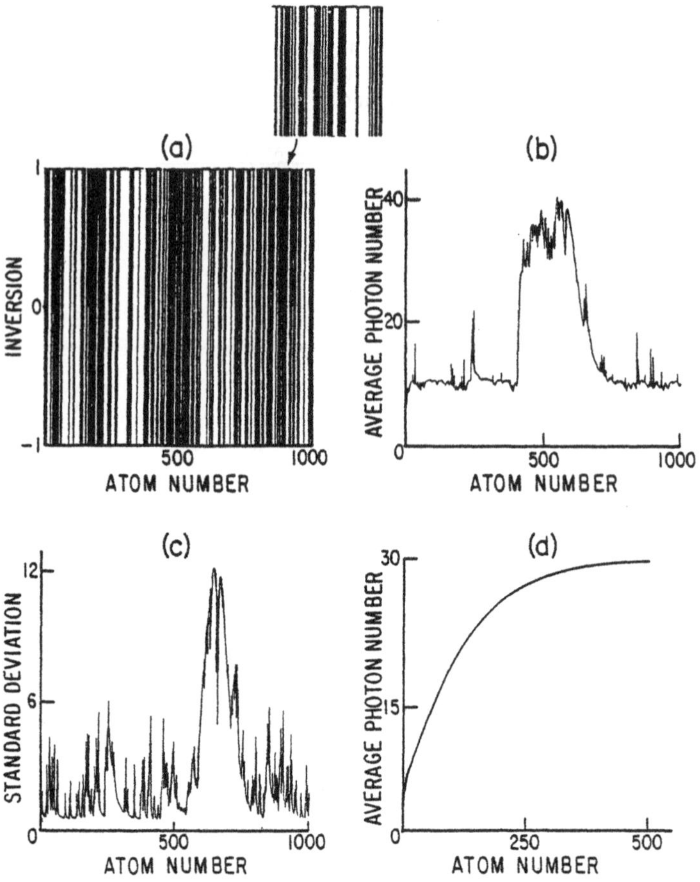

Figure 3: (*a*) Raw data from repeated measurements of the states of successive two-level atoms exiting the micromaser cavity. The value "+1" corresponds to atoms measured in the upper state and "-1" to atoms measured in the lower state. The vertical lines are for visual help only. (*b*) Average intracavity photon number $\langle n \rangle$ and (*c*) standard deviation σ^2 infered from the raw measurements as a function of the number of atoms injected and measured. (*d*) Conventional ensemble average $\langle n \rangle$ for the same parameters $N_{ex} = 50$, $\Theta = 2.116\pi$ and $n_b = 5$. Fig. 4b clearly shows measurements-induced diffusion between the two minima of the effective potential.

single realization of the maser is concerned, the double-peaked photon statistics should be interpreted in terms of transitions between the two competing minima of the potential well due to quantum diffusion above the potential barrier. Repeated measurements on the system show that this is precisely what happens.

Fig. 3a gives the raw results of a typical sequence of measurements for the case (2) of Fig. 1, i.e. $N_{ex} = 50$, $n_b = 5$, $\Theta = 2.116\pi$, for which $n_\ell = 0.18N_{ex} = 9$ and $n_u = 0.68N_{ex} = 34$. They are labeled as +1 or -1 for the atom measured in the upper or lower state, respectively. The vertical lines are for visual help only. In contrast to the situation in real experiments, we assume for simplicity that all atoms exiting the resonator are detected in their state $|a\rangle$ or $|b\rangle$ with 100% quantum efficiency. For clarity, the insert of Fig. 3a shows the same results on an expanded horizontal scale. From this raw data and Eqs. (22), one can infer back the cavity mode photon statistics and $\langle n \rangle = \Sigma\, np_n$.

The results of this reconstruction are shown in Fig. 3b, where diffusion between the two minima of the effective potential becomes quite clear. Note the good agreement between the numerical and effective potential predictions for n_ℓ and n_u. Fig. 3c gives the normalized standard deviation $\sigma^2 = \langle(n-\langle n\rangle)^2\rangle/\langle n\rangle$ of the photon statistics. It exhibits a broad peak during the slow down-switching of $\langle n\rangle$ from n_u to n_ℓ. The upswitching, in contrast, is very fast and does not show any significant change in standard deviation in this example. For comparison, Fig. 3d gives the (ensemble-averaged) mean photon number $\langle n(t)\rangle$ as obtained for the same parameters from the standard approach of Section III. The difference between the two results is striking: in one case, one can truly follow the dynamics of the micromaser for all times, while the other only gives the transient approach to (ensemble average) steady state.

Quantum relaxation-oscillations

For a reservoir at zero-temperature, $n_b = 0$, the steady-state photon statistics (16) reduces to

$$p_n = p_0 \frac{N_{ex}^n}{n!} \prod_{k=1}^{\infty} \sin^2(\theta\sqrt{n/N_{ex}}), \tag{25}$$

where p_0 provides normalization and we have introduced the parameters N_{ex} and θ. A direct consequence of (25) is that for values of the pump parameter

$$\theta = q\pi\sqrt{N_{ex}}, \quad q = 1,2,3\ldots\,, \tag{26}$$

the (ensemble average) steady-state of the micromaser photon statistics is $p_n = \delta_{n,0}$, independent of the initial conditions. That is, the cavity field is in the vacuum state. This is because the vacuum field acts precisely as a $2q\pi$-pulse for atoms spending a time t_{int} inside the cavity, as readily seen by combining Eqs. (26) and the definition of the pump parameter θ to give

$$\theta/\sqrt{N_{ex}} = \kappa t_{int}/2 = q\pi\ . \tag{27}$$

The vacuum state, as any number state, does not exhibit any intensity fluctuations and can act as a true $2q\pi$-pulse. In most other cases, however, the inherent intensity fluctuations lead to the impossibility of achieving such a 'perfect' interaction. This is the case e.g. if the micromaser cavity has a finite temperature, $n_b \neq 0$.

This situation is illustrated in Fig. 4. Here, condition (27) is fulfilled ($N_{ex} = 5$, $\theta \cong 35.$) but $n_b = 10^{-3}$. For the corresponding thermal initial field, the first atom experiences almost, but not exactly a 10π-pulse. Consequently the probability of measuring it in the upper state at the exit of the resonator is almost unity. Because the initial photon statistics is exceedingly narrow, there is almost exact conservation of the mean photon energy and the resonator field remains practically unchanged. But as further atoms are injected, there is a small but finite probability that one of them will eventually be measured to exit in its ground state. This happens first in our example for atom $\cong 310$. In this case, the back-action on the cavity mode is particularly drastic: to a very good approximation, the average intracavity photon number is increased by one, and the field becomes almost, but not exactly, the number state $|1\rangle$.

For the parameters of this example, the probability for the next atom to exit in the upper state is about $p_a \cong .98$, see Fig. 4b, so that it is also very likely for it to be measured in that state. Hence the cavity mode simply relaxes back at rate γ to a situation close to thermal equilibrium. As further atoms are injected, there is however a finite probability that eventually another atom will also exit in its lower state. In Fig. 4b, this happens next for atom $\cong 400$. The same process then starts again, resulting in the dynamics of $\langle n\rangle$ shown in Fig. 4a. This is an example of quantum measurements induced relaxation-oscillations.

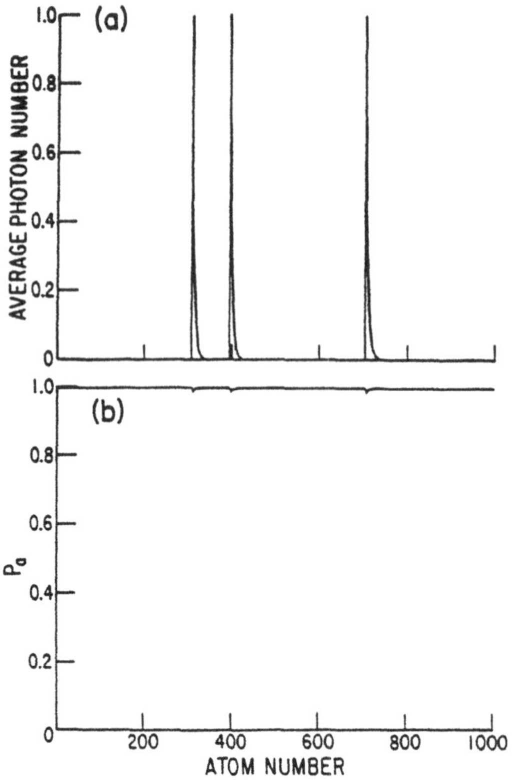

Figure 4: (*a*) Infered $\langle n \rangle$ for $N_{ex} = 5$, $\Theta = 35$, $n_b = 10^{-3}$ as a function of the number of atoms injected and measured. (*b*) Probability for the successive atoms to be measured in the upper state as they exit the cavity.

VI. SUMMARY

We have reviewed some aspects of the theory of a micromaser and showed how its dynamics is governed by the coherent interaction between the atoms and the quantized radiation mode. We showed that the major distinction between this device and conventional lasers and masers is the quantum mechanical phases characteristic of the coherent atom-field interaction are smeared out by the fluctuations associated with the laser pump and loss mechanisms. As a consequence, micromasers are truly quantum devices that typically emit non-classical fields. In contrast, the electrons in the maser or laser act then very much like a classical current, and generate in the best case a poissonian (coherent) field. Devices based on the micromaser principle could provide alternate routes towards the general goal of producing radiation sources with reduced fluctuations. A major difficulty in the microwave regime is however the absence of good photon detectors. We have shown how atoms can be used to probe the state of the field. However, they represent by no means an ideal measurement. In the case of micromaser the atoms act both as pump and detectors and have considerable back-action on the field to be measured. We are presently investigating alternate measurement schemes closer to ideal quantum non-demolition measurements.

Acknowledgements

This work is supported by NSF grant PHY-8603368 and by the Joint Services Optics Program. I have benefited from numerous discussions with J. H. Eberly, P. Filipowicz, J. Javanainen, T. A. B. Kennedy, C. M. Savage, M. O. Scully, E. M. Wright, D. F. Walls and H. Walther.

REFERENCES

1. R. Short and L. Mandel, Phys. Rev. Lett. **51**, 384 (1983).
2. H. J. Carmichael and D. F. Walls, J. Phys. **B9**, L43 and 1199 (1976).
3. M. C. Teich and B. E. A. Saleh, J. Opt. Soc. Am. **B2**, 275 (1985).
4. J. G. Walker and E. Jakeman, Optica Acta 32, 1303 (1985)
5. S. Machida, Y. Yamamoto and Y. Itaya, Phy. Rev. Lett. **58**, 1000 (1987).
6. P. Filipowicz, J. Javanainen, and P. Meystre, Phys. Rev. **A34**, 3077 (1986).
7. M. A. M. Marte and D. F. Walls, preprint (1987).
8. H. Walther, private communication (1987).
9. P. Filipowicz, J. Javanainen and P. Meystre, J. Opt. Soc. Am. **B3**, 906 (1986).
10. D. Meschede, H. Walther and G. Müller, Phys. Rev. Lett. **54**, 551 (1985).
11. J. Krause, M. O. Scully and H. Walther, Phys. Rev. **A34**, 2032 (1986).
12. M. Brune, J. M. Raimond and S. Haroche, Phys. Rev. **A35**, 154 (1987).
13. L. Davidovich, J. M. Raimond, M. Brune and S. Haroche, Phys. Rev. **A36**, 3771 (1987).
14. M. Brune, J. M. Raimond, P. Goy, L. Davidovich and S. Haroche, Phys. Rev. Lett. **59**, 1899 (1987).
15. P. Meystre, Opt. Letters **12**, 669 (1987).
16. P. Meystre and E. M. Wright, Phys. Rev. **A**, to be published.
17. W. E. Lamb, Jr., *"Quantum Mechanical Amplifiers"*, Vol. 2 of *Lectures in Theoretical Physics*, W. Brittin and D. W. Downs, eds. (Interscience, New York, 1960).
18. M. Sargent III, M. O. Scully and W. E. Lamb, *Jr*, *Laser Physics* (Addison Wesley, Reading, Mass 1974), Chap. 17.
19. P. Filipowicz, J. Javanainen, and P. Meystre, *"Why is laser light coherent ? Photon statistics in coherently driven oscillators"* in *Coherence, Cooperation and Fluctuations*, F. Haake, L. M. Narducci, and D. F. Walls, eds. (Cambridge University Press, Cambridge 1986), *p*. 206.
20. E. T. Jaynes and F. W. Cummings, Proc. IEEE **51**, 89 (1963).
21. A. O. Caldeira and A. J. Leggett, Phys. Rev. **A31**, 1059 (1985); D. F. Walls and G. J. Milburn, Phys. Rev. **A31**, 2403 (1985); C. M. Savage and D. F. Walls, Phys. Rev. **A32**, 2316 (1985)
22. G. Milburn, *"Quantum coherences in randomly kicked quantum systems"*, submitted to Phys. Rev. **A**.

PHASE SPACE, CORRESPONDENCE PRINCIPLE AND DYNAMICAL PHASES: PHOTON COUNT PROBABILITIES OF COHERENT AND SQUEEZED STATES VIA INTERFERING AREAS IN PHASE SPACE

Wolfgang P. Schleich

Max-Planck Institut für Quantenoptik,
D-8046 Garching bei München, W. Germany
and
Center for Advanced Studies and
Department of Physics and Astronomy
University of New Mexico
Albuquerque, New Mexico, USA, 87131

I. INTRODUCTION AND OVERVIEW

Motion of an electron around a nucleus or, in its most elementary version, vibratory motion of a harmonic oscillator viewed in Planck-Bohr-Sommerfeld quantized phase space;[1-3] and matching the *discrete*, microscopic world with the *continuous*, macroscopic world via Bohr's correspondence principle,[4-5] these are the essential ingredients of "Atommechanik".[4] Combined with the concept of interference – expressed in the familiar double-slit experiment[6] – these central ideas of early quantum mechanics provide in the present paper the most vivid sources of insight into the photon count probability, W_m, of a coherent state[7-9] shown in Fig. 1 and into the oscillatory[10-15] photon statistics[16] of a highly squeezed state[17] of a single mode of the electromagnetic field depicted in Fig. 2. Throughout these notes we employ the analogy[8,9] of the electromagnetic field to a harmonic oscillator described by the coordinate variable q and momentum $\tilde{p}$.

The paper is organized as follows: In Sec. II we "create" the most elementary examples of a coherent and squeezed state from the ground state, ψ_0, of a harmonic oscillator by a sudden displacement and sudden change in steepness of the potential. The original ground state, ψ_0, is no longer an energy eigenstate of the new potential and consequently shows a spread in energy. This spread in energy, that is, the probability, W_m, of finding m quanta of energy or m photons, for a coherent state is a Poisson distribution, in contrast to the oscillatory photon count probability, W_m, of a highly squeezed state.

In Sec. III we relate these photon distributions, W_m, to the area of overlap, A_m, in phase space between the m-th Bohr-Sommerfeld band representing the m-th energy eigenstate of the new oscillator and the Gaussian bell-shaped Wigner function,[18] $P^{(W)}_{\text{coh}}$, of the coherent state or the Gaussian cigar, $P^{(W)}_{sq}$, of the squeezed state. In

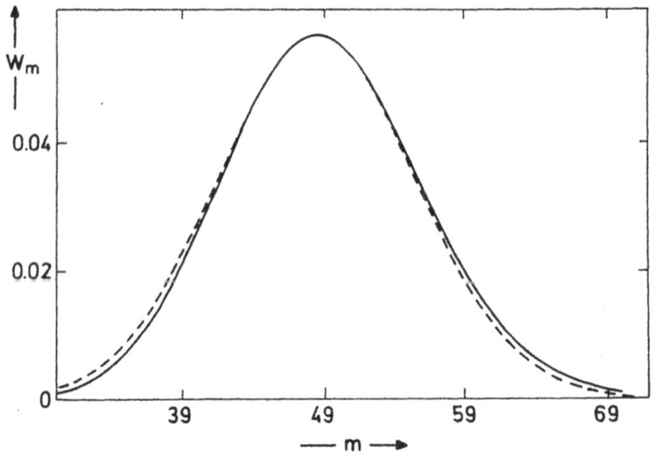

Fig. 1. The probability, W_m, of finding m photons in a coherent state is given by a Poisson distribution, (solid line). This exact distribution and its asymptotic limit for large displacements α (broken line) are almost indistinguishable in the neighborhood of the maximum $m \cong \alpha^2 - 1/2$. (We have chosen a displacement $\alpha = 7$.)

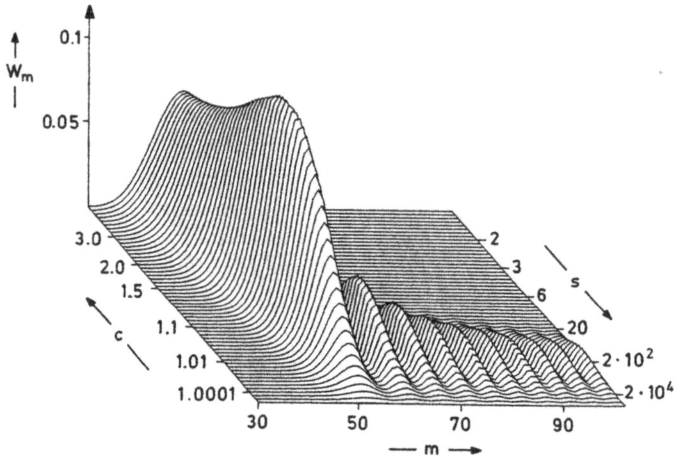

Fig. 2. Probability, W_m, of finding m photons in a squeezed state for different choices of the squeeze parameter s (or $c = (s+1)/(s-1)$). All curves are plotted for the same value $\alpha = 7$ of the displacement parameter. The rearmost curve (no squeeze at all, $s = 1$ or $c = \infty$) shows the ideal photon count Poisson distribution associated with a coherent state. Curves that are further forward display oscillations in the probability distribution of excitation. As the squeezing becomes extreme ($s \to \infty, c \to 1$) there are more and more of these oscillations at ever higher values of m and the percentage of probability under any one peak goes to zero (foremost curve, a mere line, $W_m \cong 0$).

the large-m limit, this area A_m is identical to the phase space "path" integral of the relevant Wigner function, $P^{(W)}_{\rm coh}$ or $P^{(W)}_{sq}$, evaluated along the semiclassical Bohr-Sommerfeld trajectory. In the case of a coherent state this line integral is governed by the neighborhood of a *single* point in phase space and yields $W_m \cong A_m$, that is, the photon count probability W_m is given by the area of overlap, A_m. In contrast, for a highly squeezed state there exist *two* symmetrically located points and thus *two* distinct areas, $\mathcal{A}_m$, of cross-over in phase space. These two zones corresponding to different phases of vibratory motion of the field oscillators represent two contributing probability amplitudes of identical absolute values, $\mathcal{A}_m^{1/2}$, but have a phase difference, $2\phi_m$, governed by the area caught between the center lines of the squeezed state and the Bohr-Sommerfeld band. As a result of these interfering areas in phase space the photon distribution of a highly squeezed state exhibits the striking oscillations shown in Fig. 2.

Section IV summarizes the main results and briefly compares and contrasts the area-of-overlap-and-interference-in-phase-space principle[11,19] to alternative phase space approaches such as the Wigner function concept[18] with its various semiclassical approximation schemes.[20–22] For the ingenious application of Q-function techniques and coherent phase integrals providing further insight into W_m we refer to Refs. 23 and 24.

II. COHERENT STATES AND SQUEEZED STATES: PHOTON COUNT PROBABILITY

In this section we briefly review basic properties of coherent and squeezed states of a single mode of the electromagnetic field in the language of its mechanical analogue – the harmonic oscillator. In particular, we emphasize the notion of sudden transitions[25] known from the physics of Franck-Condon transitions in diatomic molecules.[26] We "create" such states by a sudden displacement and, or squeeze of the harmonic oscillator potential in the spirit of Refs. 11, 13 and 27. We then derive the energy spread – in the field concept, the photon count probability, W_m – of a coherent state and a squeezed state and focus on the striking oscillations in W_m in the limit of strong squeezing.

The harmonic oscillator provides a description of the amplitude of one mode of the electromagnetic field[8,9] and is a powerful source of insight on what is measurable and what is not.[28] A cylinder of mass μ rolling under the influence of gravity on a metal ruler bent into the shape of a parabola can serve as a model. To begin with the oscillator is in its lowest quantum state[25]

$$\psi_0(q) = \left(\frac{\mu\omega'}{\pi\hbar}\right)^{1/4} \exp\left[-\frac{\mu\omega'}{2\hbar}q^2\right],$$

where q denotes the position and ω' the frequency of the oscillator.

A coherent state[7–9] is obtained from this ground state by a displacement. In the mechanical model this is achieved by suddenly displacing the origin of the harmonic oscillator by an amount $q_0 = (2\hbar/\mu\omega')^{1/2}\alpha$ and lowering the potential energy by $\frac{1}{2}\mu\omega'^2 q_0^2 = \hbar\omega'\alpha^2$. The wave function of the state – so prepared – a coherent state reads

$$\psi_{\rm coh}(q) = \left(\frac{\mu\omega'}{\pi\hbar}\right)^{1/4} \exp\left\{-\frac{\mu\omega'}{2\hbar}(q-q_0)^2\right\}. \tag{2.1}$$

When in addition we suddenly change the frequency ω' of the oscillator to $\omega = s^{-1}\omega'$, where $s > 0$, this wave function expressed in the new frequency ω reads

$$\psi_{sq}(q) = \left[s\left(\frac{\mu\omega}{\pi\hbar}\right)\right]^{1/4} \exp\left\{-s\frac{\mu\omega}{2\hbar}(q-q_0)^2\right\}. \tag{2.2}$$

The coherent state (2.1) is thus a special case of Eq. (2.2) for $s = 1$, that is, $\omega = \omega'$.

The wave functions $\psi_{\rm coh}$ and ψ_{sq} are not eigenfunctions of this oscillator, that is, they are not stationary states and thus undergo a time development. Both Gaussian wave packets – having the potential energy $\frac{1}{2}\mu\omega'^2 q_0^2 = \hbar\omega'\alpha^2$ – bounce back and forth between the classical turning points of the vibratory motion corresponding to this energy. The coherent state has just the "right" width to keep its shape while oscillating. In contrast, the wave function ψ_{sq} continuously changes its shape – it broadens and contracts – that is, it "breathes" while performing its vibratory motion.[8]

Oscillator phase space built out of coordinate and momentum variables reveals an additional difference between these two states. In the remainder of the present article we consider all states relative to energy eigenstates of an oscillator of frequency ω. Therefore we introduce a dimensionless position variable $x = (\mu\omega/\hbar)^{1/2} q$ and a dimensionless momentum $p = (\mu\hbar\omega)^{-1/2}\,\tilde{p}$. We thus distinguish a coherent state with wave function

$$\psi_{\rm coh}(x) = \pi^{-1/4} \exp\left[-(1/2)(x-\sqrt{2}\alpha)^2\right] \tag{2.3}$$

(and its subsequent oscillatory evolution with time) from the state

$$\psi_{sq}(x) = (s/\pi)^{1/4} \exp\left[-\frac{s}{2}(x-\sqrt{2}\alpha)^2\right]. \tag{2.4}$$

From Eq. (2.3) we most easily see that for a coherent state the spread in x-values (*probability*, not probability amplitude) about the most probable x-value goes as $\exp[-(x-\sqrt{2}\alpha)^2]$ and (by Fourier analysis) the spread in momentum goes as $\exp(-p^2)$, yielding a Gaussian bell-shaped Wigner phase space distribution[18]

$$P^{(W)}_{\rm coh}(x,p) = \pi^{-1}\exp[-(x-\sqrt{2}\alpha)^2 - p^2]. \tag{2.5}$$

The half widths, $(\Delta x)_{\rm coh}$ and $(\Delta p)_{\rm coh}$ in position and momentum respectively, following from Eq. (2.5) are thus identical and equal to unity maintaining the uncertainty relation

$$(\Delta x)_{\rm coh}\cdot(\Delta p)_{\rm coh} = 1\cdot 1 = 1.$$

Analogously the Wigner function for the state ψ_{sq}, Eq. (2.4), is a Gaussian cigar

$$P^{(W)}_{sq}(x,p) = \pi^{-1}\exp[-s(x-\sqrt{2}\alpha)^2 - s^{-1}p^2] \tag{2.6}$$

and yields for the half widths $(\Delta x)_{sq} = s^{-1/2}$ and $(\Delta p)_{sq} = s^{1/2}$. Thus for $s \neq 1$ these states have asymmetrical, elliptical uncertainties. Moreover, for $s > 1$ the uncertainty in position is reduced below that of a coherent state, which has led to the name *squeezed state*.[17] However, the inward pinch in x carries with it a corresponding stretch in p as indicated by $(\Delta p)_{sq} = s^{1/2}$. The "area" of the cigar remains the same as before,

$$(\Delta x)_{sq} \cdot (\Delta p)_{sq} = s^{-1/2} s^{1/2} = 1$$

to maintain the uncertainty relation. For $0 < s < 1$, the uncertainty in momentum p is reduced at the expense of enhanced fluctuations in x.

Neither a coherent nor a squeezed state is an energy eigenstate of the new oscillator. Both show a spread in energy. How large a spread, and how different a spread for a squeezed state from the coherent state?

Even for a coherent state, Eq. (2.3), the energy distribution is not a delta function located at the classical value

$$\bar{m} = (1/2)(\text{spring constant})(\text{displacement})^2 = (1/2) \cdot 1 \cdot (\sqrt{2}\alpha)^2 = \alpha^2.$$

Instead it is a Poisson distribution, following from standard quantum mechanics[25] as

$$\begin{aligned} W_m &= \left| \int_{-\infty}^{\infty} dx u_m(x) \psi_{coh}(x) \right|^2 = \frac{(\alpha^2)^m}{m!} \exp(-\alpha^2) \\ &\cong (2\pi)^{-1/2} |\alpha|^{-1} \exp\left\{ - \left[\frac{m + \frac{1}{2} - \alpha^2}{\sqrt{2}\alpha} \right]^2 \right\} \end{aligned} \quad (2.7)$$

Here

$$u_m(x) = \pi^{-1/4} (2^m m!)^{-1/2} H_m(x) \exp[-(1/2)x^2] \quad (2.8)$$

denotes the wave function of the m-th energy eigenstate[25] with H_m being the m-th Hermite polynomial[29] and ψ_{coh} is given by Eq. (2.3). In the last step we have performed[21] an asymptotic expansion of the exact Poisson distribution shown in Fig. 1 by the solid line yielding in the limit $\alpha^2 \gg 1$ the familiar Gaussian approximation represented in Fig. 1 by the broken line.

Analogously we find for the case of a squeezed state of wave function, Eq. (2.4), the photon count probability[16,30]

$$\begin{aligned} W_m &= \left| \int_{-\infty}^{\infty} dx u_m(x) \psi_{sq}(x) \right|^2 = \frac{2\sqrt{s}}{s+1} \left(\frac{s-1}{s+1} \right)^m (2^m m!)^{-1} H_m^2 \left(\frac{s}{(s^2-1)^{1/2}} \sqrt{2}\alpha \right) \\ &\times \exp\left(-\frac{2s}{s+1} \alpha^2 \right) \end{aligned} \quad (2.9)$$

shown in Fig. 2 for selected values of the squeeze parameter s (or its complement $c = (s+1)/(s-1)$). Here the displacement parameter α has the value $\alpha = 7$.

When we increase the squeeze, that is, increase s starting from $s = 1$ (no squeeze at all) the photon count probability, W_m, experiences a transition from a Poisson distribution, ($s = 1$) to a sub-Poisson distribution, ($s \cong 6$). In the limit of strong squeezing ($s \to \infty$) the probability W_m shows oscillations[10-15] – a striking feature[31] quite different from the quasi-monotonic Poisson distribution of a coherent state. Oscillations? Yes! But why?

How little insight into the physical origin of this rapid variation in W_m is offered by the present form of Eq. (2.9)! An asymptotic expansion[14] of W_m for the range of the parameters where the oscillations make their appearance, that is, for large squeezing, $s \to \infty$, and quantum numbers m appropriately larger than $\alpha^2 \gg 1$, throws some light on this rather surprising effect.

According to Appendix A we find for $s \cong 2/\epsilon$, where $0 < \epsilon \ll 1$, and m appropriately larger than α^2

$$W_m \cong 4\mathcal{A}_m \cos^2 \phi_m \tag{2.10a}$$

where

$$\mathcal{A}_m = (\epsilon/4\pi)^{1/2} \frac{\exp[-\epsilon(m + \frac{1}{2} - \alpha^2)]}{(m + \frac{1}{2} - \alpha^2)^{1/2}} \tag{2.10b}$$

and

$$\begin{aligned} \phi_m &= S_m(\sqrt{2}\alpha) - \pi/4 = \int_{\sqrt{2}\alpha}^{[2(m+1/2)]^{1/2}} dx[2(m+1/2) - x^2]^{1/2} - \pi/4 \\ &= (m+1/2)\arctan\left[\alpha^{-1}(m+1/2-\alpha^2)^{\frac{1}{2}}\right] \\ &\quad - \alpha(m+1/2-\alpha^2)^{\frac{1}{2}} - \pi/4. \end{aligned} \tag{2.10c}$$

Now we realize that the photon count probability W_m shown in Fig. 3 takes the typical form of an interference effect. However, interference of what? Interference in which space? This phenomenon does not take place in ordinary space but in the more sophisticated arena of phase space.

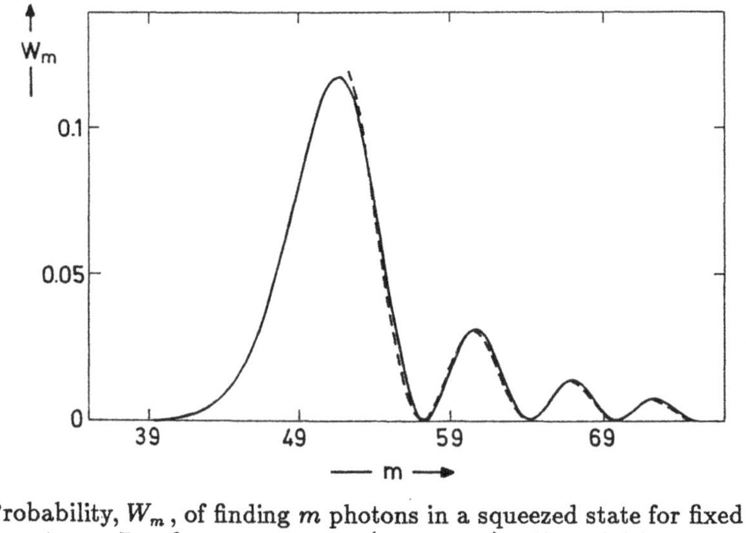

Fig. 3. Probability, W_m, of finding m photons in a squeezed state for fixed displacement $\alpha = 7$ and squeeze $s = 21$ (or $c = 1.1$). The solid line represents the exact expression for W_m, Eq. (2.9), whereas the dashed line depicts the approximation, Eq. (2.10).

III. AREA OF OVERLAP AND INTERFERENCE IN PHASE SPACE

The classical history of a harmonic oscillator in $x - p$ phase space is a circle. The quantum levels are given by the rule:[1-4] area inside ground state phase diagram equal to π (in units $\hbar$); between one orbit and the next, 2π itself. Thus the circular Bohr-Sommerfeld phase space trajectories are given by

$$m + 1/2 = (1/2)p^2 + (1/2)x^2 \tag{3.1}$$

and are traversed in clockwise direction as shown in Fig. 4.

A version[1] better suited for the present discussion associates with each quantum state not a single trajectory (3.1), but a circular band of area 2π (in units $\hbar$) with inner radius $(2m)^{1/2}$ and outer radius $[2(m+1)]^{1/2}$ as indicated in the inset of Fig. 4. The trajectory (3.1) runs in the middle of the band.

A first try at figuring the probability, W_m, of occupation of the m-th energy level in a coherent or squeezed state associates with W_m the area of overlap, A_m, in phase space between the m-th level represented by the m-th Bohr-Sommerfeld band and the Gaussian bell or the Gaussian cigar, that is, the Wigner function $P^{(W)}_{coh}$ or $P^{(W)}_{sq}$,

$$A_m \equiv \underset{\text{m} - \text{th band}}{\int dx \int dp P^{(W)}_{(j)}(x,p)} \tag{3.2}$$

as shown in Figs. 5 and 6 ($j = \text{coh}, sq$).

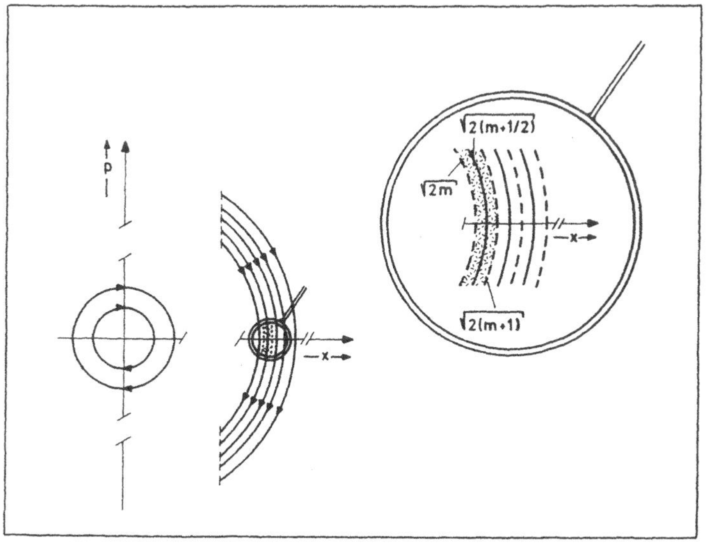

Fig. 4. A single mode of the electromagnetic field in a number state is equivalent to a harmonic oscillator with dimensionless variables x and p. The trajectories in phase space are circles of radius $[2(m+1/2)]^{1/2}$. With each state we associate a band of area 2π (in units $\hbar$) defined by its inner radius $(2m)^{1/2}$ and outer radius $[2(m+1)]^{1/2}$.

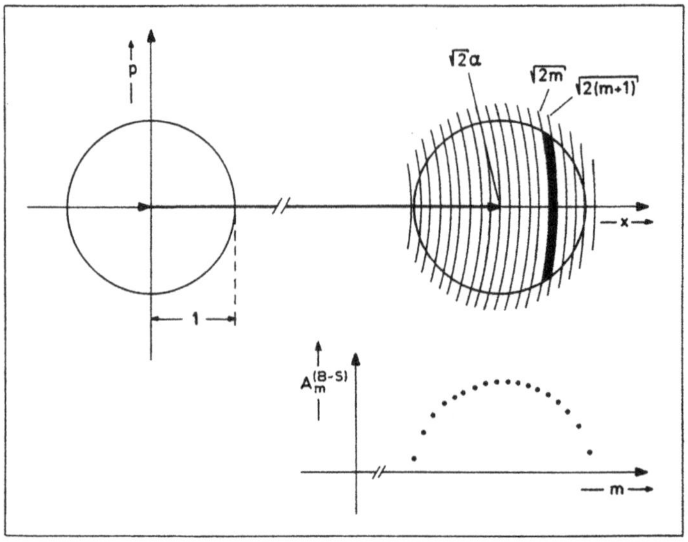

Fig. 5. The ground state of a harmonic oscillator – visualized in phase space as a Gaussian bell (depicted here by its circular contour line of exponential fall off) – when displaced from the origin by an amount $\sqrt{2}\alpha$ represents a coherent state. The area of overlap, A_m, between the m-th band (representing the m-th number state) and the Gaussian bell – the simplest algorithm for determining the photon distribution of a coherent state – yields a distribution almost indistinguishable from the correct Poissonian result as indicated in Fig. 1 by a broken line. An even simpler version of this algorithm employs the area of overlap, $A_m^{(B-S)}$, between the m-th annulus and the circle itself as suggested by the ground state Bohr-Sommerfeld trajectory $m = 0$. This yields a "semi-ellipse shaped" distribution (Appendix B) reproducing the qualitative features of the exact Poisson result.

The power of Bohr's correspondence principle[4] stands out most clearly when we express the weighted annulus, A_m, of Eq. (3.2) as the difference

$$A_m = a(m+1) - a(m) \tag{3.3}$$

between the circular areas of radii $(2m)^{1/2}$ and $[2(m+1)]^{1/2}$ weighted by the distribution function $P_{(j)}^{(W)}$, that is,

$$a(m') = \underset{\substack{\text{circle of radius} \\ (2m')^{1/2}}}{\int dx \int dp} P_{(j)}^{(W)}(x,p) \tag{3.4}$$

as indicated in Fig. 7. In the large-m limit, that is, in Bohr's correspondence limit, we replace[4] the difference (3.3) by the differential, that is,

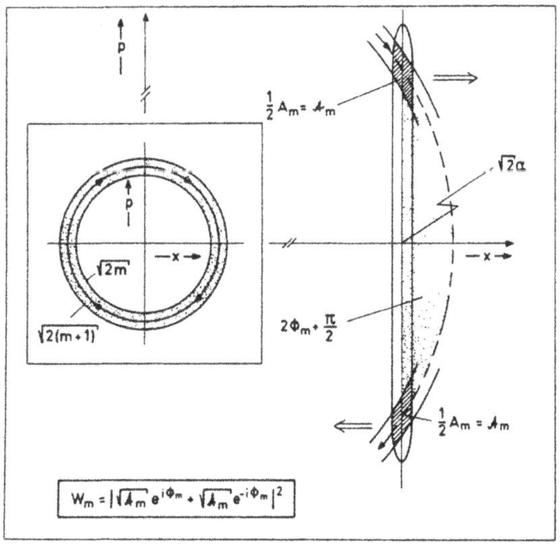

Fig. 6. Oscillations in photon distribution, W_m, of a highly squeezed state as a consequence of *interference in phase space*. For excitations m appropriately larger than α^2 the bands of inner radius $(2m)^{1/2}$ and outer radius $[2(m+1)]^{1/2}$ shown in the inset and representing the m-th number state intersect the elliptical contour line of a highly squeezed state in *two* symmetrically located diamond-shaped zones of weighted area, $\mathcal{A}_m$, at different phases of the oscillatory motion: in one zone the oscillator moves to the "right"; in the other to the "left". The total probability amplitude, $W_m^{1/2}$, is thus the sum of contributions $\mathcal{A}_m^{1/2} \exp(\pm i\phi_m)$ from the shaded areas. The phase difference, $2\phi_m$, is fixed by the dotted domain caught between the center lines of the two states. As a result of this *interference in phase space* the photon distribution oscillates for m values appropriately larger than α^2.

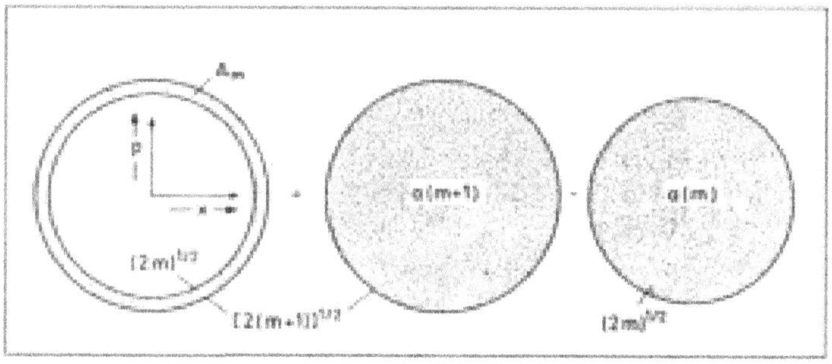

Fig. 7. The area, A_m, of the weighted m-th Bohr-Sommerfeld annulus defined in $(x-p)$ oscillator phase space by the inner radius $(2m)^{1/2}$ and outer radius $[2(m+1)]^{1/2}$ indicated by the solid band is given by the difference between the weighted circular areas of radius $[2(m+1)]^{1/2}$ and $(2m)^{1/2}$.

$$A_m \cong \left.\frac{\partial a}{\partial m'}\right|_{m'=m+1/2} . \tag{3.5}$$

When we rewrite Eq. (3.4) in polar coordinates, $x = r\cos\varphi$ and $p = r\sin\varphi$, that is,

$$a(m') = \int_0^{(2m')^{1/2}} dr r \int_{-\pi}^{\pi} d\varphi P_{(j)}^{(W)}(x = r\cos\varphi, p = r\sin\varphi)$$

the differentiation in Eq. (3.5) with respect to m' can easily be performed to yield

$$A_m \cong \int_{-\pi}^{\pi} d\varphi P_{(j)}^{(W)}(x = [2(m+1/2)]^{1/2}\cos\varphi, p = [2(m+1/2)]^{1/2}\sin\varphi). \tag{3.6}$$

Thus in the semiclassical limit the area of overlap in phase space, A_m, between the m-th Bohr-Sommerfeld band and a coherent or squeezed state described by a (Wigner) distribution $P_{(j)}^{(W)}$ shown in Fig. 8, respectively, is identical to the line integral of $P_{(j)}^{(W)}$ evaluated along the m-th Bohr-Sommerfeld trajectory, Eq. (3.1), that is, along a circle of radius $r_{m+1/2} \equiv [2(m+1/2])^{1/2}$. In the Appendices B and C we evaluate the line integral, Eq. (3.6), for the two cases of a coherent and a squeezed state using the method of steepest descents.[32]

For a coherent state the main contribution to the integral arises from the neighborhood $\Delta\varphi \cong 2^{-1/2}\alpha^{-1}$ of a *single* phase angle, $\varphi = 0$ as shown in Fig. 8a. We identify $\Delta\varphi$ as the angle under which the coherent state crudely depicted as a circle of radius unity is seen from the origin. A detailed calculation shown in Appendix B yields

$$A_m \cong (2\pi)^{-1/2}|\alpha|^{-1}\exp\left\{-\left[\frac{m+1/2-\alpha^2}{\sqrt{2}\alpha}\right]^2\right\} \cong W_m . \tag{3.7}$$

Hence in the semiclassical limit the area of overlap, A_m, in phase space between the m-th Bohr-Sommerfeld band and the coherent state is identical to the probability, W_m, of finding m-photons in a coherent state.

We now turn to the case of a state highly squeezed in the x variable described by the Wigner function (2.6) for $s \cong 2/\epsilon$, Eq. (A2), where $0 < \epsilon \ll 1$. According to Appendix C and Fig. 8b there exist *two* distinct areas of overlap corresponding to the two shaded diamond-shaped zones of Fig. 6, each of area $\mathcal{A}_m$. The total probability, W_m, of finding m photons however, is not the sum, $2\mathcal{A}_m$ of the areas of the two diamonds. Neither is the intensity on the photographic plate in the familiar double-slit experiment equal to the sum of the intensities that would arrive through the slits separately.

Quantum mechanics instructs us to add not probabilities but probability amplitudes, *before* any evaluation of probability, thus

$$W_m = \left|\mathcal{A}_m^{1/2}e^{i\phi_m} + \mathcal{A}_m^{1/2}e^{-i\phi_m}\right|^2 . \tag{3.8}$$

The physics of this result is clear: The two diamond-shaped areas $\mathcal{A}_m$ correspond to two different phases of vibratory motion of the harmonic oscillator in its m-th state of excitation and the "squeezed state oscillator" at $x = \sqrt{2}\alpha$. In one zone the

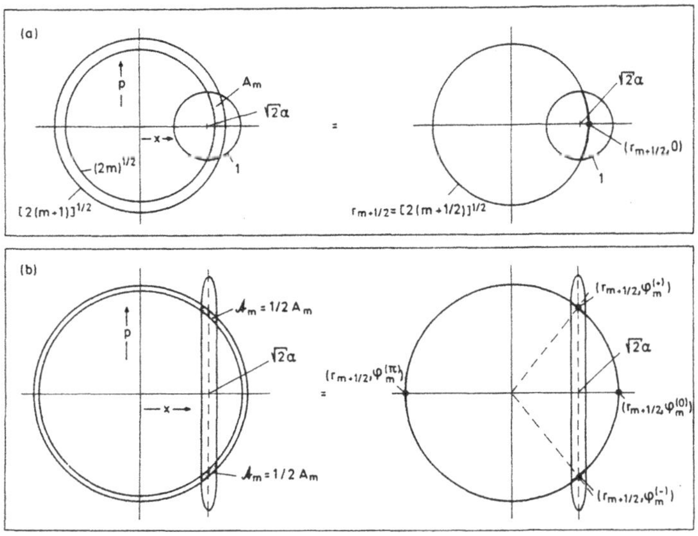

Fig. 8. In the semiclassical limit the area of overlap in phase space, A_m, between the m-th Bohr-Sommerfeld band defined by inner radius $(2m)^{1/2}$ and outer radius $[2(m+1)]^{1/2}$ and a coherent state (a) represented by a Gaussian bell-shaped Wigner function, $P_{\mathrm{coh}}^{(W)}$, depicted here in its simplest version as a circle, or a highly squeezed state (b) shown by a Gaussian cigar type Wigner function, $P_{sq}^{(W)}$, indicated by elliptical contour lines, is identical to the line integral of the corresponding phase space distributions evaluated along the semiclassical Bohr-Sommerfeld trajectory, Eq. (3.1), that is, along a circle of radius $r_{m+1/2} = [2(m+1/2)]^{1/2}$. In the case of a coherent state thc main contribution to this line integral arises from a *single* phase space point, $(r_{m+1/2}, 0)$ whereas in the case of a highly squeezed state the value of this phase space "path" integral is governed by the *two* symmetrically located, distinct points of intersection $(r_{m+1/2}, \varphi_m^{(+)})$ and $(r_{m+1/2}, \varphi_m^{(-)})$ between the circle and the center (dashed) line of the squeezed state. In the phase space points $(r_{m+1/2}, \varphi_m^{(0)})$ and $(r_{m+1/2}, \varphi_m^{(\pi)})$ the contributions to Eq. (3.6), assume minima.

momentum is positive corresponding to motion to the "right", whereas in the other, negative, as appropriate for motion to the "left". According to Eq. (2.10c) the phase difference, $2\phi_m$, between the two contributing zones of cross-over is fixed by the area in phase space caught between the center lines of the two states. This *interference-determining phase space area*, $2\phi_m + \pi/2$, shown in Figs. 6 and 9 by the dotted zone provides further insight. The area

$$(1/2)r_{m+1/2}^2\left(\varphi_m^{(+)} - \varphi_m^{(-)}\right) = 2(m+1/2)\arctan\left[(m+1/2-\alpha^2)^{1/2}/\alpha\right]$$

of the phase space segment defined by the m-th Bohr-Sommerfeld circle of radius $r_{m+1/2} = [2(m+1/2)]^{1/2}$ and the angle $\varphi_m^{(+)} - \varphi_m^{(-)} = 2\arctan[p_m(\sqrt{2}\alpha)/\sqrt{2}\alpha]$ corresponds to the phase difference of the oscillator in its m-th state of excitation at the two points $(r_{m+1/2}, \varphi_m^{(\pm)})$. On the other hand the area

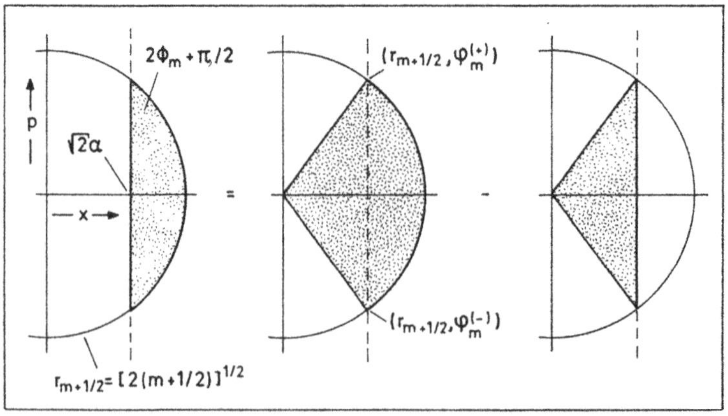

Fig. 9. The phase space area, $2\phi_m + \pi/2$, governing the phase difference between the two contributing probability amplitudes and defined by the dotted zone caught between the two center lines of the two states is the difference between the areas associated with the phases of the field oscillator in its m-th state and the squeezed oscillator confined to narrow neighborhood of $x = \sqrt{2}\alpha$ and represented by a straight line.

$$\sqrt{2}\alpha p_m\,(x = \sqrt{2}\alpha) = 2\alpha(m + 1/2 - \alpha^2)^{1/2}$$

of the triangle defined by the origin and the phase space points $(r_{m+1/2}, \varphi_m^{(+)})$ and $(r_{m+1/2}, \varphi_m^{(-)})$ represents the phase difference of the "squeezed oscillator" confined to a narrow regime around $x = \sqrt{2}\alpha$ as shown in Fig. 9. The total phase difference $2\phi_m + \pi/2$ is thus the difference of the two areas as indicated in Fig. 9 and by Eq. (2.10c).

We conclude this section by explaining the rapid oscillations in W_m of Fig. 2 via this phase fixing zone. The area between consecutive bands is 2π. Thus the dotted area governing the phase difference $2\phi_m$ rapidly increases for increasing quantum numbers. As a result the photon distribution of a highly squeezed state develops the oscillations shown in the foreground of Fig. 2.

IV. SUMMARY AND DISCUSSION

Interfering areas of cross-over in phase space as a measure of the semiclassical quantum mechanical scalar product between two states – this is the central idea of the present article. In the limit of large quantum numbers the Poissonian photon distribution, W_m, of a coherent state results from, and is identical to the area of overlap, A_m, in phase space between the m-th Bohr-Sommerfeld band and the Gaussian bell of the coherent state. In the case of a highly squeezed state, represented by a tall, thin Gaussian cigar, the cross-over with the m-th band consists of *two* symmetrically-located diamond-shaped weighted zones. The two areas of value $\mathcal{A}_m$ correspond to two distinct phases of vibratory motion of the field oscillators. The total probability amplitude $W_m^{1/2}$ to find m photons in a highly squeezed state is thus the sum of probability amplitudes $\mathcal{A}_m^{1/2}\exp(\pm i\phi_m)$ with the phase difference,

$2\phi_m$, given by the area caught between the center lines of the Bohr-Sommerfeld band and the squeezed state. As a result of this *interference in phase space* the photon distribution W_m of a highly squeezed state exhibits oscillations.

Bohr's correspondence principle identifies the area of overlap, A_m, with the "path"-integral of the corresponding (Wigner-Cohen) phase space distribution function – Gaussian bell and Gaussian cigar – taken along the semiclassical Bohr-Sommerfeld trajectory. This relates the present work to the phase space interpretation of semiclassical quantum mechanics given in Ref. 20. This approach deals in terms of Wigner functions[18] and the photon statistics W_m of the coherent state is then given[18] by the phase space integral

$$W_m = 2\pi \int_{-\infty}^{\infty} dx \int_{-\infty}^{\infty} dp P_m^{(W)}(x,p) P_{\mathrm{coh}}^{(W)}(x,p) \tag{4.1}$$

where $P_m^{(W)}$ denotes the Wigner function of the harmonic oscillator in its m-th state of excitation.[18,33] In the semiclassical approximation $P_m^{(W)}$ is approximated[20] by a delta-function density located at the Bohr-Sommerfeld trajectory (3.1), that is,

$$P_m^{(W)}(x,p) \cong \pi^{-1}\delta[2(m+1/2)-(x^2+p^2)]. \tag{4.2}$$

When we substitute Eq. (4.2) into (4.1) and introduce polar coordinates $x = r\cos\varphi$ and $p = r\sin\varphi$ we find from Eq. (4.1)

$$W_m \cong 2\int_0^{\infty} dr r\delta[2(m+1/2)-r^2] \int_{-\pi}^{+\pi} d\varphi P_{\mathrm{coh}}^{(W)}(x = r\cos\varphi, p = r\sin\varphi),$$

a result which after integration over r yields Eq. (3.6).

An alternative approach[21,22] within the Wigner function scheme does not replace $P_m^{(W)}$ by a delta function density but puts to use the asymptotic behavior of $P_m^{(W)}$ in the large-m limit. Then $P_m^{(W)}$ consists of spherical wave fronts[21,22,33] – crests and troughs – with the outermost front always being a crest located at the Bohr-Sommerfeld trajectory. The photon statistics of a coherent state is governed[21] by this outermost crest. In contrast, the oscillations in W_m for a highly squeezed state result[22] from the inner wave fronts and, in particular, from the regions in phase space, where $P_m^{(W)}$ assumes negative values.[33,34] Thus the semiclassical concept of interference in phase space identifies a complex probability *amplitude* with interfering areas in phase space whereas the Wigner function approach deals with "probabilities" some positive, some negative. Interference phenomena are taken into account by negative "pseudo"-probabilities.

We conclude by noting that the oscillations in the photon distribution of a highly squeezed state are analogous to the oscillations in transition probability between two different vibrational levels of two electronic states in diatomic molecules, (Franck-Condon transitions[26]) shown in Fig. 10 and found experimentally[35] as early as 1933. They also result[11,19] from interference in phase space. The experimental verification of this rapid variation in the photon statistics of a highly squeezed state would therefore constitute a test of the concept of interference in phase space on the level of second quantization.

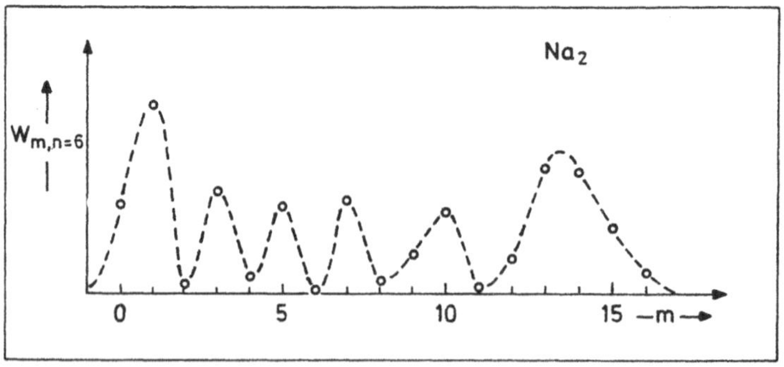

Fig. 10. Probability for a transition to occur between the $n = 6$-th vibrational level in the π-electronic state to the m-th vibrational mode of the Σ-electronic state in the diatomic molecule Na_2 (Franck-Condon effect). Between the two Franck-Condon maxima at $m = 1$ and $m = 13$ the catenary curve is modulated as a consequence of ***interference in phase space*** similar to the modulation in the decaying tail of the photon statistics of a highly squeezed state shown in Figs. 2 and 3.

ACKNOWLEDGMENTS

The author of the present article is enormously grateful to J. A. Wheeler for introducing him to the beauty and simplicity of semiclassical quantum mechanics and for the privilege of collaborating with him on these problems while being with the Center for Theoretical Physics at the University of Texas at Austin and over the last years. During the course of this work we have profited immensely from discussions with H. Carmichael, R. Y. Chiao, L. Cohen, K. Dodson, P. Drummond, R. Glauber, M. Hillery, H. J. Kimble, K. Kraus, G. Leuchs, R. F. O'Connell, G. J. Milburn, M. O. Scully, D. F. Walls and H. Walther. In particular, I would like to thank R. Pike and P. Tombesi for organizing a most splendid conference. Many thanks go to Jeanne Williams for bringing this manuscript into the present form. Support of the Office of Naval Research while being with the Center for Advanced Studies at the University of New Mexico is gratefully acknowledged. Preparation of this article was assisted by National Science Foundation.

APPENDIX A: ASYMPTOTIC EXPANSION OF PHOTON DISTRIBUTION, EQ. (2.9), OF HIGHLY SQUEEZED STATE

In this Appendix we derive an asymptotic expansion of the photon count probability, W_m, Eq. (2.9) of a highly squeezed state, that is, for $s \to \infty$, or equivalently for $c = 1 + \epsilon$, where $0 < \epsilon \ll 1$, valid for quantum numbers m appropriately larger than $\alpha^2 \gg 1$.

In the limit of $s \to \infty$, Eq. (2.9) reduces to

$$W_m \cong 2s^{-1/2} \left(1 - \frac{1}{s}\right)^{2(m+1/2)} \exp\left(\frac{2}{s}\alpha^2\right) (2^m m!)^{-1} H_m^2 \left(\sqrt{2}\alpha\right) \exp(-2\alpha^2). \quad (A1)$$

With the help of

$$\left(1-\frac{1}{s}\right)^{2(m+1/2)}=\exp\left[2\left(m+\frac{1}{2}\right)ln\left(1-\frac{1}{s}\right)\right]\cong\exp\left[-\frac{2}{s}\left(m+\frac{1}{2}\right)\right]$$

and

$$s=\frac{c+1}{c-1}=\frac{2+\epsilon}{\epsilon}\cong\frac{2}{\epsilon} \tag{A2}$$

we simplify Eq. (A1)

$$W_m\cong 2(\epsilon/2)^{1/2}\exp\left[-\epsilon\left(m+1/2-\alpha^2\right)\right](2^m m!)^{-1}H_m^2(\sqrt{2}\alpha)\exp(-2\alpha^2). \tag{A3}$$

An asymptotic expansion of the Hermite polynomials, H_m, for $\alpha^2 \gg 1$ in terms of the Airy function, Ai, has been presented in Ref. 14. According to Fig. 2 the oscillations in W_m arise for $m > \alpha^2$. In this regime we obtain an asymptotic expansion of H_m by comparing the WKB wave function[5] of the harmonic oscillator valid for $|x| < \xi_m \equiv [2(m+1/2)]^{1/2}$,

$$u_m(x)\cong(2/\pi)^{1/2}(p_m(x))^{-1/2}\cos[S_m(x)-\pi/4]$$

where

$$p_m(x)=\left[2(m+1/2)-x^2\right]^{1/2}$$

and

$$S_m(x)\equiv\int_x^{\xi_m}dx p_m(x)$$

to the exact wave function, Eq. (2.8) ,

$$H_m(x)=2^{1/2}\pi^{-1/4}(2^m m!)^{1/2}\exp[(1/2)x^2][p_m(x)]^{-1/2}\cos[S_m(x)-\pi/4].$$

When we substitute this expansion into Eq. (A3) we find for $x=\sqrt{2}\alpha < [2(m+1/2)]^{1/2}$, that is, for $\alpha^2 < m+1/2$, the result Eq. (2.10).

APPENDIX B: EVALUATION OF LINE INTEGRAL, EQ. (3.6), FOR COHERENT STATE

In this appendix we evaluate the area of overlap, A_m, between the m-th Bohr-Sommerfeld band and a coherent state, Eq. (3.2) via the line integral, Eq. (3.6), that is,

$$A_m\cong\pi^{-1}\exp\left\{-2\left[(m+1/2)+\alpha^2\right]\right\}\int_{-\pi}^{\pi}d\varphi\exp\left[4\alpha(m+1/2)^{1/2}\cos\varphi\right] \tag{B1}$$

where we have used Eq. (2.5).

The integration over φ can be performed in an exact way yielding the modified Bessel function, I_0.[36] However, in the large-m limit we apply the method of steepest descents.[32] Since

$$\cos\varphi \cong 1 - (1/2)\varphi^2 + \ldots$$

the main contribution arises from $\varphi = 0$ and Eq. (B1) reduces to

$$A_m \cong \pi^{-1/2} \exp\left\{-2\left[(m+1/2)^{1/2} - \alpha\right]^2\right\} \pi^{-1/2}$$
$$\times \int_{-\pi}^{\pi} d\varphi \exp\left\{-2\alpha\,(m+1/2)^{1/2}\,\varphi^2\right\}$$

For $2\alpha(m+1/2)^{1/2} \cong 2\alpha^2 \gg 1$ we extend the limits of integration to infinity. When we perform the resulting Gaussian integral we arrive at

$$A_m \cong (2\pi)^{-1/2}|\alpha|^{-1} \exp\left\{-2\left[(m+1/2)^{1/2} - \alpha\right]^2\right\}.$$

The relation

$$(m+1/2)^{1/2} - \alpha = \frac{m+1/2-\alpha^2}{(m+1/2)^{1/2}+\alpha} \cong \frac{1}{2\alpha}\left(m+1/2-\alpha^2\right)$$

finally yields

$$A_m \cong (2\pi)^{-1/2}|\alpha|^{-1} \exp\left\{-\left[\frac{m+1/2-\alpha^2}{\sqrt{2}\alpha}\right]^2\right\} \cong W_m\,, \qquad (B2)$$

that is, the Gaussian approximation to the Poisson distribution, Eq. (2.7).

It is interesting to compare and contrast this result, Eq. (B2), to the tentative probability $W_m \cong A_m^{(B-S)}$ obtained from Eq. (3.6) for the most elementary phase space representation

$$P_{\mathrm{coh}}^{(B-S)}(x,p) = \frac{1}{\pi}\theta[1-(x-\sqrt{2}\alpha)^2 - p^2] \qquad (B3)$$

of a coherent state suggested by the Bohr-Sommerfeld trajectory, Eq. (3.1), for $m = 0$. Here θ denotes the Heaviside step function. The area of overlap, $A_m^{(B-S)}$, between the m-th band and the circle, (B3), is thus given by

$$A_m^{(B-S)} \cong \frac{1}{\pi}\int_{-\pi}^{\pi} d\varphi\theta\left\{1 - 2\left[\frac{m+1/2-\alpha^2}{(m+1/2)^{1/2}+\alpha}\right]^2\right.$$
$$\left. - 8\alpha(m+1/2)^{1/2}\sin^2(\varphi/2)\right\}. \qquad (B4)$$

When we make use of $[2(m+1/2)]^{1/2} \cong \sqrt{2}\alpha$ and $\varphi \sim 2^{-1/2}\alpha^{-1} \ll 1$, Eq. (B4) reduces to

$$A_m^{(B-S)} \cong \frac{1}{\pi}\int_{-\pi}^{\pi} d\varphi\theta\left[1 - \frac{1}{2\alpha^2}(m+1/2-\alpha^2)^2 - 2\alpha^2\varphi^2\right],$$

that is,

$$A_m^{(B-S)} \cong \frac{1}{\pi\alpha^2}[2\alpha^2 - (m+1/2-\alpha^2)^2]^{1/2}\theta[2\alpha^2 - (m+1/2-\alpha^2)^2]. \qquad (B5)$$

This tentative "semi-ellipse shaped" probability distribution depicted schematically on the lower part of Fig. 5 is peaked at $m \cong \alpha^2 - 1/2$ and has a total width $2\cdot\sqrt{2}\alpha$, in agreement with the approximation, Eq. (B2), of the exact Poisson result. However, the distribution, $P_{coh}^{(B-S)}$, attributes to each point inside the coherent state circle equal weight resulting in the square root dependence of $A_m^{(B-S)}$, (Eq. (B5)), in contrast to the Gaussian bell, Eq. (2.5).

APPENDIX C: EVALUATION OF LINE INTEGRAL, EQ. (3.6), FOR HIGHLY SQUEEZED STATE

In this Appendix we evaluate the area of overlap, A_m, Eq. (3.2) between a highly squeezed state and the m-th Bohr-Sommerfeld band in the large-m limit. When we substitute $P_{sq}^{(W)}$, Eq. (2.6), into the line integral Eq. (3.6) we find after minor algebra

$$A_m \cong \pi^{-1/2}\exp\left(-\frac{4}{\epsilon}\alpha^2\right)\exp\left\{-\frac{2}{\epsilon}\left[1+\left(\frac{\epsilon}{2}\right)^2\right](m+1/2)\right\}\mathcal{L}_m \qquad (C1)$$

where

$$\mathcal{L}_m \equiv \pi^{-1/2}\int_{-\pi}^{\pi} d\varphi \exp\left[-\frac{2(m+1/2)}{\epsilon}h_m(\varphi)\right] \qquad (C2)$$

and

$$h_m(\varphi) = \left[1-\left(\frac{\epsilon}{2}\right)^2\right]\cos(2\varphi) - 4\alpha(m+1/2)^{-1/2}\cos\varphi. \qquad (C3)$$

The integral $\mathcal{L}_m$ is closely related to generalized Bessel functions discussed in Ref. 37. For the present discussion however, the method of steepest descents is sufficient to furnish an asymptotic expansion of $\mathcal{L}_m$.

For $[2(m+1/2)]/\epsilon \gg 1$, that is, for large squeezing and in the Bohr correspondence limit we expand h_m into a Taylor series

$$h_m(\varphi) \cong h_m\left(\varphi_m^{(j)}\right) + 1/2\left.\frac{d^2 h_m}{d\varphi^2}\right|_{\varphi=\varphi_m^{(j)}}\left(\varphi-\varphi_m^{(j)}\right)^2 + \ldots$$

around the points $\varphi_m^{(j)}$ given by the roots of

$$\left.\frac{dh_m}{d\varphi}\right|_{\varphi=\varphi_m^{(j)}} = 0. \qquad (C4)$$

When we differentiate Eq. (C3) and substitute into Eq. (C4) we arrive at the condition

$$\sin\varphi_m^{(j)}\left\{\left[1-\left(\frac{\epsilon}{2}\right)^2\right]\cos\varphi_m^{(j)} - \alpha(m+1/2)^{-1/2}\right\} = 0 \qquad (C5)$$

which yields

$$\varphi_m^{(0)} = 0 \qquad (C6)$$

and

$$\varphi_m^{(\pi)} = \pi, \qquad (C7)$$

and

$$\left[1 - \left(\frac{\epsilon}{2}\right)^2\right] [2(m + 1/2)]^{1/2} \cos\varphi_m^{(j)} = \sqrt{2}\alpha. \qquad (C8)$$

The angles $\varphi_m^{(0)}, \varphi_m^{(\pi)}$ and $\varphi_m^{(\pm)}$ are thus determined by the intersection of the center line of the m-th Bohr-Sommerfeld band, Eq. (3.1), with the x-axis and the center line of the Gaussian cigar, respectively, as indicated in Fig. 8b.

From this geometrical interpretation and Eq. (C8) follows that the solutions $\varphi_m^{(\pm)}$ exist only for

$$\frac{\sqrt{2}\alpha}{[2(m+1/2)]^{1/2}} \frac{1}{1-(\epsilon/2)^2} \leq 1.$$

In the remainder of this appendix we consider high squeezing, $0 < \epsilon \ll 1$, such that Eq. (C8) reads

$$\cos\varphi_m^{(\pm)} \cong \alpha(m + 1/2)^{-1/2}.$$

Note also that the Taylor expansion of h_m implies a *nonuniform* asymptotic expansion[38] which is only valid if the contributing saddle points, $\varphi_m^{(j)}$, are well separated from each other. Otherwise a *uniform* asymptotic expansion[38] in terms of Airy functions has to be invoked. In the remainder of this appendix we thus consider only quantum numbers m appropriately larger than α^2 such that this situation applies.

From Eq. (C3) we find

$$h_m''(0) \cong -4[1 - \alpha(m + 1/2)^{-1/2}]$$

and

$$h_m''(\pi) \cong -4[1 + \alpha(m + 1/2)^{-1/2}]$$

and

$$h_m''(\varphi_m^{(\pm)}) \cong 4[1 - \alpha^2(m + 1/2)^{-1}],$$

where we have neglected terms of the order $(\epsilon/2)^2 \ll 1$ and double prime denotes second derivative with respect to φ. For quantum numbers m such that $\alpha(m + 1/2)^{-1/2} < 1$, we thus find $h_m''(0) < 0$ and $h_m''(\pi) < 0$, whereas $h_m''(\varphi_m^{(\pm)}) > 0$. Thus $\varphi_m^{(0)}$ and $\varphi_m^{(\pi)}$ represent minima whereas $\varphi_m^{(\pm)}$ are the maxima contributing to the integral $\mathcal{L}_m$. Hence the integral $\mathcal{L}_m$ follows from Eq. (C2)

$$\mathcal{L}_m \cong \sum_{j=\pm 1} \left\{\epsilon \Big/ \left[(m+1/2)\, h_m''\left(\varphi = \varphi_m^{(j)}\right)\right]\right\}^{1/2} \exp\left[-\frac{2(m+1/2)}{\epsilon} h_m\left(\varphi_m^{(j)}\right)\right].$$

Equations (C3) and (C8) yield

$$h_m\left(\varphi_m^{(\pm)}\right) \cong -\left\{2\alpha^2(m+1/2)^{-1}\left[1+\left(\frac{\epsilon}{2}\right)^2\right]+\left[1-\left(\frac{\epsilon}{2}\right)\right]^2\right\},$$

and thus

$$\mathcal{L}_m = 2\left(\frac{\epsilon}{4}\right)^{1/2}\left[m+1/2-\alpha^2\right]^{-1/2}$$
$$\exp\left\{\frac{2}{\epsilon}\left[2\alpha^2\left(1+\left(\frac{\epsilon}{2}\right)^2\right)+\left(1-\left(\frac{\epsilon}{2}\right)^2\right)(m+1/2)\right]\right\}.$$

The contributions from $\varphi_m^{(+)}$ and $\varphi_m^{(-)}$ are identical giving rise to the factor 2. When we substitute this result into Eq. (C1) we finally arrive at

$$A_m = 2\mathcal{A}_m$$

where

$$\mathcal{A}_m = \left(\frac{\epsilon}{4\pi}\right)^{1/2}\left[m+1/2-\alpha^2\right]^{-1/2}\exp[-\epsilon(m+1/2-\alpha^2)].$$

REFERENCES

1. M. Planck, "Vorlesungen über die Theorie der Wärmestrahlung", J. A. Barth, Leipzig (1906), p. 154 and M. Planck, Die physikalische Struktur des Phasenraumes, Ann. Phys. (Leipzig) 50: 385 (1916).
2. N. Bohr, "Collected Works," L. Rosenfeld, ed., North-Holland, New York (1976), Vol. 3.
3. A. Sommerfeld, Zur Theorie der Balmerschen Serie, Sitzungsber. d. kgl. bayr. Akad. d. Wiss.: 425 (1915); Die Feinstruktur der Wasserstoff-und der Wasserstoff-ähnlichen Linien, ibid. 459 (1915); Zur Quantentheorie der Spektrallinien, Ann. Phys. (Leipzig) 51: 1 (1916).
4. M. Born, Vorlesungen über Atommechanik, in: "Struktur der Materie in Einzeldarstellungen," M. Born and J. Franck, eds., Springer, Berlin (1925).
5. P. Debye, Wellenmechanik and Korrespondenzprinzip, Physik. Zeitschr. 28: 170 (1927); H. A. Kramers, Quantentheorie des Elektrons und der Strahlung, Vol. 2 in: "Hand-und Jahrbuch der Chemischen Physik," Eucken-Wolf, Leipzig, (1938). W. Pauli, Die allgemeinen Prinzipien der Wellenmechanik, in: "Handbuch der Physik," Vol. 24, H. Geiger and K. Scheel, eds., Springer, Berlin (1933); R. L. Liboff, The correspondence principle revisited, Physics Today 37(2): 50 (1984).
6. J. A. Wheeler and W. H. Zurek, "Quantum Theory and Measurement," Princeton University Press, Princeton (1983); R. P. Feynman, R. B. Leighton and M. Sands, "The Feynman Lectures on Physics," Addison-Wesley, Reading (1964), Vol. 3.
7. R. Glauber, Coherent and incoherent states of the radiation field, Phys. Rev. 131: 2766 (1963).
8. M. Sargent, M. O. Scully, W. E. Lamb, "Laser Physics," Addison-Wesley, Reading (1974), Appendix H.
9. W. H. Louisell, "Quantum Statistical Properties of Radiation," Wiley, New York (1973).

10. R. S. Bondurant (B. S. thesis, MIT, unpublished).
11. J. A. Wheeler, Franck-Condon effect and squeezed state physics as double-source interference phenomena, Lett. Math. Phys. 10:201 (1985).
12. W. Schleich and J. A. Wheeler, Interference in phase space, in: "The Physics of Phase Space," Y. S. Kim and W. W. Zachary, eds., Springer, New York (1987).
13. W. Schleich and J. A. Wheeler, Oscillations in photon distribution of squeezed states and interference in phase space, Nature 326: 574 (1987).
14. W. Schleich and J. A. Wheeler, Oscillations in photon distribution of squeezed states, JOSA B4: 1715 (1987).
15. A. Vourdas and R. M. Weiner, Photon-counting distribution in squeezed states, Phys. Rev. A36: 5866 (1987).
16. H. P. Yuen, Two-photon coherent states of the radiation field, Phys. Rev. A13: 2226 (1976); G. J. Milburn and D. F. Walls, Squeezed states and intensity fluctuations in degenerate parametric oscillators, Phys. Rev. A27: 392 (1983).
17. D. F. Walls, Squeezed states of light, Nature 306:141 (1983); see also special issues on squeezed states in: JOSA B4(10): (1987) and J. of Mod. Opt. 34 (6-7): (1987); see also: G. Leuchs, Photon statistics, anti-bunching and squeezed states, in: "Frontiers of Nonequilibrium, Statistical Physics," G. T. Moore and M. O. Scully, eds., Plenum, New York, (1986).
18. M. Hillery, R. F. O'Connell, M. O. Scully and E. P. Wigner, Distribution functions in physics: fundamentals, Phys. Rep. 106: 121 (1984); V. I. Tatarskii, The Wigner representation of quantum mechanics, Usp. Fiz. Nauk. 139: 587 (1983); L. Cohen, Positive and negative joint quantum distributions, in: "Frontiers of Nonequilibrium Statistical Physics," G. T. Moore and M. O. Scully, eds., Plenum, New York (1986).
19. W. Schleich and J. A. Wheeler, Interference in phase space, Ann. Phys. (New York), to be published; J. A. C. Gallas, W. Schleich and J. A. Wheeler, Beyond interference in phase space, Ann. Phys. (New York), to be published.
20. E. J. Heller, Phase space interpretation of semiclassical theory, J. Chem. Phys. 67: 3339 (1977).
21. W. Schleich, H. Walther and J. A. Wheeler, Area in phase space as determiner of transition probability: Bohr-Sommerfeld bands, Wigner ripples and Fresnel zones, Found. Phys. (to be published).
22. W. Schleich, D. F. Walls and J. A. Wheeler, Area of overlap and interference in phase space versus Wigner pseudo-probabilities, Phys. Rev. A (to be published).
23. G. J. Milburn and D. F. Walls, Effect of dissipation on interference in phase space, Phys. Rev. A (to be published), G. J. Milburn, this volume.
24. P. D. Drummond, Interference in squeezed field measurements using coherent phase integrals, Phys. Rev. A (to be published).
25. D. Bohm, "Quantum Theory," Prentice Hall, Englewood Cliffs, 1951.
26. E. U. Condon, The Franck-Condon principle and related topics, Am. J. Phys. 15: 365 (1947).
27. J. Janszky and Y. Y. Yushin, Squeezing via frequency jump, Optics Comm. 59: 151 (1986); R. Graham, Squeezing and frequency changes in harmonic oscillations, J. Mod. Opt. 34: 873 (1987).
28. L. Landau and R. Peierls, Erweiterung des Unbestimmtheitsprinzips für die relativistische Quantentheorie, Z. Phys. 69: 56 (1931); N. Bohr and L. Rosenfeld, Zur Frage der Messbarkeit der elektromagnetischen Feldgrössen, Mat.-fys. Medd. Dan. Vid. Selsk. 12(8):(1933); N. Bohr and L. Rosenfeld, Field

and charge measurements in quantum electrodynamics, Phys. Rev. 78: 794 (1950); (These papers are reprinted and commented on in Ref. 6).
29. G. Szegö, "Orthogonal polynomials," American Mathematical Society, New York (1939).
30. D. Stoler, Equivalence classes of minimum-uncertainty packets, Phys. Rev. D 1: 3217 (1970); Equivalence classes of minimum-uncertainty packets, II, Phys. Rev. D 4: 1925 (1971); M. Nieto, What are squeezed states really like?, in: "Frontiers in Nonequilibrium Statistical Physics," G. T. Moore and M. O. Scully, eds., Plenum Press, New York (1986).
31. Oscillations in photon distributions of electromagnetic fields have also been found in the context of the Jaynes-Cummings model and the Rydberg maser, see for example P. Meystre, E. Geneux, A. Quattropani and A. Faist, Long time behavior of a two level atom in interaction with an electromagnetic field, Nuovo Cimento B25: 521 (1975); P. Filipowicz, P. Meystre, G. Rempe and H. Walther, A testing ground for quantum electrodynamics, Opt. Act. 32: 1105 (1985) and also in various other nonlinear optical systems, see for example G. S. Agarwal and G. Adam, Photon number distributions for quantum fields generated in nonlinear optical processes, Phys. Rev. A (to be published).
32. B. Friedman, "Lectures On Application-Oriented Mathematics," University of Chicago Press, Chicago (1957).
33. J. R. Klauder, The design of radar signals having both high range resolution and high velocity resolution, Bell Sys. Tech. 39: 809 (1960).
34. M. V. Berry, Semiclassical mechanics in phase space: A study of Wigner's function, Phil. Trans. Roy. Soc. 287: 237 (1977).
35. W. G. Brown, Intensitätsveränderungen in einigen Fluoreszenzserien von Natrium, Zs. f. Physik 82: 768 (1933); W. Demtröder, "Laser Spectroscopy," Springer, Berlin (1981) p. 418; M. Trautmann, J. Wanner, S. K. Zhou and C. R. Vidal, Dynamics of selected rovibronic B $^3\pi(0^+)$ states of IF: Variation of the electronic transition moment with internuclear distance, J. Chem. Phys. 82: 693 (1985).
36. M. Abramowitz and I. E. Stegun, "Handbook of Mathematical Functions, National Bureau of Standards, Washington, D.C., (1964).
37. C. Leubner, Uniform asymptotic expansion of a class of generalized Bessel functions occurring in the study of fundamental scattering processes in intense laser fields, Phys. Rev. A 23: 2877 (1981).
38. C. Chester, B. Friedman and F. Ursell, An extension of the method of steepest descents, Proc. Camb. Phil. Soc. 53: 599 (1957). N. Bleistein and R. A. Handelsman, "Asymptotic Expansions of Integrals," Holt, Rineheart and Winston, New York (1975).

THE EFFECT OF MEASUREMENT ON INTERFERENCE IN PHASE-SPACE

Gerard J. Milburn

Department of Physics & Theoretical Physics
Australian National University
Canberra A.C.T. 2601 Australia

In a recent paper Schleich and Wheeler have shown that the oscillations in the tail of the photon number distribution for squeezed light may be explained in terms of "interference in phase-space".(1) In this paper the concept of phase-space interference will be developed in terms of the Q-function of quantum optics. While this approach is quite different to that of Schleich and Wheeler, the result is of course the same. The use of the Q-function permits a direct comparison of the quantum result with that expected classically.

In classical mechanics the state of a system is given as a joint probability density on phase-space, an even dimensional manifold with coordinates of position and momentum. Consider the system to be a one dimensional harmonic oscillator. Given $P(q,p)$, how does one compute the probability density for energy measurements $P(E)$? One easily sees that this is given by

$$P(E) = \int dp\, dq\, \delta^{(2)}\left(E - \tfrac{1}{2}(p^2 + \omega^2 q^2)\right) P(p,q) \ . \tag{1}$$

In order to facilitate a comparison with the corresponding quantum result it is convenient to define a complex phase-space by the change of variable

$$\alpha = \left(\frac{\omega}{2\hbar}\right)^{\frac{1}{2}} q + i(2\hbar\omega)^{-\frac{1}{2}} p \ . \tag{2}$$

We may further define a "rescaled" energy by $n = E/\hbar\omega$, that is we choose to measure energy in units of $\hbar\omega$. The energy probability density is given by

$$P(n) = \int \frac{d^2\alpha}{\pi}\, \delta^{(2)}(n - |\alpha|^2) Q(\alpha,\alpha^*) \ , \tag{3}$$

where the density on the complex phase-space is defined by the transformation

$$P(p,q) \rightarrow \frac{1}{2\hbar} Q(\alpha,\alpha^*) \ . \tag{4}$$

Is there a quantum analogue for the result in Eq. (3)? Let the state of the system be $|\psi\rangle$, then the probability amplitude for an energy measurement to yield a result n is $\langle n|\psi\rangle$, where $|n\rangle$ is the corresponding energy eigenstate. The probability for this result is of course just $|\langle n|\psi\rangle|^2$. If one takes Feynman's insight (2) at face value we might hope to construct the quantum analogue of Eq. (3) at the level of appropriate probability amplitudes. Such an expression is easily constructed from $\langle n|\psi\rangle$ using the coherent states $|\alpha\rangle$, which provide the following resolution of identity

$$\int \frac{d^2\alpha}{\pi} |\alpha\rangle\langle\alpha| = \hat{1} . \qquad (5)$$

Substituting Eq. (5) into the amplitude $\langle n|\psi\rangle$ gives

$$\langle n|\psi\rangle = \int \frac{d^2\alpha}{\pi} \langle n|\alpha\rangle\langle\alpha|\psi\rangle . \qquad (6)$$

To show that this is the required result we need to be able to interpret $\langle\alpha|\psi\rangle$ as the probability amplitude for simultaneous measurement of position and momentum (and thus $|\langle\alpha|\psi\rangle|^2$ would have the same interpretation as $Q(\alpha,\alpha^*)$), and to interpret $\langle n|\alpha\rangle$ as the conditional amplitude that the results of such a measurement correspond to energy n. Let us first consider $\langle n|\alpha\rangle$. The modulus square of this function is

$$|\langle n|\alpha\rangle|^2 = \frac{|\alpha|^{2n}}{n!} e^{-|\alpha|^2} . \qquad (7)$$

For large n this may be approximated by

$$|\langle n|\alpha\rangle|^2 \sim (2\pi n)^{-\frac{1}{2}} e^{-(|\alpha|^2 - n)^2/n}$$

which is a gaussian of mean $\langle |\alpha|^2\rangle = n$ and standard deviation of $\sqrt{n}$. Thus the relative width for large n is $n^{-\frac{1}{2}}$. The function $|\langle n|\alpha\rangle|^2$ is an annular density concentrated at $|\alpha|^2 = n$ and with a relative width that tends to zero as $n \to \infty$. We may thus interpret $\langle n|\alpha\rangle$ as the conditional probability amplitude analogous to the delta function in Eq. (3). Note, however, that the delta function has zero width while it is only the relative width of the quantum result which tends to zero as $n \to \infty$. In a sense the quantum result is always "broader" than one might expect.

Now consider the amplitude $\langle\alpha|\psi\rangle$. If this is a probability amplitude then $|\langle\alpha|\psi\rangle|^2$ should be a true joint probability density. In fact $|\langle\alpha|\psi\rangle|^2$ is just the Q-function of quantum optics(3). The Q-function is an everywhere positive function normalised with respect to the measure $d^2\alpha/\pi$. It is also bounded by unity. Q-functions thus belong to a subclass of classical phase-space probability densities. This said it is still not entirely clear in what sense $Q(\alpha,\alpha^*)$ is a joint probability density for simultaneous measurement of position and momentum. To clarify this point I will now describe a simple model for such measurements, the statistics of which are determined by the Q-function. This model is a simple extension of von Neumann's position measurement model (4) and was first given by Arthurs and Kelly(5). The formulation here is due to Caves(6).

Consider the system of interest to be coupled to two distinct systems referred to as meters (figure 1) in such a way that the position of the system causes a displacement of the position of one meter while the momentum of the system displaces the momentum of the other meter (labelled 2 on figure 1). This coupling may be obtained by the impulsive interaction hamiltonian at time t_r

$$H_I = \delta(t - t_r)\,(\hat{q}\,\hat{p}_1 - \hat{p}\,\hat{q}_2) , \qquad (8)$$

where $\hat{p}_1$ is the momentum of the position meter and $\hat{q}_2$ is the position of the momentum meter. At the same time a simultaneous readout of $\hat{q}_1$ and $\hat{p}_2$

is made on each of the meters. This violates no law of quantum mechanics as $\hat{q}_1$ and $\hat{p}_2$ commute. Let the system be in an arbitrary state $|\psi\rangle$. Each meter is assumed to be in the minimum uncertainty state

$$\phi(q_i) = (2\pi\Delta_i)^{-\frac{1}{4}} e^{-q_i^2/4\Delta_i} , \tag{9}$$

where $i = 1$ for the position meter and $i = 2$ for the momentum meter. We are free to adjust Δ_1 and Δ_2 to minimise the noise added by the meters. The

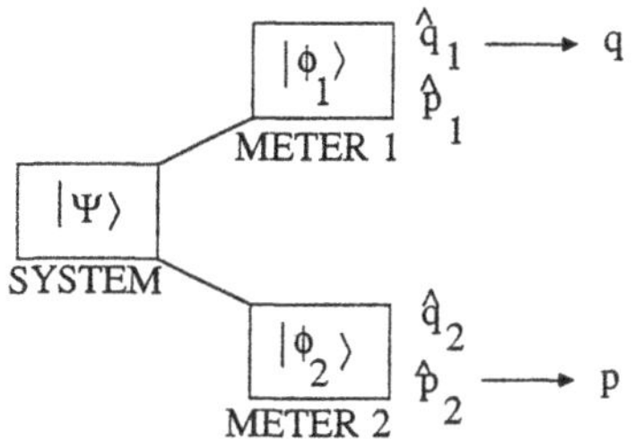

Fig. 1. Schematic outline of the model for a simultaneous measurement of position and momentum. Meter 1 is the position meter and meter 2 is the momentum meter.

joint probability for the readout of the two meters to yield the result (p,q) may be written entirely in terms of system variables as

$$P(q,p) = \langle\psi| \hat{F}(q,p) |\psi\rangle , \tag{10}$$

where $\hat{F}(q,p)$ is a bounded positive operator. In the special case that $4\Delta_1 = \Delta_2$

$$\hat{F}(q,p) = \frac{1}{2\pi\hbar} |\psi_{q,p}\rangle\langle\psi_{q,p}| , \tag{11}$$

where $|\psi_{q,p}\rangle$ is a minimum uncertainty state with position variance $2\Delta_1$ and mean values given by $\langle\hat{q}\rangle = q$, $\langle\hat{p}\rangle = p$. If we further put $\Delta_1 = \hbar/4\omega$ and make the change of variable

$$\alpha = \left(\frac{\omega}{2\hbar}\right)^{\frac{1}{2}} q + \frac{i}{\sqrt{2\hbar\omega}} p ,$$

then

$$\hat{F} = \frac{1}{\pi} |\alpha\rangle\langle\alpha| , \tag{12}$$

where $|\alpha\rangle$ is a Glauber coherent state. In this case the joint distribution becomes simply the Q-function corresponding to the state of the system:

$$Q(\alpha,\alpha^*) = |\langle\psi|\alpha\rangle|^2 . \tag{13}$$

We may thus interpret $\langle\psi|\alpha\rangle$ as the probability amplitude for this particular type of simultaneous measurement of position and momentum.

Let us now return to an analysis of Eq. 6. In general the integral on the right hand side will be complex. Further, it will only be significantly different from zero when the areas of support where $\langle n|\alpha\rangle$ and $\langle\alpha|\psi\rangle$ are sufficiently different from zero, intersect. For example, consider the case where $|\psi\rangle$ is a single mode squeezed state with non-zero amplitude α_0 and squeeze parameter r. In figure 2 we indicate schematically the contours of the two functions appearing in the integrand. Clearly for $n > |\alpha_0|^2$ there are two regions of overlap. An asymptotic analysis for large n shows that these two regions contribute equal amplitudes but opposite phases to the integral, that is

$$\langle n|\psi\rangle \approx A_n\left(e^{i\phi_n} + e^{-i\phi_n}\right) \ . \tag{14}$$

Explicit expressions for A_n and ϕ_n may be found in reference (1). Thus

$$P(n) = 4\, A_n \cos^2\phi_n \ , \tag{15}$$

and the possibility of "interference fringes" is apparent. This is interference in phase-space. Note that had we taken the modulus square of the probability amplitudes before integration, no interference fringes would have been obtained. The interference here is thus directly analogous to the interference in configuration space as demonstrated in the two slit experiment.

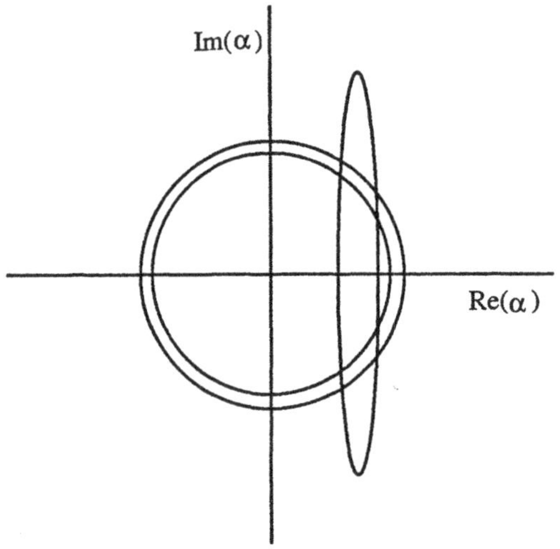

Fig. 2. Schematic representation of the contours of $\langle n|\alpha\rangle$ and $\langle\alpha|\alpha_0,r\rangle$, showing the two regions of significant overlap.

The exact energy distribution for this state is shown in figure 3(a) where the oscillations for $n > |\alpha_0|^2$ are clearly apparent.

Interference in phase-space leads to oscillations in the marginal distribution for any observable. For example consider position measurements. The classical result is

$$P(q) = \int dp'dq' \ \delta(q-q') \ P(q',p') \ . \tag{16}$$

The corresponding quantum amplitude is

$$\langle q|\psi\rangle = \int \frac{d^2\alpha}{\pi} \langle q|\alpha\rangle\langle\alpha|\psi\rangle \quad (17)$$

In this case $|\langle q|\alpha\rangle|^2$ is a gaussian centered on $q' = q$ with

$$q' = \left(\frac{\omega}{2\hbar}\right)^{-\frac{1}{2}} \mathrm{Re}(\alpha) \ .$$

In the limit $\hbar \to 0$ $|\langle q|\alpha\rangle|^2$ tends to a delta function at $q = q'$. If $\langle q|\alpha\rangle$ overlaps $\langle\alpha|\psi\rangle$ in two regions we expect interference fringes. As an example let

$$|\psi\rangle = \frac{1}{\sqrt{2}} \left(|\alpha_0\rangle + |-\alpha_0\rangle\right) ,$$

a superposition of two coherent states. When α_0 is purely imaginary $\langle\alpha|\psi\rangle$ and $\langle q|\alpha\rangle$ have significant overlap at $q = 0$, at which point there are two regions of overlap due to the fact that $|\langle\alpha|\psi\rangle|^2$ is double peaked at $\pm\alpha_0$. We thus expect to see interference fringes centered on $q = 0$. This is indeed the case(7).

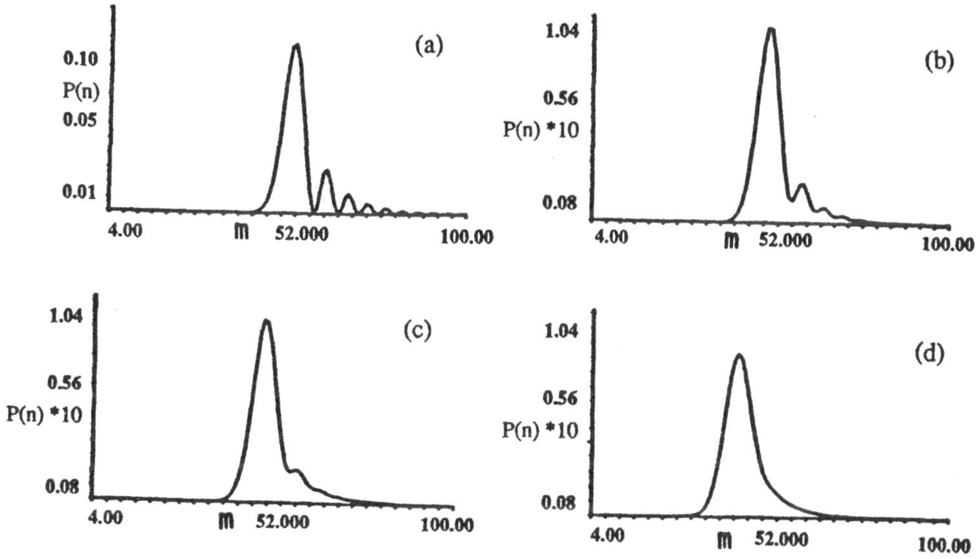

Fig. 3. Plot of $P(m,\mu)$ versus m for the squeezed state $|\beta_0, r\rangle$ with $\beta_0 = 7$ and $r = 1.52$.
(a) $\mu = 1.0$ (b) $\mu = 0.96$ (c) $\mu = 0.92$
(d) $\mu = 0.8$

Interference in phase-space may also arise as a result of the intrinsic dynamics of a system. In classical mechanics the evolution of the probability density is given by

$$Q(\alpha,\alpha^*,t) = \int d^2\beta \ J(\alpha,\alpha^*|\beta,\beta^*;t) \ Q(\beta,\beta^*;0) \ , \quad (18)$$

where $J(\alpha,\alpha^*|\beta,\beta^*;t)$ is the Greens function for the Liouville equation. In quantum mechanics the analogue of this expression is determined by the evolution of the amplitude

$$\psi(\alpha^*,t) \equiv \langle\alpha|\psi(t)\rangle \ . \quad (19)$$

The most easily interpreted way to write this amplitude is in terms of the coherent state path integral of Klauder[8];

$$\psi(\alpha^*,t) = \int \frac{d^2\beta}{\pi} K(\alpha^*|\beta;t)\ \psi(\beta,0) , \tag{20}$$

where

$$K(\alpha^*|\beta;t) \equiv \int Dz(t)\ e^{\frac{i}{\hbar} S[z(t)]} , \tag{21}$$

and S[z(t)] is the action functional over all paths in phase-space such that $z(0)=\beta$ and $z(t)=\alpha$. Then

$$Q(\alpha,\alpha^*,t) = \int \frac{d^2\beta}{\pi} \frac{d^2\beta'}{\pi} J(\alpha|\beta,\beta;t)\ \psi(\beta,0)\ \psi^*(\beta',0) \tag{22}$$

where

$$J(\alpha|\beta,\beta';t) = \int Dz(t) \int Dz'(t)\ e^{\frac{i}{\hbar}(S[z(t)] - S[z'(t)])} \tag{23}$$

and the paths z(t) and z'(t) are such that

$$z(t) = z'(t) = \alpha$$

$$z(0) = \beta$$

$$z'(0) = \beta'$$

Interference effects arise when the off-diagonal elements of the system density operator in the coherent state basis $\left(\psi(\beta,0)\psi^*(\beta',0)\right)$ are significant.

An example of this kind of interference is provided by the non-linear oscillator with hamiltonian[9]

$$H = H_0 + \frac{\mu}{\hbar\omega_0} H_0{}^2 , \tag{24}$$

where H_0 is the simple harmonic oscillator hamiltonian. Classically the dynamics of this hamiltonian correspond to rotations about the origin in phase-space with an angular frequency proportional to the energy of the initial state. This leads to a rotational sheer of an initial Gaussian probability density displaced from the origin.

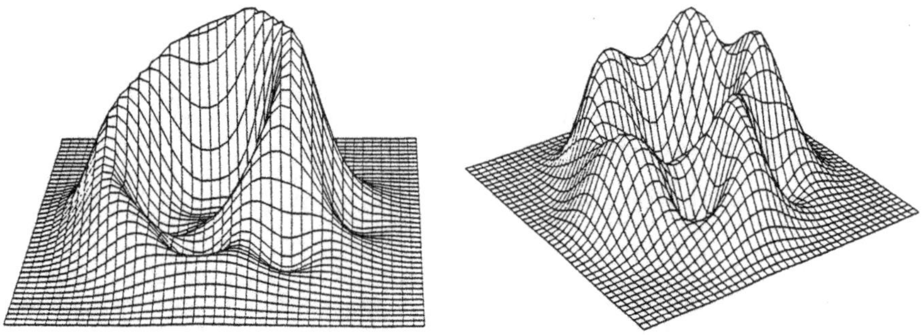

Fig. 4. Plot of the Q-function for the non-linear oscillator as the leading edge of the density overlaps the trailing tail, for an initial coherent state with $\alpha_0 = 2.0$.

The evolution of the Q-function, however, is quite different. For short times the density does indeed undergo a rotational sheer. However, as the leading edge of the sheered density begins to overlap the trailing tail phase-space interference sets in. Points on the leading edge of the density move along paths with different action to those on the trailing edge leading to a phase difference when they return to the same region of phase-space. This interference is seen in figures 4(a), (b). Eventually the system evolves towards a quantum superposition of two coherent states(7) and periodically reconstructs the initial state.

I will now consider the effect of continual measurement on phase-space interference. Suppose one is monitoring the evolution of the Q-function by making simultaneous measurements of position and momentum of the kind discussed earlier. We will suppose that this is done by making a sequence of such measurements at regular intervals over the period [0,t). The theory of continuous measurement developed by Barchielli and coworkers(10) (see also reference 11) is ideally suited to this problem. Given suitable approximations(11) the evolution of the system subjected to such measurements is given by

$$\frac{d\hat{\rho}}{dt} = -\frac{i}{\hbar}[\hat{H},\hat{\rho}] - \frac{Dq}{2\hbar^2}[\hat{p},[\hat{p},\hat{\rho}]] - \frac{Dp}{2\hbar^2}[\hat{q},[\hat{q},\hat{\rho}]] \quad , \tag{25}$$

where

$$Dq = \frac{\Delta_2}{2\hbar^2\tau} \tag{26}$$

and

$$Dp = \frac{1}{8\Delta_1\tau} \quad . \tag{27}$$

The measurements have two complementary effects. First they cause a diffusion of the system position and momentum variables with diffusion constants Dq and Dp respectively;

$$\frac{d\langle\hat{p}^2\rangle}{dt} = \begin{pmatrix}\text{free}\\ \text{motion}\end{pmatrix} + Dp \tag{28}$$

$$\frac{d\langle\hat{q}^2\rangle}{dt} = \begin{pmatrix}\text{free}\\ \text{motion}\end{pmatrix} + Dq \quad . \tag{29}$$

The second effect of the measurements is most clearly illustrated in the path-integral picture. In terms of the real phase-space variables (q,p) the propagator for the Q-function solution of Eq. (25) is

$$\int D(q,p) \int D(q',p')\, e^{\frac{i}{\hbar}\left(S[q,p] - S[q',p']\right)} I[q,p,q',p'] \quad , \tag{30}$$

where I is the Feynman-Vernon influence functional given by

$$I[q,p,q',p'] = \exp\{-\frac{t}{2}\Gamma[q,p,q',p']\} \quad , \tag{31}$$

where we have defined the coherence decay rate functional Γ by

$$\Gamma \equiv \frac{1}{t}\int_0^t dt' \left[\frac{Dp}{2\hbar^2}\left(q(t') - q'(t')\right)^2 + \frac{Dq}{2\hbar^2}\left(p(t') - p'(t')\right)^2\right] \quad . \tag{32}$$

The coherence decay rate functional causes a suppression of interference between phase-space paths at a rate determined by Dp and Dq and the average squared separation of two phase-space trajectories in time t.

This result may be of some relevance in the subject of quantum chaos. In chaotic systems the average squared separation of two initially close points would very quickly become very large.

Let us now return to the example with which we began, the photon number distribution for a squeezed state. In this case the continual measurement corresponds to photon counting over some interval of time. Consider the simplest case of a single made cavity field with a single source of loss to a photon counter. That is all photons last from the cavity are counted. The photon number distribution inside the cavity at time t is given by

$$P(m,\mu) = \sum_{n=m}^{\infty} P_0(n) \binom{n}{m} \mu^m (1-\mu)^{n-m} , \tag{33}$$

where

$$\mu = e^{-\gamma t} , \tag{34}$$

and $P_0(n)$ is the initial photon number distribution. By a redefinition of μ this is also the expression for the probability to detect m photons in some interval with a detector efficiency μ(12).

In terms of phase-space amplitudes, Eq. (33) becomes

$$P(m,\mu) = \int \frac{d^2\alpha}{\pi} \int \frac{d^2\beta}{\pi} \psi(\alpha,0)\psi^*(\beta,0)\langle m|\alpha\sqrt{\mu}\rangle \langle \beta\sqrt{\mu}|m\rangle \langle \beta|\alpha\rangle^{1-\mu} . \tag{35}$$

Thus interference between the off-diagonal elements of the system state in the coherent state basis is suppressed by the factor $\langle \beta|\alpha\rangle^{1-\mu}$. For short times the modulus of this factor is

$$|\langle \beta|\alpha\rangle|^{1-\mu} \simeq e^{-\frac{\gamma t}{2}|\alpha - \beta|^2} , \tag{36}$$

showing the expected dependence on the squared separation of the phase-space points. This interference suppression factor occurs in other models as well(14).

In figures 3(a-d) we plot $P(m,\mu)$ for various values of μ. One easily sees that as μ decreases from unity the interference fringes "fade out". This indicates the need to have very high efficiency detectors in order to observe the oscillations in the tail of the photon number distribution for squeezed light.

ACKNOWLEDGEMENTS

I would like to thank D.F. Walls, W. Schleich and C.M. Caves for useful discussions.

REFERENCES

1. W. Schleich and J.A. Wheeler, *Nature* 326, 574 (1987) and W. Schleich and J.A. Wheeler, *JOSA B* 4, 1715 (1987).
2. R.P. Feynman and A.R. Hibbs, "Quantum Mechanics and Path Integrals", McGraw-Hill, New York (1965).
3. D.F. Walls and G.J. Milburn *in* "Quantum Optics, Experimental Gravitation and Measurement Theory", edt. P. Meystre and M.O. Scully (New York, Plenum) p.209 (1983).

4. J. von Neumann, "Mathematische Grundlagen der Quanten mechanik" (Springer, Berlin, 1932) (English translation: "Mathematical Foundations of Quantum Mechanics", Princeton University, Princeton, New Jersey, 1955).
5. E. Arthurs and J.L. Kelly Jr., *Bell Syst. Tech. J.* 725 (1965).
6. C.M. Caves, private communication (1986).
7. B. Yurke and D. Stoler, *Phys. Rev. Letts.* 57, 13 (1986).
8. J.R. Klauder, *in* Path Integrals, proceedings of the NATO Advanced Summer Institute, edt. G.J. Papadopoulos and J.T. Devreese (Plenum, New York, 1978).
9. G.J. Milburn, *Phys. Rev. A* 33, 674 (1986).
10. A. Barchielli, L. Lanz, G.M. Prosperi, *Nuovo Cimento B* 72, 79 (1982).
11. C.M. Caves and G.J. Milburn, *Phys. Rev. A* 36, 5543 (1987).
12. M. Sargeant III, M.O. Scully and W.E. Lamb Jr., "Laser Physics", Addison-Wesley, Reading Massachusetts (1974).
13. G.J. Milburn and D.F. Walls, submitted to *Phys. Rev. A*.
14. D.F. Walls and G.J. Milburn, *Phys. Rev. A* 35, 3546 (1985).

SQUEEZING IN OPTICAL BISTABILITY

L.A. Lugiato*, M. Vadacchino* and F. Castelli**

*Dipartimento di Fisica, Politecnico Turin, Italy
**Dipartimento di Fisica, Università, Milan, Italy

INTRODUCTION

Some recent relevant experiments of the Austin group[1-3] attained the generation of squeezed states by a system of two-level atoms contained in an optical cavity and driven by a coherent incident field. This type of system is fundamental in quantum optics, and since the mid seventies has been the object of the extensive literature on *optical bistability* (OB) (see e.g. Ref.4). The squeezing problem in this system is studied in Refs. 5-14.

The possibility of squeezing in two-level OB was first predicted in Ref.5 as a straightforward consequence of a previous prediction of photon antibunching[15]. Reference 5 analyzed the squeezing effect in the internal cavity field, and considered the conditions which allow for the simplest theoretical treatment: exact resonance between the input field, the atoms and the cavity and adiabatic elimination limit.The latter means that the cavity relaxation rate κ is much smaller than the atomic relaxation rates $\gamma_\perp$ and $\gamma_\parallel$, or vice versa. In the first case, which is called good cavity limit, the dynamics of the system is governed by the field whereas in the opposite bad cavity limit it is controlled by the atomic variables. In both limits, the squeezing effect arises when the ratio $f=\gamma_\parallel/2\gamma_\perp$ is unity (purely radiative damping) or slightly smaller than unity. One has squeezing in the intracavity field over a portion of the lower transmission branch; the effect is small in the good cavity limit and even smaller in the bad cavity limits in which it is proportional to the small ratio $\gamma_\perp/\kappa$ (ref.5). If, instead, one considers the spectrum of squeezing in the *output* field, as first introduced by Refs. (16-17), the factor $\gamma_\perp/\kappa$ drops and the amount of squeezing in the bad-cavity limit becomes larger than in the good cavity limit. In fact, there is a part of the higher transmission branch such that the bad-cavity spectrum exhibits a small amount of squeezing over two suitable frequency intervals, symmetrically placed with respect to the frequency $\omega=0$[14]. On the other hand, in the good cavity limit the output field shows squeezing only when the intracavity field is squeezed, and the maximum squeezing occurs for $\Omega=0$[14].

An essential ingredient to obtain a significant squeezing is detuning as shown in Ref. 9, which reports an extensive study of the squeezing spectrum in the good cavity limit, with arbitrary detunings between the input field, the cavity and the atoms.

For reasonable values of the parameters a large level of squeezing can be obtained, however, only by moving away from the good cavity limit. A step in this direction was made by Ref. 10, which analyzes the fluctuations in OB for arbitrary values of the ratio $\mu=\kappa/\gamma_{\|}$ in the purely radiative limit $f=1$; the treatment is however limited to resonant conditions. This paper emphasizes the effects which arise from the oscillatory exchange of excitation between the cavity mode and the atoms when μ has order unity, and influenced the selection of the parametric values in the experiments[1-3].

Reference 14 gives a general analysis of the squeezing problem in OB, and does not introduce any restriction on the detuning parameters or on the relative order of magnitude of the atomic and cavity relaxation rates. Thus, it is possible to investigate how the amount of squeezing and the main features of the spectrum change when one varies one or more of the external parameters. In such a way, one can find general indications on how to optimize the conditions in any experiment devised to observe squeezing in OB. The following note summarizes the main results of this analysis.

Our treatment is based on the quantum statistical singlemode model of OB[4] and on a suitable Fokker-Planck approximation involving all the five variables for the field and the atomic system[18]. We exploit the quantum statistical dressed-mode formalism[19-21], which allows to obtain a general explicit analytical expression for the squeezing spectrum in two-level OB.

THE SYSTEM AND THE FOKKER-PLANCK EQUATION IN FIVE VARIABLES

We consider a ring cavity of length l, or an equivalent Fabry-Perot cavity of length $l/2$, which contains a homogeneously broadened system of N two-level atoms, with transition frequency ω_a. The system is driven by a coherent stationary field of frequency ω_0.

We assume that the transitions from one atomic level to the other are induced by absorption or emission of single photons, and that the electric field is uniform in the transverse directions (plane wave approximation).

We describe the dynamics of the system by a singlemode model, which considers only the cavity mode of frequency ω_c nearest to ω_0 (resonant mode). We call A and A^+ the annihilation and creation operators of photons of the resonant mode; they obey the boson commutation rule

$$[A, A^+] = 1, \tag{1}$$

On the other hand, the atomic system is associated with three collective operators: the macroscopic polarization operators R^+ and R^- and the population inversion operator R_3; they obey the angular momentum commutation relations

$$[R^+, R^-] = 2\, R_3, \qquad [R_3, R^{\pm}] = \pm\, R^{\pm}, \tag{2}$$

In the dipole and rotating wave approximation, the atom-field interaction is governed by the Hamiltonian

$$H_{AF} = g(AR^+ + A^+ R^-) \tag{3}$$

where g is the coupling constant.

The singlemode model for OB, first formulated in[22], is a master equation for the density operator W(t) of the system atoms + cavity mode[4]. The parameters of this model are:

- The cavity damping constant

$$\kappa = \frac{c\,T}{l}, \tag{4}$$

which coincides with the cavity linewidth;
- The relaxation rate $\gamma_\perp = T_2^{-1}$ of the atomic polarization, which coincides with the atomic linewidth;
- The relaxation rate $\gamma_\parallel$ of the population inversion,
- The cavity detuning parameter

$$\theta = \frac{\omega_c - \omega_0}{\kappa}, \tag{5}$$

- The atomic detuning parameter

$$\Delta = \frac{\omega_a - \omega_0}{\gamma_\perp}, \tag{6}$$

- The bistability parameter

$$C = \frac{g^2\,N}{2\kappa\,\gamma_\perp}, \tag{7}$$

- The input field α_0 such that $\kappa|\alpha_0|^2$ is the incident flux of photons.

Using the quantum-classical correspondence[23]

$$\begin{aligned} &A \to \beta \quad , \; A^+ \to \beta^* , \\ &R^- \to v \quad ; \; R^+ \to v^* ; \quad R_3 \to m , \\ &W(t) \to P(v, v^*, m, \beta, \beta^*, t), \end{aligned} \tag{8}$$

one can translate the operator master equation into a classical-looking partial differential equation for the quasiprobability distribution of the five variables $v, v^*, m, \beta, \beta^*$, which generalizes to the atomic system the Glauber P-function[24]. The moments of P correspond to the normally ordered expectation values. The partial differential equation for the generalized P-function has derivatives of all orders with respect to the variable m; however, when the number N of atoms is very large, the terms with derivatives of order higher than second are negligible and the equation reduces to a Fokker-Planck equation (FPE). This is best formulated in terms of the normalized variables

$$\begin{aligned} &x = \beta/\sqrt{N_s} \; ; \quad x = \alpha_0/\sqrt{N_s} , \\ &\bar{m} = \left(\frac{N}{2}\right)^{-1} m \; ; \; \bar{v} = -\left(\frac{N}{2}\sqrt{\frac{\gamma_\parallel}{\gamma_\perp}}\right)^{-1} v \end{aligned} \tag{9}$$

where N_s is the saturation photon number

$$N_s = \frac{\gamma_\perp \gamma_\parallel}{4\, g^2} \tag{10}$$

For the explicit expression of the FPE, as well as for all details that are not given here, we refer the reader to Ref. 14.

THE LINEARIZED FOKKER-PLANCK EQUATION

If one neglects fluctuations, i.e. the diffusion terms of the FPE, one recovers the semiclassical theory of OB. The semiclassical stationary solution is given by[4]

$$\mathrm{v}_s = \frac{(1-i\Delta)\, x_s}{1+\Delta^2+|x_s|^2} \; ; \quad \mathrm{m}_s = \frac{1+\Delta^2}{1+\Delta^2+|x_s|^2} \tag{11}$$

where x_s obeys the steady-state equation

$$|y| = |x_s| \left\{ \left[1+ \frac{2C}{1+\Delta^2+|x_s|^2}\right]^2 + \left[\theta - \frac{2C\Delta}{1+\Delta^2+|x_s|^2}\right]^2 \right\}^{1/2} \tag{12}$$

From (12) one obtains the steady-state curve of output field $|x_s|$ as a function of input field $|y|$. For C larger than a suitable threshold value $C_{min}(\Delta,\theta)$ the steady-state curve is S-shaped[4]; the portion with negative slope is unstable and therefore the system is bistable in a suitable interval of the input field $|y|$.

Next we select a stable stationary solution, introduce the fluctuation variables

$$\begin{aligned} x' &= x - x_s \quad &;& \quad (x^*)' = x^* - x^*_s, \\ v' &= \mathrm{v} - \mathrm{v}_s \quad &;& \quad (v^*)' = \mathrm{v}^* - \mathrm{v}^*_s, \\ m' &= \mathrm{m} - \mathrm{m}_s \;, \end{aligned} \tag{13}$$

and linearize the FPE around steady state. Introducing the variables

$$\begin{aligned} x'_1 &= \frac{1}{2}\,[x' + (x^*)'] \quad ; \quad x'_2 = \frac{1}{2i}\,[x' - (x^*)'] \\ v'_1 &= \frac{1}{2}\,[v' + (v^*)'] \quad ; \quad v'_2 = \frac{1}{2i}\,[v' - (v^*)'] \end{aligned} \tag{14}$$

and the five-component vector

$$q = \begin{pmatrix} x'_1 \\ m' \\ v'_1 \\ x'_2 \\ v'_2 \end{pmatrix} \tag{15}$$

the linearized FPE reads

$$\frac{\partial}{\partial t} P(\underset{\sim}{q},t) = \sum_{i,k=1}^{5} \left\{ - \frac{\partial}{\partial q_i} M_{ik} \; q_k \; + \frac{1}{N_s} \frac{\partial^2}{\partial q_i \, \partial q_k} D_{ik} \right\} \; P \; (\underset{\sim}{q},t) \, ; \qquad (16)$$

the explicit expressions of the drift matrix M_{ik} and of the diffusion matrix D_{ik} can be found in Ref. 14.

It is convenient to introduce the eigenstates of the drift matrix and of its adjoint (j=1,2,..,5)

$$M \, \underset{\sim}{O_j} = \lambda_j \, \underset{\sim}{O_j}$$

$$M^+ \underset{\sim}{\bar{O}_j} = \lambda^*{}_j \, \underset{\sim}{\bar{O}_j} \qquad (17)$$

they obey the orthonormality relations

$$(\underset{\sim}{\bar{O}_j} , \; \underset{\sim}{O_{j'}}) = \sum_{i=1}^{5} \; \bar{O}^*{}_{ji} \; O_{j'i} = \delta_{jj'} \, , \qquad (18)$$

the vector $\underset{\sim}{q}$ can be expanded as follows:

$$\underset{\sim}{q}(t) = \sum_{i=1}^{5} \; \xi_j \; (t) \; \underset{\sim}{O_j} \, , \qquad (19)$$

where the dressed-mode amplitudes[19,21] ξ_j are given by

$$\xi_j = (\underset{\sim}{\bar{O}_j} \, , \, \underset{\sim}{q}) \equiv \sum_{i=1}^{5} \; \bar{O}^*{}_{ji} \; q_i \, , \qquad (20)$$

The linearized FPE (16) becomes

$$\frac{\partial}{\partial t} P(\underset{\sim}{\xi},t) = \left\{ - \sum_{j=1}^{5} \frac{\partial}{\partial \xi_j} \lambda_j \; \xi_j \; + \frac{1}{N_s} \sum_{j,j'=1}^{5} \frac{\partial^2}{\partial \xi_j \, \partial \xi_{j'}} D^{jj'} \right\} \; P \; (\underset{\sim}{\xi},t) \, , \qquad (21)$$

with a diagonal drift matrix and a diffusion matrix given by

$$D^{jj'} = \sum_{i,k=1}^{5} \; \bar{O}^*{}_{ji} \; \bar{O}^*{}_{j'k} \; D_{ik} \qquad (22)$$

THE SPECTRUM OF SQUEEZING

Let us consider the quadrature component

$$A_\varphi = A\, e^{-i\varphi} + A^+ e^{i\varphi} \tag{23}$$

where φ is an arbitrary phase. According to the quantum-classical correspondence (8), the operator A_φ corresponds to the variable $\beta_\varphi=\beta\, e^{-i\varphi} + \beta^* e^{i\varphi}$. Similarly to (13), we introduce the fluctuation variable

$$\beta'_\varphi=\beta_\varphi-\beta_{\varphi,s}= \sqrt{N_s}\,[x'\, e^{-i\varphi} + (x^*)'\, e^{i\varphi}] \tag{24}$$

the squeezing spectrum[16,17] for the component A_φ is given by the Fourier transform of the time correlation function of β'_φ:

$$S(\omega,\varphi) = 2\kappa \int_{-\infty}^{\infty} d\omega\; e^{-i\omega t}\, \langle \beta'_\varphi(t)\, \beta'_\varphi(o)\rangle_s \tag{25}$$

where the steady-state time correlation function must be calculated from the FPE[16] associated with normal ordering. The spectrum (25) is defined in such a way that squeezing at frequency ω corresponds to $S(\omega,\varphi) <0$, and perfect squeezing to $S(\omega,\varphi)=-1$. One finds easily that the function $S(\omega,\varphi)$ can be expressed in the following form:

$$S(\omega,\varphi) = 8\, [S_a(\omega) + \mathrm{Re}(e^{-2i\varphi}\, S_b(\omega))] \tag{26}$$

where

$$S_a(\omega) = \frac{\kappa N_s}{2} \int_0^{\infty} dt\; \cos \omega t\; \{\langle (x^*)'(t)\;\; x'(o)\rangle_s + \langle x'(t)\;\; (x^*)'(o)\rangle_s\}$$

$$S_b(\omega) = N_s\, \kappa \int_0^{\infty} dt\; \cos \omega t\; \langle x'(t)\;\; x'(o)\rangle_s \tag{27}$$

It is convenient to select the phase φ in such a way that the amount of squeezing is maximized.

Hence in the following we will consider the spectrum $\bar{S}(\omega)$ maximized over φ for all the value of ω. This means that, for each value of ω, we must select a different value of φ, namely

$$e^{-2i\varphi} = - \frac{S^*_b(\omega)}{|S_b(\omega)|} \tag{28}$$

which gives

$$\overline{S}(\omega) = 8\,[S_a(\omega) - |S_b(\omega)|] \tag{29}$$

Using the regression theorem[25], one obtains analytic expressions for the functions S_a and S_b. In terms of the scaled quantities

$$\Omega = \frac{\omega}{\gamma_{\parallel}} \;;\quad \tilde{\lambda}_j = \frac{\lambda_j}{\gamma_{\parallel}} \;;\quad \mu = \frac{\kappa}{\gamma_{\parallel}} \tag{30}$$

one has[14]

$$S_a(\Omega) = \mu \sum_{j,j'=1}^{5} [O_{j1}\;O_{j'1} + O_{j4}\;O_{j'4}\,] \times \frac{\tilde{\lambda}_j}{\tilde{\lambda}^2_j + \Omega^2}\,\frac{D^{jj'} + D^{j'j}}{\tilde{\lambda}_j + \tilde{\lambda}_{j'}} \tag{31}$$

$$S_b(\Omega) = \mu \sum_{j,j'=1}^{5} [O_{j1}\;O_{j'1} - O_{j4}\;O_{j'4} +$$

$$+\, i\,(O_{j1}\;O_{j'4} + O_{j4}\;O_{j'1})]\,\frac{\tilde{\lambda}_j}{\tilde{\lambda}^2_j + \Omega^2}\quad\frac{D^{jj'} + D^{j'j}}{\tilde{\lambda}_j + \tilde{\lambda}_{j'}} \tag{32}$$

Using the same formalism, one can calculate also other quantities which concern fluctuations in this system, as for example the second-order intensity correlation function or the parameter Q defined by Mandel[26]. If the phase of the input field y is selected in such a way that the stationary output field x_s is real, one has

$$Q = 4\,N_s\,\langle(x'_1)^2\rangle = -4 \sum_{j,j'=1}^{5} O_{j1}\;O_{j'1}\;\frac{D^{j'j} + D^{jj'}}{\lambda_j + \lambda_{j'}} \tag{33}$$

THE VARIATION OF THE SQUEEZING SPECTRUM OVER THE PARAMETER SPACE

The analysis of the squeezing spectrum is not an easy task because there are six distinct parameters: C, Δ, θ, $\mu = \kappa/\gamma_{\parallel}$ $f=\gamma_{\parallel}/2\,\gamma_{\perp}$, x_s. In the following we indicate x instead of x_s.

Some general informations can be obtained by optimizing the squeezing over the frequency Ω and, possibly, over the stationary output field x. Figure 1 shows the quantity $-\bar{S}(\Omega)$, maximized over Ω and x, for C=160, f=1, in the intervals $-12 \leq \theta \leq 12$, $0 \leq \Delta \leq 120$, with $\mu=5$ (a) and $\mu=500$ (b). The main message is that the maximum level of squeezing is obtained for $|\theta|$ of order unity, and decreases substantially when $|\theta|$ is increased. On the other hand, $|\Delta|$ must not be small and relevant degrees of squeezing can be obtained also for large value of $|\Delta|$.

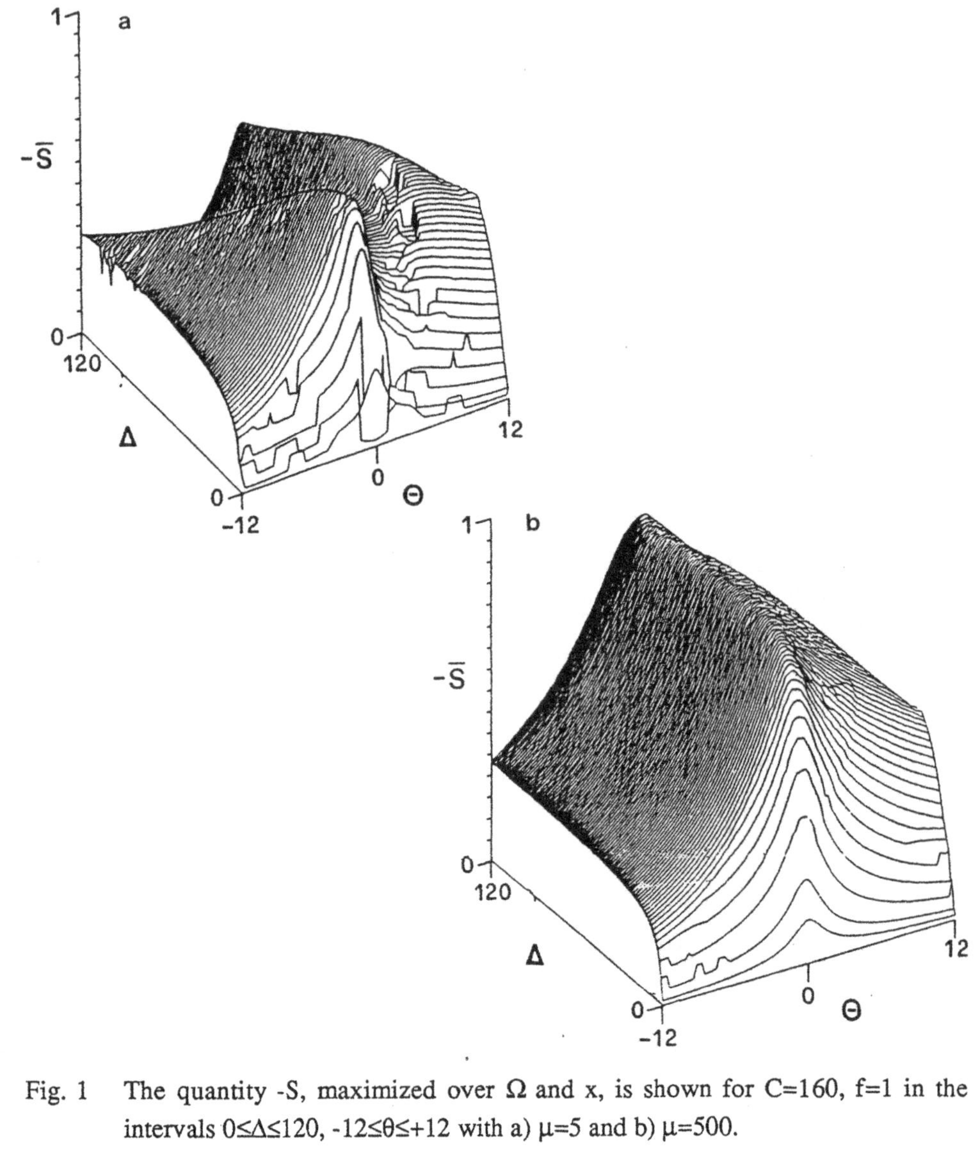

Fig. 1 The quantity -S, maximized over Ω and x, is shown for C=160, f=1 in the intervals $0 \leq \Delta \leq 120$, $-12 \leq \theta \leq +12$ with a) $\mu=5$ and b) $\mu=500$.

All the remaining diagrams in this section show the quantity $-\bar{S}$ optimized only over Ω, but not over x. These graphs illustrate the variation of $-\bar{S}$ for $0 \leq x \leq 100$ over a suitable interval of a parameter chosen among the quantities C, Δ, θ, μ, f while all the other quantities remain fixed. These diagrams exhibit occasionally some "holes" in correspondence to the intervals of the variable x in which the steady-state curve x(|y|) defined by (12) has negative slope. In fact, these unstable portions are physically meaningless and in these regions the value of $\bar{S}(\Omega)$ is arbitrarily set equal to zero.

Figure 2 illustrates the dependence of the squeezing on the parameter f. As expected, the optimal condition is the purely radiative case f=1 and the squeezing decreases, but not so dramatically, with f.

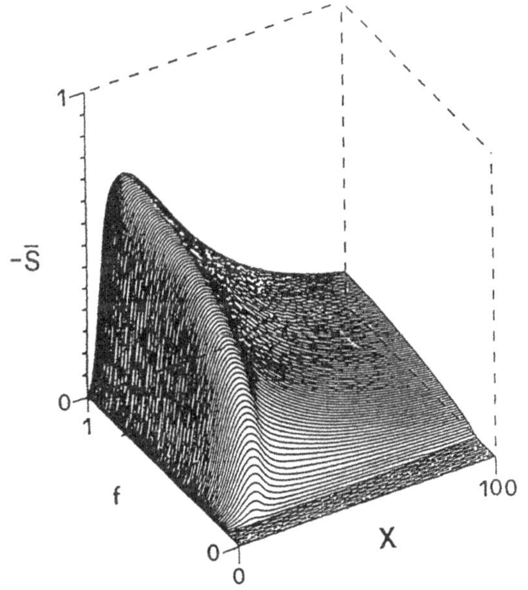

Fig. 2 The quantity -S, maximized over Ω, is shown for $C=20$, $\Delta=10$, $\theta=1$, $\mu=5$ in the intervals $0 \leq x \leq 100$, $0 \leq f \leq 1$.

The diagram 3 describes the joint dependence on Δ and x. They show a "mountain" which becomes higher and more extended when μ is increased, as we see by comparing Figs. 3a and 3b. When x is increased with all the other parameters fixed, the squeezing first increases and then decreases. In the bistable cases, the level of squeezing is usually larger in the lower branch when μ is on the order of 1-10, whereas it is larger in the lower portion of the upper branch when μ is very large. Similarly, when Δ is increased with all the other parameters fixed,the squeezing first increases and then decreases. We found generally a good accord with the "correlation rule", formulated in Refs. 1-3, which prescribes that best squeezing occurs for x~Δ. This rule implies that the optimal value for x is independent of the parameters C,θ,μ,f; this feature is usually confirmed with occasional exceptions.

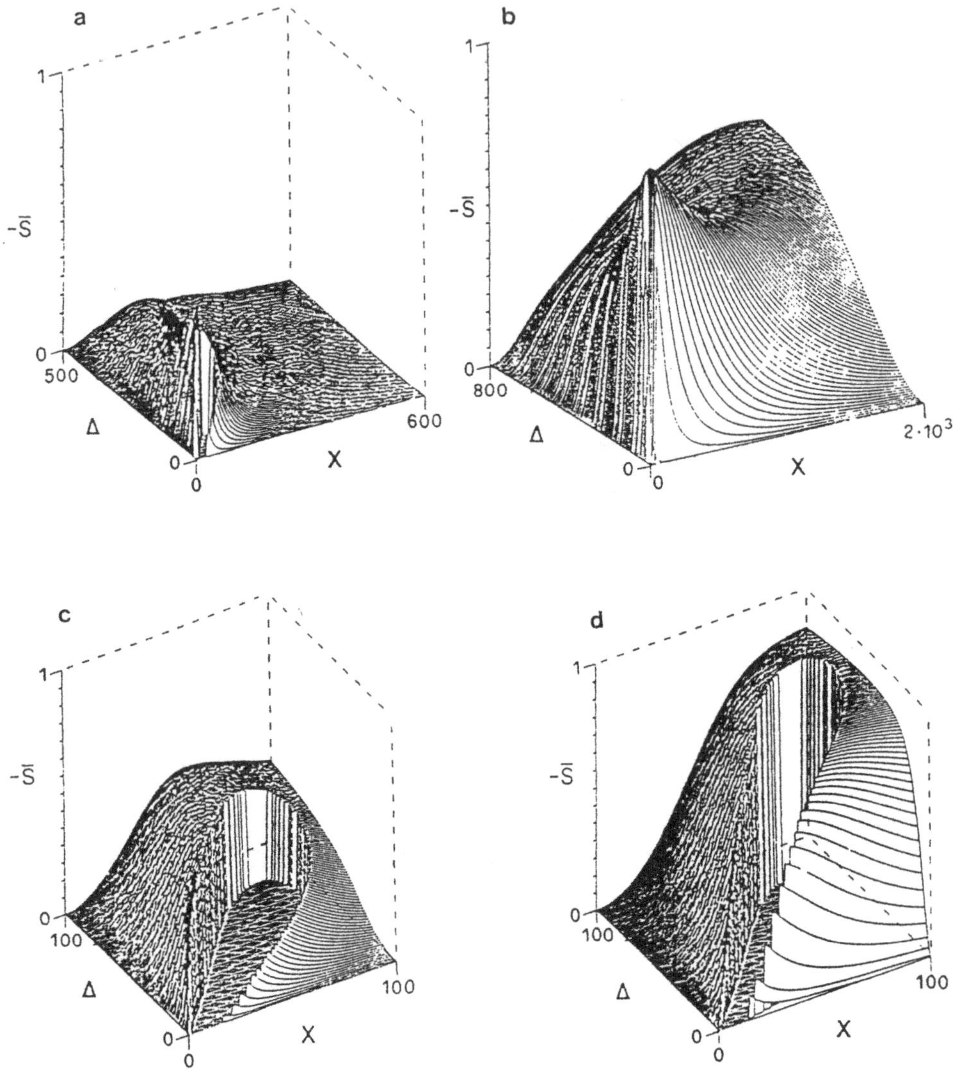

Fig. 3 The quantity -S, maximized over Ω, is shown for C=160, θ=1, f=1 and a) μ=5, 0≤x≤600, 0≤Δ≤500, b) μ=500, 0≤x≤2000, 0≤Δ≤800, c) μ=5, 0≤x≤100, 0≤Δ≤100, d) μ=500, 0≤x≤100, 0≤Δ≤100, c) and d) are a magnification of a portion of a) and b), respectively.

From Fig. 4 we see clearly that the degree of squeezing increases with the bistability parameter C.

Fig. 4
The quantity -S, maximized over Ω, is shown for f=1, μ=5, Δ=10, θ =-1.5, 0≤x≤100, 0≤C≤400.

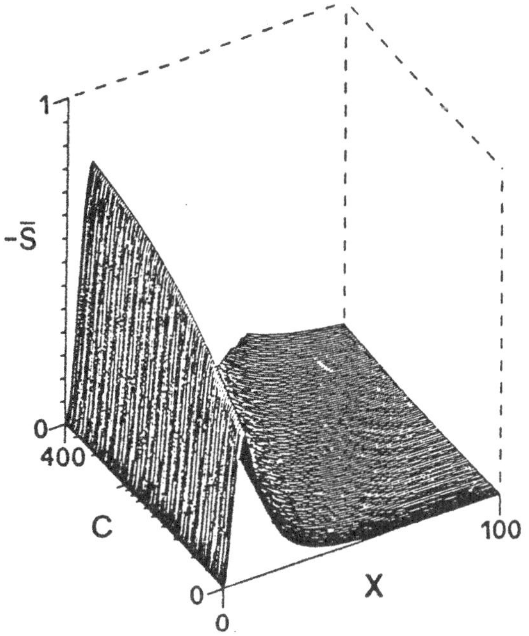

Figure 5 describes the variation with the parameter μ. Very often, an increase of μ implies an increase of squeezing, this appears also from some previous figures. In the limit $\mu \rightarrow \infty$, with fixed values of the other parameters, the spectrum approaches an asymptotic value; this point is further discussed in the following section.

Fig.5
The quantity -S, maximized over Ω, is shown for C=20, f=1, Δ=10, θ=1, 0≤x≤100, 5≤μ≤10^4. (μ on a logarithmic scale).

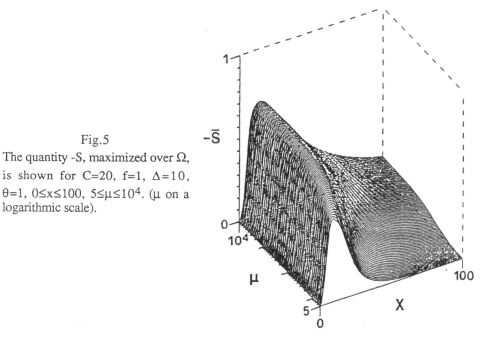

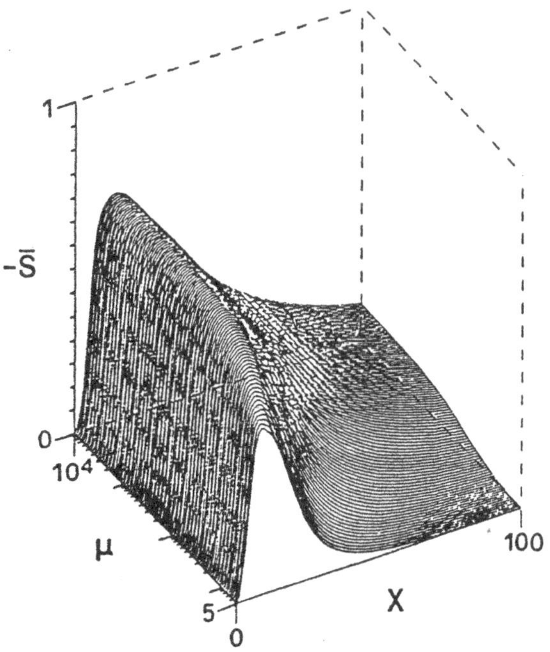

THE BAD CAVITY LIMIT

The first prediction of squeezing for two-level OB[5] concerned the good and the bad-cavity limits, in which the set of five eigenvalues split in two distinct groups of two "field" eigenvalues of order k and three "atomic" eigenvalues of order $\gamma_{\parallel}$[4]. In the bad-cavity limit $\mu \rightarrow \infty$ one can perform an adiabatic elimination of the field variables; we focus now on this situation, which allows for high levels of squeezing, and study how it is approached as μ is increased.

Figures 6 b-c describe the approach to the bad cavity limit in correspondence to the points B and C in Fig. 6a. We note that for $\mu \rightarrow \infty$ the squeezing not only persists, but extends itself over very broad regions of frequency. In the lower branch, the width of the spectrum coincides with the cooperative linewidth $\gamma_R = 2\,C\gamma_\perp$ of pure superfluorescence[27].

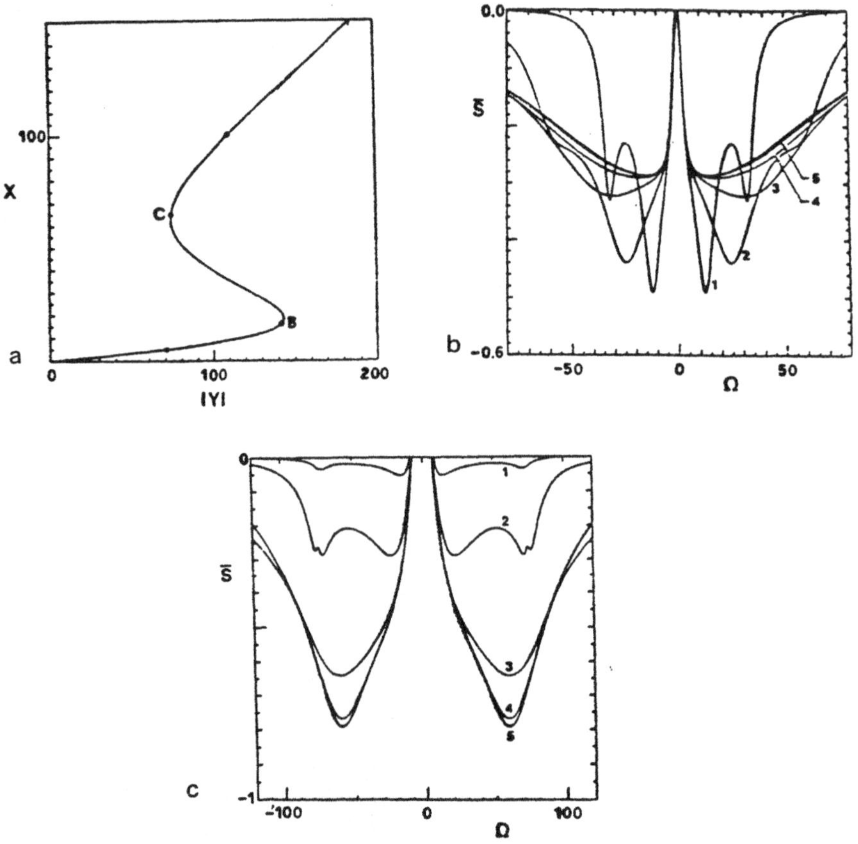

Fig.6 a) Steady-state curve of output field x as a function of input field |y| for C=160, Δ=20, θ=1. b) and c). The spectrum of squeezing $S(\Omega)$ is shown for f=1 and 1) μ=5, 2) μ=20, 3) μ=80, 4) μ=320, and 5) μ=1280. The dark curve corresponds to the limit $\mu \rightarrow \infty$ b) x=16, c) x=100. The diagrams b) and c) correspond to the points B, C in a), respectively.

Over the upper branch, the degree of squeezing grows very markedly as μ is increased, until the squeezing spectrum approaches the asymptotic configuration, which coincides with that predicted by the model after adiabatic elimination of the field variables. The frequency domain in which there is sizable squeezing is centered around the Rabi frequency of the internal cavity field $\Omega/\gamma_{||} = [(2fx^2+\Delta^2)/4f^2]^{1/2}$. The role of γ_R in the "cooperative" lower transmission branch, as well as the role of the Rabi frequency in the upper branch, are familiar from the previous studies of the spectrum of transmitted light[4-28] and of photon antibunching[15] in the bad cavity limit. The picture given by Fig.6 agrees with the numerical results for a low Q cavity given in Ref. 12.

All the cases considered up to now correspond to detuned configurations. If the detunings Δ and θ are simultaneously conveyed to zero the squeezing decreases substantially[14]. On the one hand this result confirms the importance of detuning, on the other hand it explains why the early prediction[5] displayed only a small amount of squeezing. However, on retracing the path backwards by increasing Δ and θ, one sees that the spectrum changes in a continuous and smooth fashion from the resonant to the detuned configuration. This proves that the seed of the large squeezing, which emerges in appropriate regions of the parameter space, is just the small squeezing predicted in Ref. 5.

Our paper concerns a system contained in an optical cavity. Recent works[29,30] develop quantum theories of four wave mixing in cavityless configurations. In particular, a significant squeezing is found in correspondence to the Rabi sidebands of the pump field, when the absorption parameter αL of the medium and the pump field are large enough. Our picture in the bad cavity limit, higher transmission branch, is presumably correlated with these results. In fact, in the bad cavity limit the features of the spectrum are determined by the atoms whereas the field eigenvalues, which are governed by the cavity linewidth, are ininfluential. For C large the intracavity field is large in the upper branch, and we remind in this connection the alternative expression of the bistability parameter $C=\alpha L/2T$, where T is the transmissivity coefficient of the mirrors.

We observe finally that, using the dressed-atom approach[31], the Rabi-sidebands squeezing in the upper branch, bad-cavity limit, can be explained as a four-wave mixing process in which the absorption of two photons of the input field is followed by the emission of two photons with frequencies $\omega_0-\Omega$ and $\omega_0+\Omega$, where Ω is the Rabi frequency of the intracavity field. This argument holds strictly in the resonant case $\Delta=0$, but provides a useful analogy between the squeezing in OB under these conditions and the four-wave mixing mechanism.

ACKNOWLEDGEMENTS

Work in the framework of the EEC twinning project on Nonclassical States of the Electromagnetic Field.

REFERENCES

1. H.J. Kimble, M.J. Raizen, L.A. Orozco, Mim Xiao and T.L. Boyd, in "Fundamentals of Quantum Optics Ii", ed. by F. Ehlotzky, Springer-Verlag, Berlin 1987.
2. M.G. Raizen, L.A. Orozco, Min Xiao, T.L. Boyd and H.J. Kimble, Phys. Rev. Lett. 59: 198 (1987).

3. L.A. Orozco, M.G. Raizen, Min Xiao, R.J. Brecha, and H.J. Kimble, Journ. Opt. Soc. Am. B 4: 1490 (1987).
4. L.A. Lugiato, "Theory of Optical Bistability", in Progress in Optics, Vol. XXI, ed. by E. Wolf, North-Holland, Amsterdam 1984 and references quoted therein.
5. L.A. Lugiato and G. Strini, Opt. Commun. 41, 67 (1982).
6. L.A. Lugiato and G. Strini, Opt. Commun. 41, 374 (1982).
7. D.F. Walls and G.J. Milburn, in "Quantum Optics, Gravitation and Measurements Theory", ed. by P. Meystre and M.O. Scully, Plenum Press, New York 1983.
8. M.D. Reid and D.F. Walls, Phys. Rev. A 28: 332 (1983).
9. M.D. Reid and D.F. Walls, Phys. Rev. A 32: 396 (1985).
10. H.J. Carmichael, Phys. Rev. A 33: 3262 (1985).
11. L.A. Lugiato, Phys. Rev. A 33, 4079 (1986).
12. M.D. Reid, A. Lane and D.F. Walls, in "Quantum Optics IV", ed. by J.D. Harvey and D.F. Walls, Springer Proceedings in Physics n. 12, Springer-Verlag,Berlin 1986, p. 31.
13. D.A. Holm and M. Sargent III, Phys. Rev. A 35: 2150 (1987).
14. F. Castelli, L.A. Lugiato and M. Vadacchino, Nuovo Cimento D, in press.
15. F. Casagrande and A.L. Lugiato, Nuovo Cimento 55B: 173 (1980).
16. B. Yurke, Phys. Rev. A 29: 408 (1984).
17. M.J. Collett and C.W. Gardiner, Phys. Rev. A 30: 1386 (1984).
18. M. Gronchi and L.A. Lugiato, Lett. Nuovo Cimento 23: 593 (1978).
19. V. Benza and L.A. Lugiato, in "Optical Bistability", ed. by C.R. Bowden, H.R. Robl and M. Ciftan, Plenum, New York 1981, p. 9.
20. L.A. Lugiato, Z. Phys. B 41: 85 (1981).
21. L.A. Lugiato and F. Castelli, Z. Phys. B 64: 375 (1986).
22. R. Bonifacio and L.A. Lugiato, Phys. Rev. A 18: 1129 (1978).
23. H. Haken, "Laser Theory", Encyclopedia der Physik vol. XXV/2c, Springer Verlag, Berlin 1970.
24. R.J. Glauber, Phys. Rev. 130: 2529 (1963);131:2766 (1963).
25. M. Lax, Phys. Rev. 157: 213 (1967).
26. L. Mandel, Opt. Commun. 42: 437 (1982); Phys. Rev. Lett. 49: 136 (1982).
27. R. Bonifacio and L.A. Lugiato, Phys. Rev. A 11: 1507 (1975).
28. L.A. Lugiato, Nuovo Cimento B 50: 89 (1979).
29. S-T. Ho, P. Kumar and J.H. Shapiro, Phys. Rev. A 35: 3892 (1987) and this volume.
30. G.S. Agarwal and R.W. Boyd, submitted for publication and this volume.
31. C. Cohen - Tannoudij and S. Reynaud, J. Phys. B 10: 345 (1977).

SQUEEZED-LIGHT GENERATION IN OPTICAL WAVEGUIDES[1]

Prem Kumar

Department of Electrical Engineering and Computer Science
The Technological Institute, Northwestern University
Evanston, Illinois 60208, USA

1. Introduction

Within the past few years, squeezed states of light have been generated in a number of distinct physical systems [1]. All of them can be categorized into the following two groups: i) those exploiting the resonant nonlinear interaction of light with two-level atoms and ii) those involving the nonresonant nonlinearity of the interaction of high-intensity light with transparent media. The experiments in the first group have included intracavity four-wave mixing in an atomic beam [2], forward four-wave mixing in a Doppler-broadened gaseous medium [3], and the strong interaction of a small-volume high-finesse optical cavity with a beam of two-level atoms [4]. The experiments in the second group, namely forward four-wave mixing in a single-mode optical fiber [5] and intracavity parametric down-conversion in a nonlinear crystal [6], have been more interesting from an applications point of view and indeed the latter has emerged to be a prototypical system for the generation of squeezed light. Over 60% squeezing has been measured in light emitted by a cavity containing the $MgO{:}LiNbO_3$ down-converter [6].

A major reason for the success of a parametric down-converter as a state squeezer is the absence of uncorrelated excess noise generating mechanisms that almost invariably degrade squeezing in other systems. For example, in resonant systems, the spontaneous emission generated by the pump field degrades squeezing because of proximity to the atomic resonance [2,3] and in the fiber experiment, guided acoustic-wave Brillouin scattering (GAWBS) of the pump beam introduces excess noise in

[1]Supported in part by the National Science Foundation under Grant No. EET-8715275.

the squeezed quadrature of interest [5]. On the other hand, and only due to the absence of any excess-noise generating mechanism, the parametric down-converter generates a minimum uncertainty squeezed state which can have certain practical applications not offered by the squeezing generated in other systems.

Speaking of applications, table top demonstrations of the enhancement in sensitivity of an optical interferometer using squeezed-light have already been reported [7,8]. Optical communication applications (incidently, for which squeezed states were proposed in the first place) are yet to be demonstrated. For such applications, both in fiber-optic and integrated-optic systems, it is desirable that squeezing be generated in the guiding medium itself. Moreover, larger coupling constants are obtained in nonlinear materials at lower pump powers due to confinement of the various beams to the small cross-section of the waveguide. In fact, these advantages were realized by Levenson et al. [5] very early on in their demonstration of squeezing in an optical fiber. Unfortunately, as pointed out earlier, their experiment was marred by GAWBS which created significant excess noise at frequencies where squeezing was generated by the $\chi^{(3)}$ nonlinearity of the fiber. Although they were able to reduce GAWBS by immersing the fiber in liquid helium, for most practical applications, it is cumbersome and undesirable.

As the acronym suggests, GAWBS is Brillouin light-scattering of the strong pump beam into the neighboring sidebands by the guided acoustic waves. In a $\chi^{(3)}$ medium, the sidebands that receive scattered light are the same which are squeezed by the nonlinear interaction. In a $\chi^{(2)}$ medium, however, *the strong pump beam is at twice the frequency of the optical modes that are squeezed.* Therefore, a process like GAWBS, which scatters light into the neighboring sidebands of the strong pump beam, *will not mask squeezing* being generated at the fundamental frequency by the parametric interaction. It is due to this reason, and the fact that parametric down-converter has proven to be an excellent squeezer in bulk nonlinear media, that I have chosen to explore the possibility of squeezed-light generation in optical waveguides in which the dominant nonlinearity is of second order. Also, such waveguide based squeezed-light sources are potentially integrable with fiber-optic communication systems and local-area networks. Before discussing squeezed-light generation in optical waveguides, let me review some properties of the commonly employed linear and nonlinear waveguides [9].

2. Nonlinear Optics in Waveguides

A typical planar waveguide is shown in Fig. 1. It consists of a thin film of refractive index n_1 and thickness d sandwiched between a substrate of refractive index n_2 and a cladding of refractive index n_3 ($n_1 > n_2, n_3$). For nonlinear optical interactions, any or all of the three media can be nonlinear. In the absence of any nonlinearity, an electromagnetic wave of angular frequency ω propagating in such

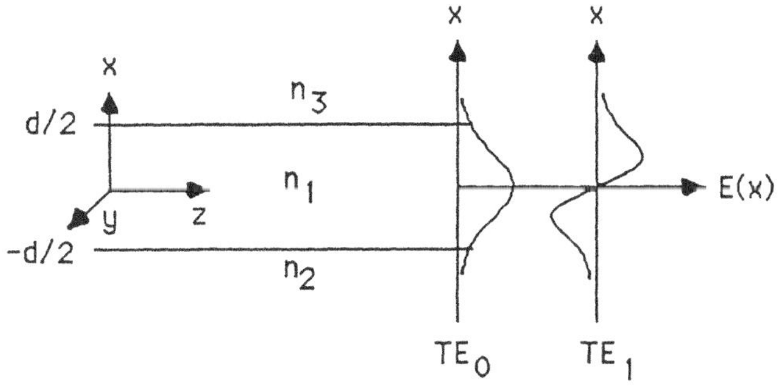

Figure 1: *A typical planar waveguide geometry. At right are shown the representative field patterns of the two lowest order modes.*

a structure must satisfy the following Helmholtz equation:

$$\nabla^2 \vec{E}(\vec{r}) + \frac{n_i^2 \omega^2}{c^2} \vec{E}(\vec{r}) = 0 \tag{2.1}$$

with n_i $(i = 1, 2, 3)$ chosen appropriately in each medium. Imposition of the boundary conditions at the $\pm d/2$ interfaces leads to guided-wave modes which are either transverse electric (TE_m) or transverse magnetic (TM_m). Representative field patterns for the lowest order modes, TE_0 and TE_1, are also shown in Fig. 1.

The electric field associated with, for example, a TE_m mode of the above geometry can be written as

$$e_y^m(\vec{r}, t) = \text{Re}\left[E_y^m(x)\ a^m(z) \exp\left\{-i(\omega t - n_e^m kz)\right\}\right], \tag{2.2}$$

where the modal patterns have been so chosen as to satisfy the orthonormality condition

$$\frac{n_e^m}{2\mu_0 c} \int E_y^m(x) E_y^{n*}(x) dx = \delta_{n,m} , \tag{2.3}$$

and $n_e^m = n_e^m(n_i, kd)$, for $i = 1, 2, 3$, is the effective refractive index for propagation along z. In general, $n_e^m(\omega)$ is a solution of coupled transidental equations and numerical techniques are employed to generate dispersion curves of the kind shown in Fig. 2. As will be discussed later, these curves play an important role in determining the phase-matching possibilities for nonlinear optical processes in optical waveguides.

For second-order processes, the nonlinear polarization generated by two waveguide modes $\vec{E}^{m_1}(x)$ and $\vec{E}^{n_2}(x)$ of frequecies ω_1 and ω_2, respectively, can be written as

$$\vec{p}^{NL}(\vec{r}, t) = \text{Re}\left[\vec{P}^{NL}(x, z, \omega) \exp\{-i(\omega t - nkz)\}\right] \tag{2.4}$$

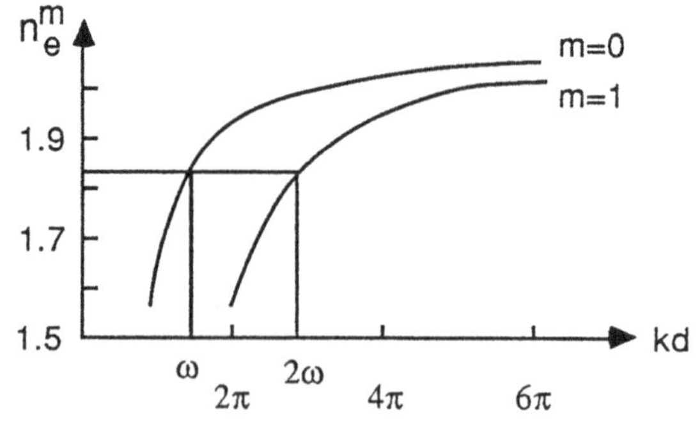

Figure 2: *Representative dispersion curves for the two lowest order modes.*

with

$$\vec{P}^{NL}(x,z,\omega) = \epsilon_0 \vec{E}^{m_1}(x)a^{m_1}(z) \cdot \overset{\rightleftarrows}{\chi}^{(2)}(-\omega;\omega_1,\omega_2) \cdot \vec{E}^{n_2}(x)a^{n_2}(z), \tag{2.5}$$

where $nk = n_e^{m_1}k_1 + n_e^{n_2}k_2 \neq n_e^m k$ for n_e^m – the effective refractive index for the mth mode of frequency ω. The presence of the nonlinear polarization results in radiation at frequency ω which, under slowly varying envelope approximation, leads to the following growth (or decay) equation for any mode m:

$$\frac{d}{dz}a^m(z) = \frac{i\omega}{4}\int_{-\infty}^{\infty} dx\, \vec{P}^{NL}(x,z,\omega) \cdot \vec{E}^{m*}(x) \exp\{-i(n_e^m - n)kz\}. \tag{2.6}$$

Since, for squeezed-state generation, we are interested in parametric down conversion of a pump field mode m_2 at frequency 2ω, the following coupled-mode equations are easily obtained for degenerate interaction (generalization to nondegenerate parametric amplification is straightforward) of signal and idler modes m_1 and n_1 at frequency ω:

$$\frac{d}{dz}a^{m_1}(z) = \kappa a^{m_2}(z)a^{n_1 *}(z)\exp\{-i(n - n_e^{m_2})kz\}\,, \tag{2.7}$$

$$\frac{d}{dz}a^{n_1}(z) = \kappa a^{m_2}(z)a^{m_1 *}(z)\exp\{-i(n - n_e^{m_2})kz\}\,. \tag{2.8}$$

Here, $n = n_e^{n_1} + n_e^{m_1}$ and κ defines the interaction efficiency in terms of the overlap integral

$$\kappa = \frac{i\omega\epsilon_0}{2}\int_{-\infty}^{\infty} dx\; d_{ijk}(x)E_i^{m_2}(x)E_j^{m_1 *}(x)E_k^{n_1 *}(x). \tag{2.9}$$

The indices i, j, and k are determined by the polarizations of the m_2, m_1, and n_1 modes respectively (whether they are TE or TM etc.), and $d_{ijk}(x)$ is the relevant coefficient of the nonlinear susceptibility tensor $\overset{\rightleftarrows}{\chi}^{(2)}$. Note that the x dependence of d_{ijk} in the overlap integral can be exploited to substantially improve the interaction efficiency of different order modes [10].

The exponential factors in Eqs. (2.7) and (2.8) arise because of differing phase speeds of the interacting modes, m_2, m_1, and n_1 resulting in reduced parametric interaction due to the accompanying phase mismatch. With appropriate choice of the two frequencies ω and 2ω, the material refractive indices, n_1, n_2, and n_3, and the waveguide thickness d, these exponential factors can be set equal to 1 for optimum phase-matched interaction. An example is shown in Fig. 2 for type I phase matching.

3. Quantum Optics in Waveguides

It has been demonstrated experimentally as well as through theoretical investigations that squeezed-state generation in resonant systems is inherently less efficient [1]. Mainly it is due to saturation of the associated resonance, causing spontaneous emission into modes that are being squeezed by the resonant nonlinear coupling. Because of this reason my interest lies only in nonresonant optical processes in waveguides. Since the field of a waveguide mode extends well into the substrate as well as the cladding regions, neither of them should be composed of materials that resonate with either of the two frequencies taking part in the parametric down-conversion process.

For such nonresonant processes, the transition from a classical description to a quantum mechanical one is fairly straightforward. It is obtained by replacing the dimensionless modal variables a^m by the annihilation operators $\hat{a}^m$. Furthermore, assuming type II phase matching, the coupled-mode equations (2.7) and (2.8) can be easily derived, under constant classical pump approximation, using the Heisenberg equations of motion obtained from the following effective Hamiltonian:

$$H_I = i\hbar\, \bar{\kappa}\, \hat{a}^{m_1\dagger} \hat{a}^{n_1\dagger} + H.c. \,, \tag{3.1}$$

where $\bar{\kappa} = cn_e\kappa a^{m_2}$. Thus the quantum mechanical description of parametric processes in optical waveguides is not very much different from that in the bulk optics so long as one takes into account the appropriate effective refractive indices of the various waveguide modes. Furthermore, parametric interactions in waveguides should produce identical quantum effects as those observed in bulk optics.

With type II phase matching, as assumed in writing the interaction Hamiltonian of Eq. (3.1), neither of the two modes, m_1 or n_1, are squeezed individually. Squeezing, however, occurs in a 50-50 linear-combination mode whose spatial pattern is complicated and determined by the participating modes m_1 and n_1. Although squeezing would be present in this linear combination mode, it would be difficult to detect because of the requirement of a local oscillator beam with a matching spatial profile. Thus from a practical standpoint, squeezed-light generation in an optical waveguide using type II phase matching of the waveguide modes is not very useful. Type I phase matching, however, is much more interesting. In this case, both of the parametrically down-converted photons are emitted into the same waveguide mode, say n_1, at frequency ω. The interaction Hamiltonian then becomes

$$H_I = i\hbar\,\bar{\kappa}\left(\hat{a}^{n_1\dagger}\right)^2 + H.c. \,, \tag{3.2}$$

and the coupling constant is given by

$$\kappa = \frac{i\omega\epsilon_0}{2}\int_{-\infty}^{\infty} dx\; d_{ijj}(x)E_i^{m_2}(x)\left(E_j^{n_1 *}(x)\right)^2 . \tag{3.3}$$

Squeezing occurs in mode n_1 which can be easily detected by employing a local oscillator beam with the same spatial profile as that of mode n_1. Such a beam can be obtained, for example, by passing a Gaussian beam of frequency ω through an identical waveguide as that used for the generation of squeezing. For pulsed squeezed-light generation [11], a temporal interlacing of the pump and local-oscillator pulses can be employed.

4. Effect of Linear Losses

The above transition from classical physics to quantum physics does not take into account the effect of linear losses incurred by the modes undergoing parametric amplification. In optical waveguides, the dominant linear loss mechanism is the surface scattering of the guided modes due to imperfections in the core-cladding and core-substrate interfaces. According to the fluctuation dissipation theorem, any such loss adds uncorrelated fluctuation to the modes that are being squeezed. Physically speaking, vacuum fluctuations from the modes which receive the scattered light sneak back into the squeezed modes. For nonresonant systems, the effect of these linear losses is easily taken into acccount either by the straightforward application of the fluctuation-dissipation theorem or by coupling the squeezed system to a thermal reservoir [12]. Either way, the following expression for the variance in the noise of an appropriate squeezed quadrature is easily obtained:

$$< (\Delta\hat{a}_1^{n_1})^2 >= \frac{1}{4}\left[\frac{\gamma}{\gamma+\bar{\kappa}} + \left(1 - \frac{\gamma}{\gamma+\bar{\kappa}}\right)\exp\{-2(\gamma+\bar{\kappa})L\}\right], \tag{4.1}$$

where γ is the linear loss coefficient per unit length and L is the interaction length. As can be easily seen from the above equation that, for lossy systems, there is a lower bound of $\gamma/4(\gamma+\bar{\kappa})$ on the achievable squeezing. Nevertheless, this limit is pushed to zero as long as the nonlinear coupling coefficient far exceeds the linear loss coefficient. In optical waveguides, the coupling coefficient is directly proportional to the overlap integral of Eq. (6). Thus, the prescription for efficient squeezed-state generation in optical waveguides is to *maximize the overlap integral and to minimize the linear losses.*

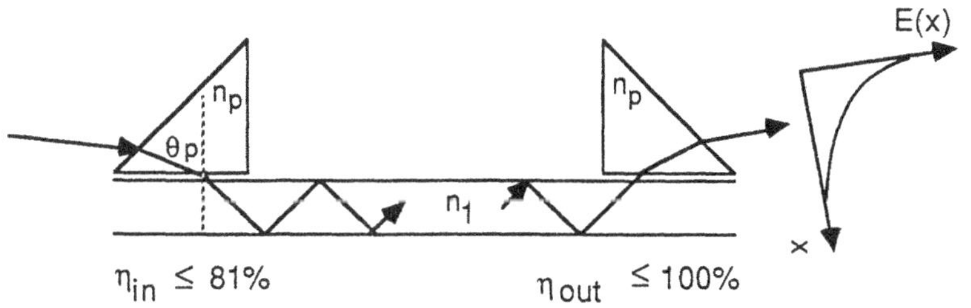

Figure 3: *Prism coupling technique for exciting specific waveguide modes.*

5. Coupling Issues

From the discussion in Sec. 3, it is clear that to generate squeezing, one needs to excite specific modes of the optical waveguide. For coupling light into and out of such a structure, there are two techniques that are commonly employed. In the first, called 'end-fire coupling', mode-matching lenses are used at each end of the waveguide. This technique is not very useful for intra-waveguide squeezed-light generation because with it, it is very difficult to excite specific modes of the waveguide.

In the second technique, prisms placed adjacent to the waveguide as shown in Fig. 3, are used to couple light into and out of the waveguide [13]. If a prism of refractive index $n_p > n_1$ is used, then the evanescent field penetrating the waveguide causes coupling between the incident beam and that waveguide mode for which $n_e^m = n_p \sin\theta_p$, where θ_p is the angle of incidence. Thus by varying θ_p, any mode m of the waveguide can be selectively excited. Using this technique, a maximum input coupling efficiency η_{in} of 81% is obtained. This is because, as the field in the waveguide builds up, it starts leaking back into the prism. The output coupling efficiency η_{out}, on the other hand, can be as high as 100%. As shown in Fig. 3, the output field spatial profile, however, is not the same as that of the guided mode.

For the purpose of an initial demonstration of intra-waveguide squeezed vacuum light generation, only the pump beam needs to be launched into the waveguide. The constraint on η_{in} then results only in some loss of the input pump power. The exponential spatial profile of the output beam will require a mode-matched local-oscillator beam. Techniques mentioned at the end of Sec. 3 can be employed for such a purpose.

In practical applications, where an incoming signal may need to be parametrically amplified [14], the loss associated with non-ideal η_{in} will cause degradation of the quantum properties of the input signal beam. For these applications, highly efficient, integrated, end-coupled, and mode matched techniques will need to be developed for coupling the signal beam into and out of the waveguide.

6. Potential for Squeezed-Light Generation in Optical Waveguides

In recent years, various nonlinear phenomena have been investigated in optical waveguides [9]. To date, the best results for parametric down-conversion have been obtained by Sohler and Suche [15] using y-cut Ti–in-diffused $LiNbO_3$ waveguides [16]. In their experiments, the pump beam, consisting of 200W, 60ns long pulses obtained from a tunable dye-laser operating in the vicinity 650nm, was prismatically coupled into the waveguide to excite the TE_{10} mode. Parametric down-conversion into TM_{00} modes (signal at 1.15μm and idler at 1.499μm) of the waveguide was observed with a gain equivalent to a coupling coefficient κ of 0.35cm^{-1}. By temperature tuning the Ti:$LiNbO_3$ waveguide, degenerate operation was also possible with comparable interaction strength. The 3.2cm long high-quality waveguide used by Sohler and Suche [15] had a linear loss coefficient γ of 0.07cm^{-1}. Substituting their values of κ and γ in Eq. (4.1), loss limited squeezing of 6.5 dB is possible.

The above result is very encouraging for intra-waveguide squeezed-light generation. Sohler and Suche [15] also observed degenerate parametric oscillation in their waveguides. Therefore, a waveguide version of the Wu et al. [6] experiment to generate squeezed-vacuum light is also possible. A potential disadvantage of using $LiNbO_3$ based waveguides is their low damage threshold resulting from the photorefractive effect. In the early days of nonlinear optics in waveguides, second harmonic generation was observed in waveguides fabricated from a number of nonlinear materials [9]. Conversion efficiencies, however, were very low due to poor quality of the waveguides. Since then, modern epitaxial growth techniques have become available and it should be possible to fabricate nonlinear waveguides from high damage-threshold materials for intra-waveguide squeezed-light generation [17].

In conclusion, I have shown that it is possible to generate squeezed-light by parametric down-conversion in nonlinear optical waveguides. Experiments conducted in Ti–in-diffused $LiNbO_3$ waveguides suggest that at least 6dB of squeezing should be observable.

References

1. *J. Opt. Soc. Am. B* **4**, No. 10 (1987); special issue on squeezed states of light.
2. R. E. Slusher, L. W. Hollberg, B. Yurke, J. C. Mertz, and J. F. Valley, *Phys. Rev. Lett.* **55**, 2409 (1985).
3. M. W. Maeda, P. Kumar, and J. H. Shapiro, *Opt. Lett.* **12**, 161 (1987); *J. Opt. Soc. Am. B* **4**, 1501 (1987).

4. M. G. Raizen, L. A. Orozco, M. Xiao, T. L. Boyd, and H. J. Kimble, *Phys. Rev. Lett.* **59**, 198 (1987).

5. R. M. Shelby, M. D. Levenson, S. H. Perlmutter, R. G. DeVoe, and D. F. Walls, *Phys. Rev. Lett.* **57**, 691 (1986).

6. L.-A. Wu, H. J. Kimble, J. L. Hall, and H. Wu, *Phys. Rev. Lett.* **57**, 2520 (1986).

7. M. Xiao, L.-A. Wu, and H. J. Kimble, *Phys. Rev. Lett.* **59**, 278 (1987).

8. P. Grangier, R. E. Slusher, B. Yurke, and A. LaPorta, *Phys. Rev. Lett.* **59**, 2153 (1987).

9. G. I. Stegeman, C. T. Seaton, W. M. Hetherington III, A. D. Boardman, and P. Egan, in *Nonlinear Optics: Materials and Devices*, Flytzanis and Oudar, Eds. (Springer, Berlin, heidelberg, New York, 1985) pp. 31-64.

10. H. Ito and H. Inaba, *Opt. Lett.* **2**, 139 (1978).

11. R. E. Slusher, P. Grangier, A. LaPorta, B. Yurke, and M. J. Potasek, *Phys. Rev. Lett.* **59**, 2566 (1987).

12. W. H. Louisell, *Quantum Statistical Properties of Radiation* (Wiley, New York, 1973).

13. P. K. Tien, *Appl. Opt.* **10**, 2395 (1971).

14. H. P. Yuen, *Opt. Lett.* **12**, 789 (1987).

15. W. Sohler and H. Suche, *Appl. Phys. Lett.* **37**, 255 (1980); *Proc. SPIE* **408**, 163 (1983).

16. R. V. Schmidt and I. P. Kaminov, *Appl. Phys. Lett.* **25**, 458 (1974).

17. B. W. Wessels, private communication.

GENERATION OF NUMBER-PHASE SQUEEZED STATES

Gunnar Björk and Olle Nilsson
Department of Microwave Engineering
The Royal Institute of Technology, S-10044 Stockholm

Yoshihisa Yamamoto and Susumu Machida
Basic Research Laboratories, Nippon Telegraph and Telephone Corporation
Musashino-shi, Tokyo 180,Japan

1. INTRODUCTION

Squeezed states have attracted quite a lot of attention the last few years. In such states the operators between which the quantum noise is redistributed are the quadrature components $\hat{a}_1$ and $\hat{a}_2$. While squeezed states may lead to substantial improvements of the sensitivity of precision measurements, specifically in interferometry [1], they may not be optimal for information transmission and processing. Another type of non-classical states of the electromagnetic field are number-phase minimum uncertainty states (NUS). In these, the quantum noise is instead shared between the photon number operator $\hat{n}$, and the sine operator $\hat{S}$ [2]. For high mean photon numbers the latter can be interpreted as the quantity we in classical physics call phase. In the extreme limit of quantum noise reduction the photon number has no fluctuations, while the phase is completly random. These states are called photon number states or Foch states. NUS are particularly interesting in information transmission systems for a variety of reasons. This paper will describe two different generation principles of states where the photon number fluctuations are reduced resulting in sub-Poissonian photon counting statistics. Neither scheme will generate a true minimum uncertainty state, however, except in the limits of infinite pumping or infinite feedback gain. We will also present a generation scheme for quadrature-phase squeezed states.

One of the reasons NUS are interesting for communication purposes is that if the photon number fluctuations are reduced, the corresponding increase of the phase fluctuations doesn't have to alter the mean photon number of the state. Thus, there is no trade-off relation between the photon number in a NUS and the degree of noise reduction possible, so the quantum signal to noise ratio can in principle be arbitrarily high. This is contrary to a squeezed state where the maximum signal to noise ratio for a given average photon number n is 4n(n+1) [3].

A second reason NUS are attractive in information transmission systems is that if the information is coded on the transmitted photon number, the detection is achieved by simple photon counting. Using squeezed states the detection is phase sensitive. Phase sensitive detection requires that both the carrier frequency and the carrier phase must be reconstructed by the receiver. We feel that such detectors will be more difficult to operate than simple photon counters. In both cases it is necessary that the detector quantum efficiency is close to unity.

Photon number states also utilize the channel capacity optimally [4]. Finally, localized single photon number states can be used in optical computers. With nonlinear interferometers it is at least in theory possible to construct logical gates which switches states using single photon inputs [5].

2. PUMP NOISE SUPPRESSED SEMICONDUCTOR LASERS

The most successful scheme so far to generate sub-Poissonian light is the pump noise suppressed semiconductor laser [6]. It's fundamental working principle is easy to understand. Since the pumping process is phase-insensitive, the pump phase information can be sacrificed in favor of reduced pump amplitude noise. In an optically pumped laser this is clearly irrelevant, but in a semiconductor laser it is a relatively straight-forward thing to do. Coupling a large source impedance in series with the diode, the voltage over the diode junction will be modulated. This will effectively limit the current fluctuations in the diode PN-junction. The thermal Johnson noise associated with the impedance can be made much lower than the shot noise level if the impedance is sufficiently large. This technique has been also been used by Tapster, Rarity and Satchell [7] to generate sub-Poissonian light form a light emitting diode, and by Teich, Capasso and Saleh who used a pump noise suppressed electron-beam to pump mercury vapor [8, 9].

The calculated external field spectra for a Poisson process pumped laser with a single ended output [10] is shown in Fig. 1. For frequencies higher than the cavity bandwidth the driving noise source is the vacuum field fluctuations reflected by the output mirror. Since we are off-resonance, the vacuum-field fluctuations cannot penetrate the cavity mirror. The high frequency pump fluctuations do not contribute to the high frequency external field fluctuations. Since the cavity lifetime of the photons is much longer than the fluctuation period at these frequencies, the photon field averages out these rapid fluctuations. The external amplitude above the cavity cut-off bandwidth is thus equal to the vacuum field fluctuation level. The driving noise sources below the cavity cutoff bandwidth are the vacuum field fluctuations transmitted through the mirror and the pump fluctuations. The former can be suppressed by the gain saturation mechanism, the harder the pumping the more effective suppression. The pump fluctuations, however, will always couple to the photon field. At high pumping rates the output spectrum below the cavity cut-off frequency will be determined by the pump fluctuations. If the pumping process is Poissonian, the signal amplitude spectrum will be essentially flat and equal to the standard quantum limit (SQL). This is well known from experiments.

In Fig. 2 the corresponding calculated spectrum for a pump noise suppressed laser is shown [6]. For frequencies higher than the cavity decay rate nothing has changed. Within the cavity bandwidth, however, the amplitude fluctuation spectral density will drop below the SQL for pump rates a few times above threshold pumping. At infinite pumping the external amplitude fluctuation spectral density drops proportionally to Ω^2 below the cavity bandwidth.

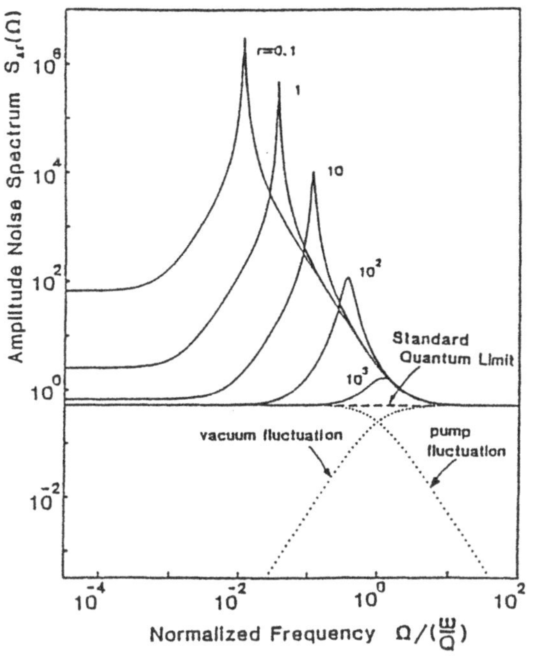

Fig. 1 The amplitude fluctuation spectrum for an ordinary (Poisson process pumped) laser. The parameter r is the normalized pumping level $P/P_{th}-1$. The origins of the fluctuations at $r = \infty$ are indicated.

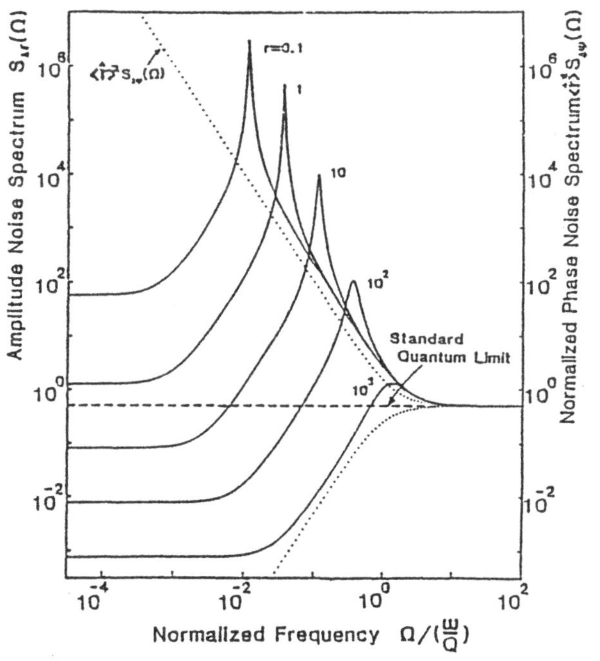

Fig. 2 Quadrature amplitude fluctuation spectra for a pump noise suppressed laser.

At the same time the phase spectral density is proportional to Ω^{-2}. The output state will be very close to a NUS at infinite pumping, and for low frequencies (photon flux integration times much longer than the photon cavity lifetime) it approaches a number state.

A schematical experimental configuration to confirm the theoretical predictions [11] is shown in Fig. 3. The sub-Poissonian light from the pump noise suppressed laser is collimated, passed through an isolator and then focused on the photodiode D1. After preamplification the photocurrent spectrum is measured by a spectrum analyzer. The SQL is calibrated using a highly attenuated reference laser beam detected by a carefully balanced reciever. (The pump noise suppressed laser beam must of course be blocked during this calibration.) When the photocurrents from the detectors D1 and D2 are balanced, the spectrum analyzer will measure only the vacuum field fluctuations incident on the unused port of the beamsplitter.

In the experiment [12] the pump noise suppressed laser was a commercially available, unmodified AlGaAs/GaAs transverse junction stripe laser (Mitsubishi ML-2308, I_{th}=1mA, λ=0.81 μm). The laser was connected to a low noise voltage source in series with a 1 kΩ source resistance. This resistance should be compared with the laser diode's differential resistance which was about 20 Ω at threshold injection current and is inversly proportional to the injection current. The laser had a high reflectivity rear facet and an anti reflection coated front facet, and was operated at 77 K to minimize the internal optical losses. The output was detected by a Si-photodiode with a 90% quantum efficiency.

The reference laser was an AlGaAs/GaAs channeled-substrate planar semiconductor laser (Hitachi HLP-1400, λ=0.84 μm) with a single longitudinal mode output power exceeding 15 mW. By attenuating the output by more than 10 dB, the amplitude noise of the reference beam could be brought down to within 1.4 dB of the SQL (see Fig. 11 of Ref. [13]). This means that the excess amplitude noise is smaller than 40% of the quantum noise. Since the balanced reciever was measured to suppress excess amplitude noise by more than 30 dB, the SQL could be calibrated within 0.04%.

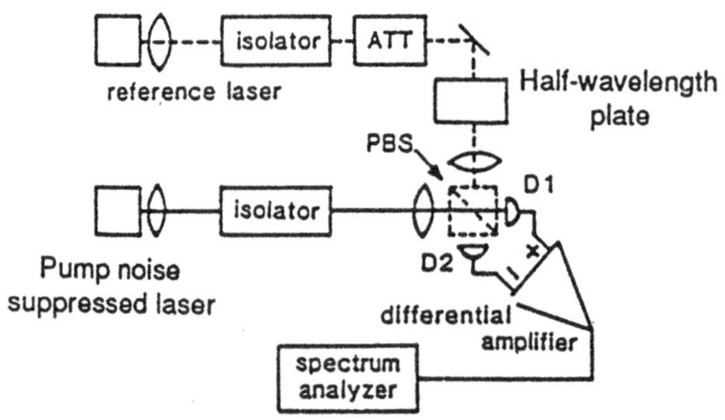

Fig. 3 Schematic measurement setup. PBS is a polarization beamsplitter. By rotating the polarization of the laser the reciever can be accurately balanced.

In Fig. 4 the measured spectral densities are shown, calibrated to the SQL-level obtained from the reference laser. The measured noise reduction is between 0.4 to 0.9 dB below the SQL in the entire frequency region from about 10 MHz to over 1.1 GHz. This is the broadest squeezing bandwidth for cw light reported so far. If the detection quantum efficiency is compensated for, the inferred noise level is about 1.7 dB below the SQL. The measurement of the squeezing spectrum below 10 MHz was prevented by the 1/f-noise of the FET preamplifier, and above 1.1 GHz it was prevented by the cut-off of the Si-photodiode.

As can be seen there is a discrepancy between the SQL calibrated with the reference laser (trace B) and that calibrated with the pump noise suppressed laser (trace D). The reason for this is unclear, but may be due to a detector quantum efficiency difference at the slightly different test laser- and reference laser-wavelengths. If that is the case, the trace D should be taken as the SQL. In the frequency range above 0.5 GHz this reduces the observed noise reduction by between 0.2 to 0.4 dB. The discrepancy may also be due to saturation effects in the detection photo diode. Trace A,B and C are all determined at a detector photocurrent level of 1.87 mA. Trace D is taken at roughly half that level, 0.98 mA. It should be stressed, however, that repeated measurements reproduced the traces in Fig. 4 and that similar results were obtained when measuring other lasers with similar structure.

The theoretical and experimental values for the measured noise level versus the estimated optical loss is shown in Fig. 5. As the external optical attenuation is increased, the noise approaches the SQL. As can be seen, there is a good agreement between the theoretical curve and the measurements. The finite noise reduction at unity quantum efficiency depends on the

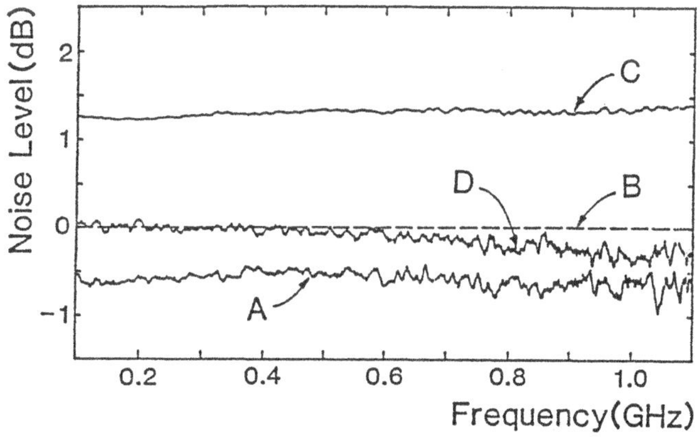

Fig. 4 Amplitude fluctuation spectra normalized to the SQL. Traces B and D are the balanced reciever output for the reference laser and the pump noise suppressed laser respectively. Traces C and A are the respective single detector outputs. The pump noise suppressed laser was pumped at 9.5 times the threshold current.

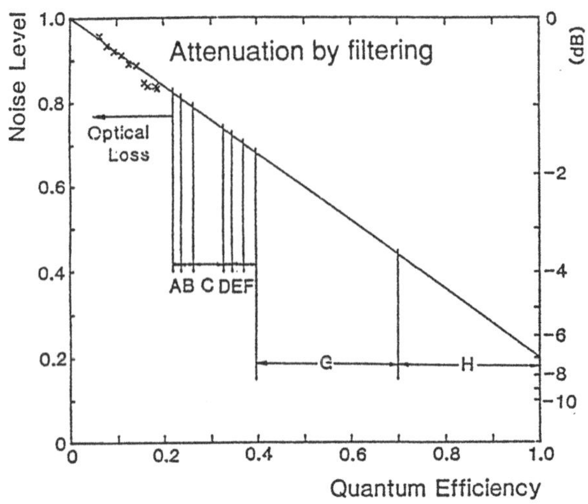

Fig. 5 Normalized amplitude noise level vs. optical attenuation at f = 800 MHz and at 11.4 times the injection current threshold. Inferred noise levels (corrected for detection quantum efficiency and laser oytput coupling efficiency) are also shown. The points A-H indicate, (A) photodetector quantum efficiency of 0.93, (B) focussing lens loss of 0.90, (C) isolator loss of 0.81, (D) mirror reflection loss of 0.95, (E) cryostat window loss of 0.93, (F) collimating lens loss of 0.93, (G) laser rear facet output loss of 0.57 and (H) laser internal absorption loss of 0.70.

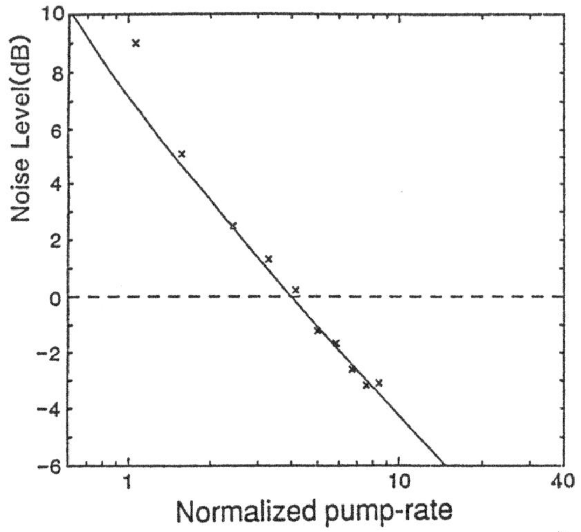

Fig. 6 Theoretical and experimental amplitude noise levels normalized to the SQL vs. laser pumping level. The experimental results are corrected for detection and laser output coupling quantum efficiency.

finite pumping as can be seen in Fig. 6 (and Fig. 2). In the former figure the theoretical curve and experimental values for the noise level versus pump rate are drawn. The experimental values are corrected for the non-ideal detection and the laser output coupling quantum efficiency beacause the theoretical curve is drawn under the assumption that the laser has single ended output and that the photons are ideally detected.

There are several reasons that the pump noise suppressed semiconductor laser is a promising squeezing device. One is that the pumping process is a first order process. The coupling between the pump-field and the signal-field is thus not limited by the relatively small nonlinear optical material constants. The semiconductor technology is also becoming a mature technology. To our knowledge, the highest overall quantum efficiency reported so far [14] is 79% but in the not to distant future it is not overoptimistic to expect devices with a 90% quantum efficiency. This paves the way for devices with substantial squeezing.

The second factor in favor of the semiconductor lasers is the short photon lifetimes achievable. Since the devices are intrinsically small and tolerate large mirror coupling losses, photon lifetimes as short as 1ps should be feasible. This means that squeezing bandwidths of the order of 10 GHz may be attainable for cw-oscillation.

Another factor of practical importance is the wavelength of the noise suppressed light. In an interferometer the wavelength may not be very important but if transmission of the non-classical light is involved, the wavelength must match the transparent "window" of the transmission media. Semiconductor lasers can be made in the wavelength region between 0.7 to 10 μm.

Lastly the pump noise suppressed semiconductor lasers are simple compared to other generation schemes. They require no stabilization, and no feedback or feedforward is involved.

3. GENERATION OF SUB-POISSONIAN LIGHT USING NON-DEGENERATE OPTICAL PARAMETRIC OSCILLATORS

A rather different way of producing sub-Poissonian light is to use the photon number correlation between the photon twins produced in a non-degenerate parametric process. The fundamental equation that ensures that there is a correlation, is the operator Manley-Rowe relation. For a travelling wave type parametric amplifier it reads [15]

$$\hat{n}_s(\text{out}) - \hat{n}_s(\text{in}) = \hat{n}_i(\text{out}) - \hat{n}_i(\text{in}) \tag{1}$$

where $\hat{n}_s$ and $\hat{n}_i$ are the photon number operators for the signal and the idler respectively. If the input states are vaccum states, that is $\hat{n}_s(\text{in}) = \hat{n}_i(\text{in}) = 0$, the output signal- and idler- photon fluxes are perfectly correlated. If the nonlinear crystal is put in a cavity and the device is brought to oscillate, there will be a perfect correlation between the produced signal and idler photons [16]. Due to the fact that the mirror coupling is a random process, the photon number correlation for such a device will be partially destroyed. For photon flux integration times much smaller than the photon cavity lifetime the correlation is completely lost, but for integration

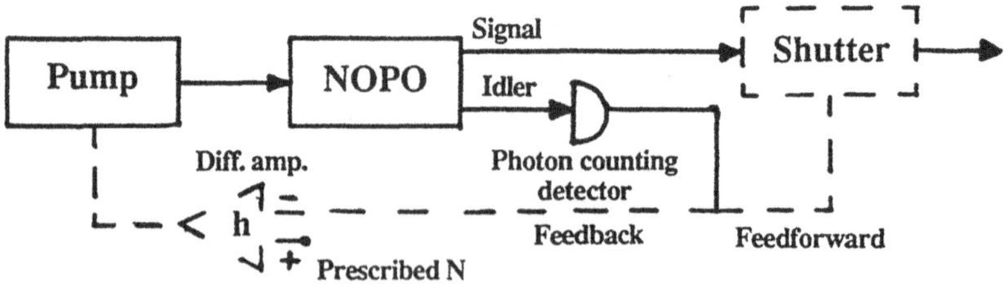

Fig. 7 Proposed schemes to generate sub-Poissonian light using the correlated photon pairs generated in a non-degenerate OPO.

times much longer than the photon lifetime there is a substantial correlation. This has recently been shown experimentally by Professor Giacobino's group at ENS in Paris [17, 18].

By measuring the idler photon flux with a photon counting detector and using the information to manipulate the signal flux, light with reduced photon number noise can be produced. Essentially two methods have been proposed. One is regulating the pump intensity with feedback [19], the other is using an optical shutter in a feedforward [20] configuration (Fig. 7). Both can in theory suppress the noise to the level imposed by the degree of correlation between the signal- and idler- photon flux.

Fig. 8 shows the calculated signal spectrum of a non-degenerate optical parametric oscillator (NOPO) with pump intensity feedback [16]. The phase noise spectrum is similar to that of a laser, so the signal phase will diffuse. The photon number noise will decrease below the SQL with increasing feedback strength. When the feedback-loop gain h approaches infinity, the output state approaches a NUS. Furthermore, for long photon flux integration times and high feedback gain, the output state is close to a number state.

There are practical problems associated with these schemes, however. As for feedback, the finite feedback delay time will impose a limited squeezing bandwidth . Feedforward can solve this problem, but it is still difficult to find an optical shutter that satisfies all the requirements. If the output photons are few and far between, like in low intensity parametric flourecense, suitable shutters can be found. If one is interested in high intensity squeezed light, however, the requirements for the speed and transmission are very severe indeed. The shutter must have very small losses when it is open, it must have a high extinction ratio and it must have a closing time in the sub-picosecond range.

One way of solving this problem is using linear manipulation of the signal beam, that is attenuation, amplification or simple photon adding. All these schemes introduces additional inevitable fluctuations, but it can be shown that they can still produce states with substantially reduced photon number fluctuations [21].

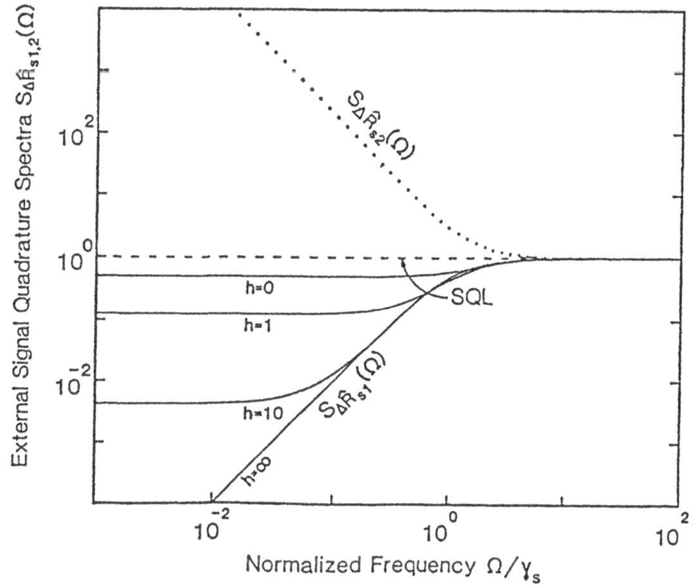

Fig. 8 Signal in-phase (solid) and quadrature-phase (dotted) amplitude fluctuation spectra for a singly resonant, feedback controled non-degenerate OPO. At high feedback gain h, the output spectra are similar to those in Fig. 4.

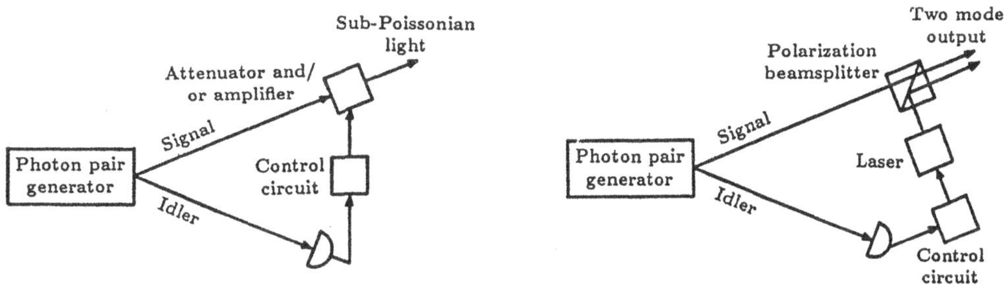

Fig. 9 (a) Generation scheme for sub-Poissonian light using linear photon attenuation and/or amplification. (b) Generation scheme for sub-Poissonian light using linear photon adding. The signal and the laser have mutually orthogonal polarizations.

Let's first assume that one has at ones disposition two photon beams with identical Poissonian photon counting statistics. One of the beams is detected with a photon counting detector. Each time the photon count exceeds the mean photon number, the signal beam is attenuated by an appropriate amount. If the photon count is smaller than the mean value the signal is amplified by an amount determined by the ratio between the mean and the observed photon number (Fig. 9a). Due to the dissipative nature of the attenuator it is not possible to bring the signal to exactly the mean signal photon number. The same is true for the amplifier. Therefore the photon variance will not vanish, but it can be substantially reduced as can be seen in the example in Fig. 10.

Low noise linear amplifiers are difficult to realize it practice, however. A question that comes naturally is if one can reduce the signal photon number fluctuations by *only* attenuating the signal selectively. Since it is only possible to reduce the signal photon number with an attenuator it is fair to assume that one should try to bring the mean signal photon number down to some photon number smaller than the original mean. Assume that we have decided to bring the photon number down to N_D photons. Each time the idler photon count exceeds N_D we attenuate the signal so that on the average N_D photons leave the attenuator. If the idler photon count is smaller than N_D the attenuator is completely transparent. The lower tail of the original

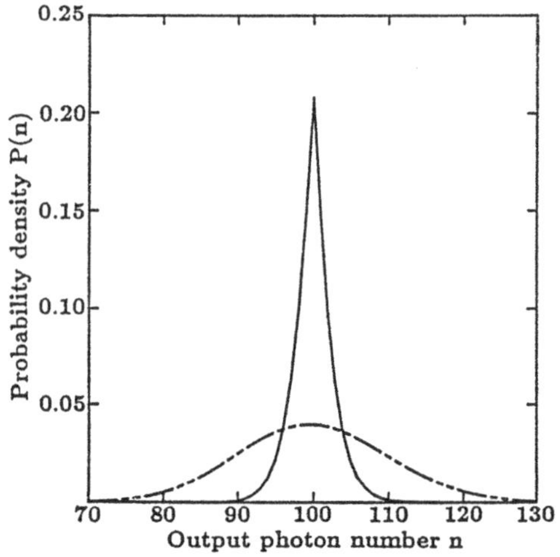

Fig. 10 The signal photon number distributions before (dash-dotted) and after the attenuation and amplification (solid). The mean signal photon count remains the same.

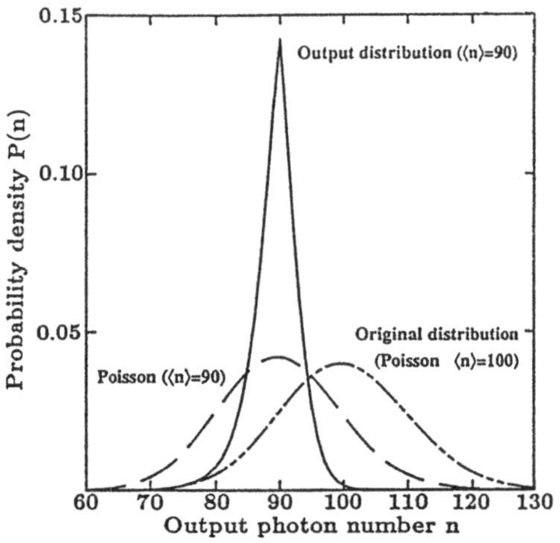

Fig. 11 The signal photon number distributions before and after the attenuation. The mean signal photon number has been reduced from 100 to 90 after the attenuation. For comparasion a Poisson distribution with the mean photon number 90 has been drawn.

photon counting statistics is thus left unchanged. It is obvious that N_D must be optimized. In Fig. 11 an optimized signal output distribution is shown together with the input distribution and a Poisson distribution with the same mean photon number as in the attenuator output. The output distribution is sub-Poissonian but comparing with Fig. 10 it is clear that it is slightly wider than the distribution obtained with an attenuator and amplifier in tandem.

The simplest configuration to test linear feedforward noise suppression is photon adding (Fig. 9b). Attenuating the controled light of a laser diode whose polarization direction is orthogonal to the signal's polarization, and letting the two beams combine spatially in a polarization beamsplitter, the sum of the two photon fluxes will have reduced photon number noise. This should be relatively easy to implement in the lab.

Figs. 12a and 12b show the Fano-factor and the optimum N_D versus the mean signal output photon number for the different feedforward schemes. The limit for the attainable variance is [21]

$$\langle \Delta\hat{n}^2 \rangle = \sqrt{2n/\pi} \qquad (2)$$

where n is the mean output photon number, if attenuation is used in tandem with amplification or adding. If the signal photon distribution is manipulated uni-directionally, the minimum Fano factor is approximately proportional to $n^{-0.466}$. A numerical example gives an idea about the

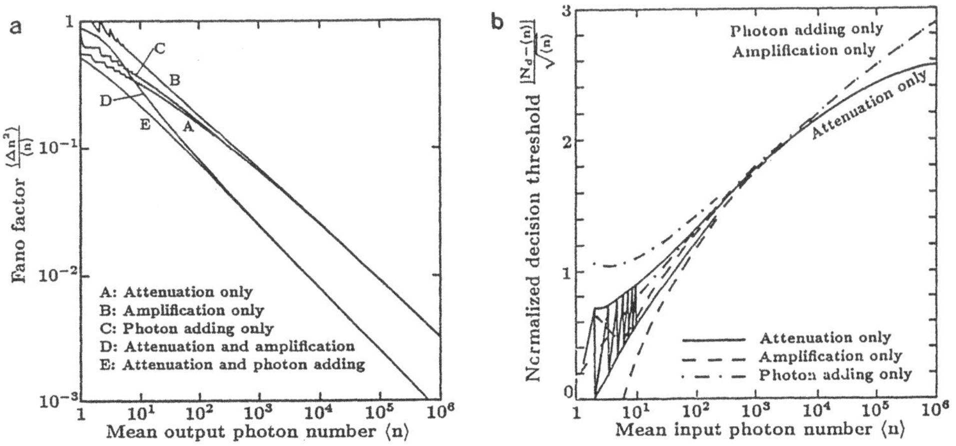

Fig. 12 (a) Fano-factor vs. mean output photon number for the different feedforward schemes. (b) The normalized optimum decision threshold N_D vs. the mean input photon number. The normalized decision threshold oscillates between a lower and a upper limit due to the fact that N_D has an integer value wheras the mean photon number is continuous. The oscillations are shown only for the attenuator scheme and only for photon numbers up to ten. For the other schemes only the limits are shown.

magnitudes of the achievable photon number fluctuations reduction. At a signal wavelength of λ=1 μm, a signal power of 1 mW and a photon flux integration time of 100 ns the photon count will be around $5 \cdot 10^8$ and using simply photon adding or attenuation (or linear amplification) the photon number fluctuation variance can be brought down by a factor of $2 \cdot 10^{-4}$ (37 dB) below the Poisson limit. It is evident that yet for some time to come, this limit is orders of magnitude smaller than the noise suppression limits imposed by technical imperfections in a real measurement setup.

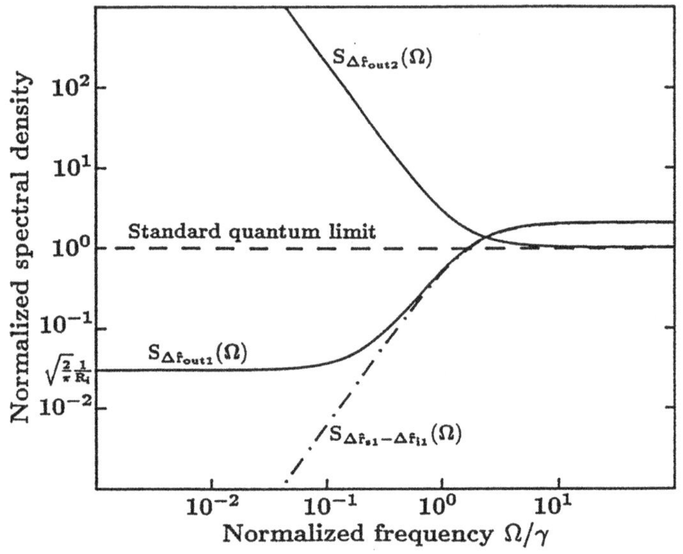

Fig. 13 Signal in-phase ($\Delta\hat{r}_{out1}$) and quadrature-phase ($\Delta\hat{r}_{out2}$) amplitude fluctuation spectra for a singly resonant non-degenerate OPO with linear feedforward controled attenuation and amplification. The dash-dotted line shows the signal and idler in-phase correlation. The pumpig rate is well above threshold.

The reader is reminded that Figs. 10, 11, and 12 were calculated from the assumption that two identical photon beams were available. How close can this assumption be realized in practice? The answer is that it can be very closely approximated. Fig. 13 shows the calculated signal output spectrum from an attenuator and amplifier in tandem, assuming that a NOPO supplies the correlated photons. At low frequencies the limit (2) can be attained while the feedforward configuration assures that the squeezing bandwidth is only limited by the NOPO cavity bandwidth. It should be noted, however, that the output state is not even close to a NUS.

4. QUADRATURE-PHASE SQUEEZED STATE GENERATION USING NON-DEGENERATE OPTICAL PARAMETRIC AMPLIFIERS

Another alternative when it comes generating non-classical photon states based on photon twin generation is using phase correlation instead of photon number correlation [22]. The signal and the idler phases in a parametric amplifier have a strong negative phase correlation. This can immediately be seen from the evolution equations for the output states of a nondegenerate parametric amplifier.

$$\begin{cases} \hat{b}_s = \sqrt{G}\,\hat{a}_s + \sqrt{G-1}\,\hat{a}_i^\dagger & (3a) \\ \hat{b}_i = \sqrt{G}\,\hat{a}_i + \sqrt{G-1}\,\hat{a}_s^\dagger & (3b) \end{cases}$$

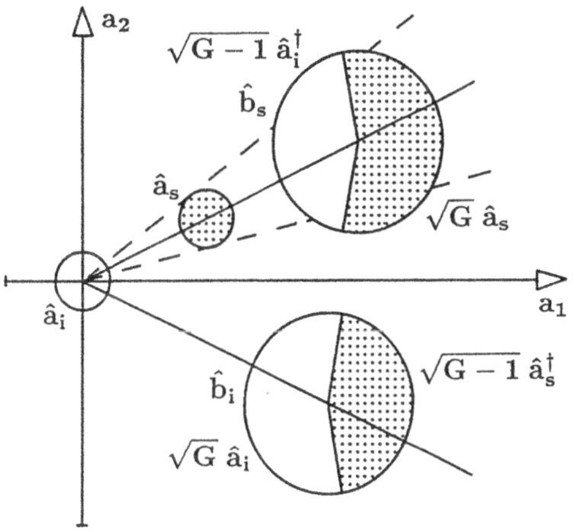

Fig. 14 Quantum pie-chart illustrating the evolution equations for the signal and idler of a travelling wave non-degenerate OPA. The pump coherent exitation axis is taken as the reference for the a_1 axis. Pump depletion is neglected. The signal power gain is 10.

where $\hat{a}_s$ is the input state of the signal, $\hat{b}_s$ is the output state of the signal (correspondingly for the idler) and G is the signal power gain of the amplifier. Rewriting Eq. (3)

$$\hat{b}_s = \hat{b}_i + \exp(-r)(\hat{a}_s - \hat{a}_i^\dagger)\,, \qquad \text{where } r = \operatorname{arccosh}(\sqrt{G}) \qquad (4)$$

it becomes clear that the output signal and idler states are nearly conjugate. Illustrating the evolution equations for a parametric amplifier with a "quantum pie-chart" (Fig. 14), the correlation is evident. In the figure the idler input state, $\hat{a}_i$, is assumed to be in a vacuum state, while the signal input, $\hat{a}_s$, is assumed to be in a coherent state. The circles represent quasiprobability contours ("uncertainty fuzzballs"). The dashed lines represent noiseless amplification (which is prohibited by the laws of quantum mechanics for coherent states). The insides of the circles have been used to draw pie-charts showing the relative "weights" of the states in the output superposition state. It is obvious that at high gain (in the figure the gain is equal to 10) the signal- and idler- output states are very nearly conjugate states, that is, they have a positive in-phase amplitude correlation but a negative quadrature-phase amplitude correlation. This correlation is slowly lost if the amplifier is saturated (Eq. (3) assumes a non-depleted pump).

The easiest way to utilize the phase correlation seems to be in conjunction with a feedforward controled linear phase modulator [21]. We belive that an ideal phase modulator need not add any additional phase uncertainty, since it is a dissipation free device. No thermal noise needs to be coupled from the driver circuits either, since the modulation voltage in available phase modulators is macroscopic. This means that the full signal and idler quantum correlation can be utilized, in contrast to the linear photon number manipulation schemes.

A possible configuration is shown in Fig. 15. In this example the input signal is in a vacuum state, whereas the input idler is in a coherent state. Most of the idler photons incident on the OPA are bounced off a semitransparent mirror to homodyne detect the output idler quantum fluctuations using a balanced reciever configuration. The obtained idler quadrature-phase fluctuations are imposed on the signal quadrature-phase using a phase modulator. The negative correlation quenches the output signal quadrature-phase quantum fluctuations. Since a homodyne reciever cannot measure phase fluctuations near dc, it is not possible to reduce the signal quadrature-phase fluctuations all the way down to zero frequency. Squeezing is only possible in a frequency band below the cut-off frequency of the parametric amplifier.

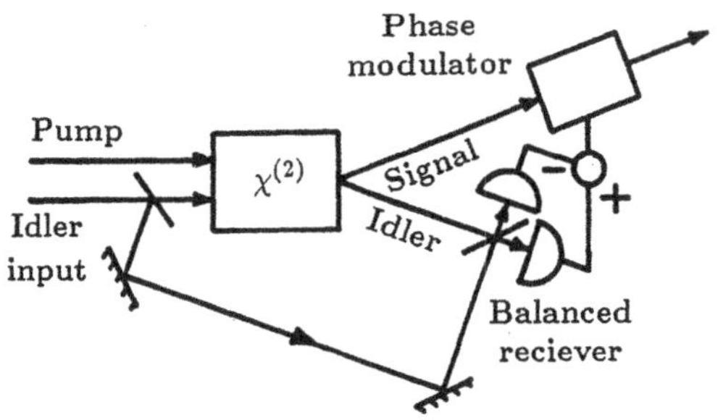

Fig. 15 Generation scheme for quadrature-phase squeezed states.

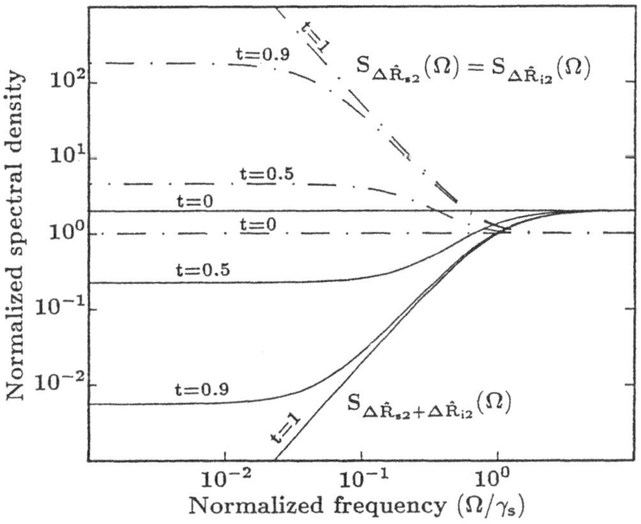

Fig.16 Signal quadrature amplitude fluctuation spectra. The normalized pumping parameter t is given by $t^2 = P/P_{th}$. Before the phase modulator (dash-dotted) both quadratures have identical spectra (the NOPA is phase insensitive). After the modulator the quadrature-phase fluctuations (solid) are reduced below the SQL.

In Fig. 16 the spectra for the the original signal and the phase-modulated signal are calculated using the normalized pumping rate t as the parameter. The quadrature-phase noise of the signal mode leaving the parametric amplifier increases with increasing pumping, or gain. The correlation increases faster, however, so the signal leaving the phase modulator has a quadrature-phase fluctuations lower than the SQL, provided that the feedforward loop-gain is properly adjusted. The output state is in a minimum uncertainty state, and for each additional 10 dB increase in gain, the fluctuations can be reduced by the same amount.

The reason the quadrature-phase correlation is finite is the finite signal photon number. The correlation can never be better than Heisenberg's uncertainty principle allows. Perfect quadrature-phase correlation would mean that the quadrature-phase fluctuations could in principle be reduced to zero. Zero quadrature-phase fluctuation variance would require an infinite in-phase fluctuation variance, and thus an infinite photon number, to preserve the uncertainty relation.

It must be pointed out that the quadrature-phase fluctuation reduction has another, less obvious limit. Since any pump phase diffusion is imposed with the same sign on the signal- and the idler-phases, the quadrature-phase fluctuations of the pump laser will be doubly imposed by

Table 1 Acheivable noise levels (low frequency spectral density) normalized to the SQL for ideal non-degenerate parametric systems. The NOPO's have been assumed to be singly resonant. Ω is the measurement angular frequency, γ is the OPO cavity decay rate, R_i is the idler photon flux amplitude expressed in square root photons per second and G is the amplifier gain.

Device	Squeezing direction	Feedback	Linear feedforward
NOPO	Photon number	$2\Omega^2/\gamma^2$	$\sqrt{2/\pi R_i^2}$
	Phase	-	(p-1)/p, p=P/Pth
NOPA	Photon number	1/G	No or little squeezing
	Phase	-	1/2G

the idler measurement feedforward. The pump laser phase-diffusion will thus set a limit to the degree of squeezing possible. Since the pump laser in practice is a high power laser, and since phase diffusion is inversly proportional to the output power, this is not a serious limitation.

Table 1 sums up the noise reduction properties of NOPO's and NOPA's. The low frequency spectral densities normalized to the SQL are shown for several configurations. In the case of NOPO's, the table assumes singly resonant devices. (For a triply resonant NOPO it is possible to squeeze the quadrature-phase amplitude fluctuations at an arbitrary pumping rate, albeit only in a narrow frequency range [21].) The reason quadrature-phase measurement feedback schemes are not even considered in the table is that nothing is gained by feedback when manipulating the signal quadrature-phase amplitude. Feedforward can utilize the full quantum correlation of the signal and idler quadrature-phases while the delay-time problem is avoided. From Table 1 it also becomes clear that linear feedforward photon number manipulation is not suitable if the correlated photon pair generator is a non-degenerate OPA. The reason is that with no signal or idler input (parametric flourecence) the photon number correlation is perfect but the photon counting statistics is exponential and not Poissonian. If the signal input is gradually increased, the output photon counting distribution narrows but the photon correlation is also lost.

5. CONCLUSIONS

The first NUS generation principle presented was the pump noise suppressed semiconductor laser. At high pumping and for photon flux integration times much longer than the photon cavity lifetime an ideal such laser will produce a near numberstate. In reality the largest noise reduction observed so far is 1.7 dB, but over a 1.1 GHz frequency band. In the near future it is not unrealistic to expect devices with 10 dB squeezing over a 10 GHz bandwidth.

It has also been shown that there is a strong correlation between the signal- and idler-photons produced by a NOPO or a NOPA. The photon number correlation is positive, whereas the phase correlation is negative. With feedback or feedforward it is thus possible to produce states with either reduced photon number noise or with reduced phase noise.

The most promising generator at present seems to be the linear feedforward NOPO. While it cannot reduce the photon-number fluctuations to zero, it can beat the SQL noise level with a factor $n^{1/2}$, where n is the mean signal photon number.

A NOPA with linear phase modulation feedforward can generate quadrature-phase squeezed states. The amount of squeezing possible is proportional to the amplifier gain. Attention must be paid to the pump laser frequency stability, however, since the pump phase fluctuations set a limit to the achievable squeezing.

REFERENCES

[1] D. F. Walls, Nature **306**, 141 (1983)
[2] R. Jackiw, J. Math. Phys. **9**, 339 (1968)
[3] H. P. Yuen, Phys. Rev. **A13**, 2226 (1976)
[4] Y. Yamamoto and H. A. Haus, Rev. Mod. Phys. **56**, (1986)
[5] K. Igeta and Y. Yamamoto, to be published.
[6] Y. Yamamoto, S. Machida and O. Nilsson, Phys. Rev. **A34**, 4025 (1986)
[7] P. R. Tapster, J. G. Rarity and J. S. Satchell, Europhys. Lett. **4**, 293 (1987)
[8] M. C. Teich and B.E.A. Saleh, J. Opt. Soc. Am. **B2**, 275 (1985)
[9] M. C. Teich, F. Capasso and B.E.A. Saleh, J. Opt. Soc. Am. **B4**, 1663 (1987)
[10] Y. Yamamoto and N. Imoto, IEEE J. Quantum Electron. **QE-22**, 2032 (1986)
[11] Y. Yamamoto, S. Machida and Y. Itaya, Phys. Rev. Lett. **58**, 1000 (1987)
[12] Y. Yamamoto and S. Machida, submitted for publication.
[13] S. Machida and Y. Yamamoto, IEEE J. Quantum Electron. QE-22, 617 (1986)
[14] A. Larsson, M. Mittelstein, A. Arakawa and A. Yariv, Electron. Lett. **22**, 79 (1986)
[15] W. H. Louisell, A. Yariv and A. E. Siegman, Phys. Rev. **124**, 1646 (1961)
[16] G. Björk and Y. Yamamoto, Phys. Rev. **A37**, 125 (1988)
[17] A. Heidmann, R. J. Horowicz, S. Reynaud, E. Giacobino, C. Fabre and G. Camy, Phys. Rev. Lett. **59**, 2555 (1987)
[18] S. Reynaud, C. Fabre and E. Giacobino, J. Opt. Soc. Am. **B4**, 1520 (1987)
[19] E. Jakeman and J. G. Walker, Opt. Commun. **55**, 219 (1985)
[20] D. Stoler and B. Yurke, Phys. Rev. **A34**, 3143 (1986)
[21] G. Björk and Y. Yamamoto, accepted for publication in Phys. Rev. **A.**
[22] G. Björk and Y. Yamamoto, submitted for publication.

QUANTUM OPTICS OF DIELECTRIC MEDIA

Roy J. Glauber and M. Lewenstein*

Harvard University, Physics Department
Cambridge, Massachusets 02138

Because quantum fluctuations impose fundamental limits on the accuracy of measurements, much attention is now being devoted to the problem of moderating or supressing their effect. If a quantity to be measured can be regarded as one of a pair of conjugate variables, for example, then its variance can usually be made arbitrarily small, but only at the expense of increasing the variance of the unmeasured variable. This technique requires putting the system being observed in a special sort of quantum state, referred to as "squeezed"[1].

The vacuum state of the electromagnetic field, and indeed all the coherent states are ones in which the variances of the electric and magnetic fields are equal. Squeezed states alter this balance by giving a smaller value to one variance than to the other. Nonlinear interactions frequently leave the field in such squeezed states and so all of the proposals for the practical generation of such states have been based on the techniques of non-linear optics[2]. It is worth emphasizing therefore that linear interactions of the sort associated with simple polarizable media can likewise break the vacuum symmetry of the electric and magnetic fields, and in this way they can lead much more simply to certain of the effects of squeezing.

Indeed, photons created in free space change substantially in nature when they enter polarizable media such as dielectrics. There are several senses, we shall see, in which this transformation characteristically introduces the effects of squeezing. But does that mean, for example, that a laser beam entering a dielectric medium will exhibit non-Poissonian photon counting statistics? To answer this question we must note that placing a photoabsorption counter within a dielectric medium likewise changes the nature of the photons it detects. When that change is taken into account, the photon counting statistics are found to remain Poissonian. Other detection mechanisms however, ones that respond more directly to field fluctuations, can still reveal the effects of squeezing.

We shall confine ourselves, in the present note, to discussing some elementary examples of squeezing in finite media, and then to outlining our general approach to the treatment of inhomogeneous dielectric media in unbounded space.

Let us begin by recalling how the notion of squeezing is defined in the current literature. In the standard approach[1,2] one considers a harmonic oscillator of unit mass (or a single mode of radiation field in a cavity). This oscillator can be described in terms of anihilation and creation operators a and $a^\dagger$ which fulfill the canonical commutation relation,

$$[a, a^\dagger] = 1. \tag{1}$$

---

* permanent adress: Institute for Theoretical Physics, Polish Academy of Sciences, 02-668 Warsaw, Poland

The position of the oscillator and its canonical momentum are expressed in terms of a and $a^\dagger$ as

$$q = \sqrt{\frac{\hbar}{2\omega}}(a^\dagger + a), \tag{2a}$$

$$p = i\sqrt{\frac{\hbar\omega}{2}}(a^\dagger - a), \tag{2b}$$

while the hamiltonian of the system is

$$H = \frac{1}{2}(p^2 + \omega^2 q^2) = \hbar\omega(a^\dagger a + \frac{1}{2}). \tag{3}$$

If we introduce the quantum mechanical variances of the canonical variables (2) as the mean values

$$\langle(\Delta q)^2\rangle = \langle q^2\rangle - \langle q\rangle^2, \tag{4a}$$

$$\langle(\Delta p)^2\rangle = \langle p^2\rangle - \langle p\rangle^2, \tag{4b}$$

then the uncertainity principle implies

$$\langle(\Delta q)^2\rangle\langle(\Delta p)^2\rangle \geq \frac{\hbar^2}{4}. \tag{5}$$

It is easy to check that in any stationary state of the hamiltonian (3)

$$\langle q\rangle = 0, \qquad \langle p\rangle = 0, \tag{6a}$$

and

$$\langle(\Delta p)^2\rangle = \omega^2\langle(\Delta q)^2\rangle. \tag{6b}$$

More generally, we can perform a rotation in phase space by means of the unitary operator $U(\Theta) = \exp(ia^\dagger a\Theta)$,

$$p \to p(\Theta) = U(\Theta)pU^{-1}(\Theta) = p\cos\Theta - \omega q\sin\Theta, \tag{7a}$$

$$\omega q \to \omega q(\Theta) = U(\Theta)\omega qU^{-1}(\Theta) = p\sin\Theta + \omega q\cos\Theta, \tag{7b}$$

which evidently leaves the hamiltonian (3) invariant. It follows then that $\langle\Delta p(\Theta)^2\rangle$ is independent of Θ in any stationary state. The variances, in other words, are rotationally symmetric in the $p - \omega q$ phase space.

Obviously the ground state of (3), $|0\rangle$ is stationary. The coherent states $|\alpha\rangle$ are defined as "displaced" ground states[3]

$$|\alpha\rangle = D(\alpha)|0\rangle, \tag{8}$$

where the unitary displacement operator is defined by the relation

$$D^{-1}(\alpha)aD(\alpha) = a + \alpha, \tag{9}$$

for any complex number α. Adding a constant α to a (or the corresponding constants to $a^\dagger$, q or p) does not change their variances. Hence the variance $\langle\Delta p(\Theta)^2\rangle$ is independent of Θ in coherent states too, even though the state as a whole does not in general share this rotational symmetry.

The minimum uncertainty states $|\ \rangle$ (Kennard states[4]) are defined by the requirement

$$(p - \langle p\rangle)|\ \rangle = i\mu(q - \langle q\rangle)|\ \rangle \tag{10}$$

for any real number μ. It follows then that for them

$$\langle(\Delta p)^2\rangle = \mu^2\langle(\Delta q)^2\rangle,$$

and in particular, the coherent states (8) are minimum uncertainty states with $\mu = \omega$. For $\mu \neq \omega$ the Θ-symmetry of $\langle(\Delta p(\Theta))^2\rangle$ is broken in these states and they are described as "squeezed"[1]. That symmetry can trivially be restored by the canonical scale transformation

$$q' = \lambda q, \tag{11a}$$

$$p' = \frac{1}{\lambda}p, \tag{11b}$$

with $\lambda = \sqrt{\omega/\mu}$. Alternatively, one can say that each minimum uncertainty "squeezed" state is a coherent state for a different oscillator with frequency μ.

The foregoing expression can be simplified a bit and made dimensionless by defining the operators

$$X(\Theta) = ae^{-i\Theta} + a^\dagger e^{i\Theta} = X^\dagger(\Theta), \tag{12a}$$

$$Y(\Theta) = -i(ae^{-i\Theta} - a^\dagger e^{i\Theta}) = X(\Theta + \frac{\pi}{2}). \tag{12b}$$

Their comutation relation takes the form

$$[X(\Theta), Y(\Theta)] = 2i, \tag{13}$$

while the uncertainty principle implies that

$$\langle(\Delta X)^2\rangle\langle(\Delta Y)^2\rangle \geq 1. \tag{14}$$

For coherent states $\langle(\Delta X)^2\rangle$ is equal to 1 for all values of Θ. Squeezed states, on the other hand, are characterized by the conditions $\langle(\Delta X(\Theta))^2\rangle < 1$ and $\langle(\Delta Y(\Theta))^2\rangle > 1$ for some range of angles Θ.

How can we realize squeezed states in physical terms? To do that we must break the circular symmetry in the phase plane. A simple way of achieving that, if the oscillator is initially in a stationary state, is to add, beginning at some moment of time, a term $\frac{1}{2}\beta^2 q^2$ to the hamiltonian[5], so that at all later times we have

$$H' = \hbar\omega\left(a^\dagger a + \frac{1}{2}\right) + \frac{1}{2}\beta^2 q^2. \tag{15}$$

In mechanical terms Eq. (15) means that we have stiffened the spring - i.e. raised the frequency, by replacing ω^2 by $\omega^2 + \beta^2$. The simple time dependence of $a(t)$ then changes its character, and the equation of motion becomes

$$\dot{a} = -i\omega a - \frac{i\beta}{\sqrt{2\hbar\omega}}(a + a^\dagger). \tag{16}$$

Instead of taking this slightly complicated view of the evolution of a and $a^\dagger$, a much simpler approach to the problem can be taken. We may introduce two new operators A and $A^\dagger$, by writing

$$q = \sqrt{\frac{\hbar}{2\mu}}(A^\dagger + A), \tag{17a}$$

$$p = i\sqrt{\frac{\hbar\mu}{2}}(A^\dagger - A), \tag{17b}$$

with $[A, A^\dagger] = 1$. If the parameter μ is then determined by

$$\mu^2 = \omega^2 + \beta^2,$$

the hamiltonian H' falls into the diagonal form

$$H' = \hbar\mu\left(A^\dagger A + \frac{1}{2}\right). \tag{18}$$

The parameter μ, in other words, is the new angular frequency of the oscillator, and there is a corresponding shift of the ground state energy.

The electromagnetic case can be treated similarly. If a medium with a dielectric constant ϵ is added to a cavity the Lagrangian of electromagnetic field becomes

$$L = \frac{1}{2}\int(\epsilon E^2 - B^2)d^3\mathbf{r}. \tag{19}$$

We now fix our attention on a single mode of oscillation of the cavity field. If we let the amplitude of its vector potential (suitably normalized) be A, we can let the coordinate of the corresponding oscillator to be $q = -A/c$, so that

$$\dot{q} = -\frac{1}{c}\dot{A} = E, \tag{20a}$$

and the Lagrangian for the mode becomes

$$L = \frac{1}{2}(\epsilon\dot{q}^2 - \omega^2 q^2). \tag{20b}$$

Note that the dielectric constant plays the role of the mass in a mechanical oscillator. The canonical momentum is

$$p = \frac{\partial L}{\partial \dot{q}} = \epsilon\dot{q} = D. \tag{21}$$

The hamiltonian is then

$$H' = \frac{1}{2}\left(\frac{p^2}{\epsilon} + \omega^2 q^2\right). \tag{22}$$

Under the action of the canonical scale transformation $p' = p/\sqrt{\epsilon}$, $q' = \sqrt{\epsilon}q$, this becomes

$$H' = \frac{1}{2}\left((p')^2 + \mu^2(q')^2\right). \tag{23}$$

The new frequency is $\mu = \omega/\sqrt{\epsilon} < \omega$ (for $\epsilon > 1$). The annihilation operator for the stationary states of this hamiltonian

$$A = \frac{1}{\sqrt{2\hbar\mu}}(\mu q' + ip'), \tag{24}$$

can be written in terms of the operators a and $a^\dagger$, appropriate to $\epsilon = 1$, as

$$A = \frac{\omega + \mu}{2\sqrt{\omega\mu}}a + \frac{\omega - \mu}{2\sqrt{\omega\mu}}a^\dagger \tag{25a}$$

$$= \frac{1}{2\epsilon^{1/4}}\{(\sqrt{\epsilon}+1)a + (\sqrt{\epsilon}-1)a^\dagger\} \tag{25b}$$

$$= SaS^{-1}, \tag{25c}$$

where

$$S = e^{\frac{1}{2}r(a^2-(a^\dagger)^2)}, \tag{26}$$

and

$$r = \ln(\epsilon)/4. \tag{27}$$

If the state $|0\rangle'$ is the ground state of H' given by Eq. (22), then $A|0\rangle' = 0$. It follows from Eq. (25c) then, that apart from a phase factor,

$$|0\rangle' = S|0\rangle. \tag{28}$$

where $|0\rangle$ is the ground state of the cavity with $\epsilon = 1$, i.e. the vacuum ground state.

The ground state $|0\rangle'$ for $\epsilon \neq 1$, contains an interesting distribution of bare vacuum photons. The probability $P(n)$ for the presence of n vacuum photons vanishes for odd n

$$P(n) = 0, \tag{29}$$

but for even n it has the value

$$P(n) = \frac{(2n)!}{2^{2n}(n!)^2}\frac{2\epsilon^{1/4}}{\sqrt{\epsilon}+1}\left(\frac{\sqrt{\epsilon}-1}{\sqrt{\epsilon}+1}\right)^{2n}. \tag{30}$$

This distribution is super-Poissonian with

$$\langle n\rangle = \frac{(\sqrt{\epsilon}-1)^2}{4\sqrt{\epsilon}}, \tag{31a}$$

and

$$\langle n^2\rangle - \langle n\rangle^2 = 2\langle n\rangle(1+\langle n\rangle). \tag{31b}$$

Let us consider the operator (12a) for $\Theta = 0$, that is $X = a + a^\dagger$. It is then easy to show that in the state $|0\rangle'$ the variance of X is

$$\langle(\Delta X)^2\rangle = \mu/\omega = \frac{1}{\sqrt{\epsilon}} < 1. \tag{32}$$

When the variance of an operator is less than its vacuum value we shall speak of it as *subfluctuant*. Hence X is *subfluctuant* in the state $|0\rangle'$ and the state is *squeezed* in X. Analogously the observable $Y = X(\frac{\pi}{2})$ has the variance $\omega/\mu = \sqrt{\epsilon}$ and is *superfluctuant.*

When the operator for the variance, $(\Delta X)^2$, is written in the normally ordered form, $:(\Delta X)^2:$, the inequality that expresses its subfluctuance becomes

$$\langle:(\Delta X)^2:\rangle \leq 0. \tag{33}$$

This ifality expresses concisely the non-classical character of the squeezed states.If they were pseudo-classical in character, they would have positively-valued P-representations. The mean-square, squared mean inequality would then imply

$$\langle:\Delta(a+a^\dagger)^2:\rangle =$$

$$\int P(\alpha)(\alpha+\alpha^*)^2 d^2\alpha - \langle \alpha+\alpha^* \rangle^2 \geq 0, \tag{34}$$

which contradicts the assumption of subfluctuance.

It is also interesting to note that squeezing does not survive the usual process of amplification. Any phase-insensitive quantum amplifier wipes out squeezing as soon as its gain exceeds 2. The reason for that is that the amplification adds noise in both quadratures[6].

Coherent excitations of amplitude α can be superposed on the altered ground state $|0\rangle'$. For the case of the stiffened string, for example, the resulting state

$$|\alpha\rangle' = D(\alpha)|0\rangle' \tag{35}$$

may have photon statistics very different from the Poisson statistics of the coherent state. They will be sub-Poissonian for α real, and super-Poissonian for α imaginary[7a]; they may also exhibit certain interference effects[7b] in phase space.

The foregoing calculations show clearly that the effects of squeezing can be produced without appealing to non-linear interactions of any sort (see for instance one of the early references on the subject, ref. 8). All that is neccessary is to break the symmetry between the fields E and B that is characteristic of the vacuum. We have seen how that can be done within a resonant cavity, but most optical experiments are carried out in free space rather than in cavities. Let us consider a block of a transparent dielectric medium, say glass, without the reflecting cavity walls around it. To what degree will the field modes within the glass block exhibit the same behavior they did when the walls were present?

The calculations have shown us that each cavity mode is occupied by a distribution of $n = 0, 2, 4, 6, \ldots$ "vacuum" photons. Those vacuum photons are precisely the ones detected in most photon counting experiments. If they are present in the ground state of the field in the glass block, why can't we count them with the usual sorts of photodetection equipment? Indeed, the usual description of the detection process in the rotating wave approximation would permit us to do just that.

If that free-standing glass block were equivalent to the one in a cavity, the field variances within it would be "squeezed" relative to the vacuum, and an ideal laser beam within the glass could be found to have non-Poissonian statistics. These are interesting if rather implausible conjectures, and they call for a good deal of clarification of the behavior of quantized fields within polarizable media. A number of comments on closely related problems have been given in ref. 9, 10 and 11, and these show that there are advantages to various ways of carrying out the analysis.

Our own approach to dealing with dielectric media is quite close in spirit to the traditional method of quantizing the electromagnetic field. It regards monochromatic photons as quantized excitations of the various stationary field modes. Since the modes are defined on the basis of global rather than local conditions, those photons usually can not be thought of as localized to any particular volume; they are spread throughout space. We must deal with media that have finite boundaries however, and we do that by considering the general case of a non-uniform medium with a position-dependent dielectric constant $\epsilon(\mathbf{r})$.

To carry out the quantization we begin by defining a generalized form of the transversality condition for the radiation gauge

$$\nabla \cdot [\epsilon(\mathbf{r})\mathbf{A}(\mathbf{r}, t)] = 0. \tag{36}$$

We then define a complete set of scattering wave functions $\{\mathbf{u}_k(\mathbf{r})\}$ fulfilling the gauge condition (36) and the generalized wave equation

$$\left[-\nabla \times \nabla + \frac{\omega_k^2}{c^2}\epsilon(\mathbf{r})\right]\mathbf{u}_k(\mathbf{r}) = 0, \tag{37}$$

together with appropriate boundary conditions. These satisfy the orthonormality condition

$$\int \mathbf{u}_k^*(\mathbf{r}) \cdot \mathbf{u}_{k'}(\mathbf{r})\epsilon(\mathbf{r})d^3\mathbf{r} = \delta_{kk'}. \tag{38}$$

The vector potential is then shown to have the expression

$$\mathbf{A}(\mathbf{r}, t) = c\sum_k \sqrt{\frac{\hbar}{2\omega_k}}\left[A_k(t)\mathbf{u}_k(\mathbf{r}) + A_k^\dagger(t)\mathbf{u}_k^*(\mathbf{r})\right], \tag{39}$$

in which the operators A_k and $A_k^\dagger$ obey the boson commutation rule

$$[A_k, A_{k'}^\dagger] = \delta_{kk'}. \tag{40}$$

The hamiltonian can then be shown to fall into the diagonal form

$$H = \frac{1}{2}\sum_k \hbar\omega_k(A_k^\dagger A_k + A_k A_k^\dagger). \tag{41}$$

The time evolution of the operator A_k thus reduces to the elementary form

$$A_k(t) = A_k(0)e^{-i\omega_k t}. \tag{42}$$

These anihilation and creation operators A_k and $A_k^\dagger$ are the ones that define physical photons, and those photons are precisely conserved in number, whatever may be the nature of the inhomogeneous dielectric medium.

All initially coherent states (defined in terms of the physical photon operators) remain coherent at all later times. Furthermore a coherent state with a specified set $\{\alpha_k\}$ of ingoing wave amplitudes at $t \to -\infty$ will evolve into a coherent state of the form $|\{\sum_{k'} S_{kk'}^* \alpha_{k'}\}\rangle$ in terms of the outgoing field amplitudes for $t \to \infty$. The matrix $S_{kk'}$ in this expression is the scattering matrix for the classical wave equation (37).

It was shown many years ago[3] that the rate at which photons are registered by an ideal photon counter is proportional to the expectation value

$$G^{(1)}(\mathbf{r}, t, \mathbf{r}, t) = \langle E^{(-)}(\mathbf{r}, t)E^{(+)}(\mathbf{r}, t)\rangle, \tag{43}$$

in which $E^{(+)}$ and $E^{(-)}$ are the positive and negative frequency parts, respectively, of the electric field operator $E(\mathbf{r}, t)$. When the photons are counted in free space, as they usually are, it is appropriate to expand the field $E^{(+)}$ in terms of the vacuum annihilation operators a_k and the field $E^{(-)}$ in terms of the creation operators $a_k^\dagger$. The product $E^{(-)}E^{(+)}$ is then said to be in "normally ordered form" in terms of the vacuum creation and annihilation operators.

When the photons are counted in a dielectric medium, the counting rate is still proportional to $G^{(1)}(\mathbf{r}, t, \mathbf{r}, t)$, but the expansion of the field into positive and negative frequency parts is no longer the same as it is in the vacuum. The positive and negative frequency parts are given by expansions in terms of the physical photon annihilation operators A_k and creation operators $A_k^\dagger$, respectively. The difference between these operators and the vacuum operators a_k and $a_k^\dagger$ is a substantial one. The A_k for example can not in general be expressed in terms of the vacuum annihilation operators a_k alone. They require an admixture of vacuum creation operators $a_k^\dagger$ as well in order to cancel the negative frequency parts of the a_k. The operator product $E^{(-)}E^{(+)}$ required in Eq. (43) is thus no longer "normally ordered" in terms of the vacuum operators a_k and $a_k^\dagger$. Annihilating a physical photon, for example, may mean creating a vacuum photon.

Once we agree that $E^{(+)}$ must be expanded in terms of the physical photon annihilation operators A_k all of the photon counting paradoxes mentioned earlier are removed. No photons at all can be counted in the ground state; photoabsorption measurments in the states generated by ideal lasers will always reveal Poisson statistics. Other types of measurments can be made however, and some of these can indeed reveal effects that we associate with "squeezing".

There are several ways in which it is possible in principle to measure the ground-state fluctuations of the electric field (typically over a rather limited frequency band). The variance of the field is given by

$$\langle E^2(\mathbf{r},t)\rangle = \langle E^{(+)}(\mathbf{r},t)E^{(-)}(\mathbf{r},t)\rangle \tag{44}$$

and the latter expectation value can be found, for example, by measuring the rate of spontaneous emission by excited atoms. The parts of the expectation value (44) corresponding to the appropriate frequencies may then be found to exceed their vacuum value, that is to be superfluctuant, or to be smaller than their vacuum value, i.e. subfluctuant.

It is useful to define the fluctuance of any variable $V(\mathbf{r},t)$ within the medium as

$$\mathcal{F}[V(\mathbf{r},t)] = \frac{\langle(\Delta V(\mathbf{r},t))^2\rangle_{med}}{\langle(\Delta V(\mathbf{r},t))^2\rangle_{vac}}. \tag{45}$$

This non-negative quantity can be used as a measure of the modifications of fluctuations that are brought about in cavities and waveguides[12], and in periodic dielectric structures[13]. It can also be used as a measure of squeezing effects in non-linear systems such as parametric oscillators and amplifiers.

We are happy to acknowledge some informative discussions with H. J. Kimble. This work has been supported by the Office of the Naval Research under the contract N00014-85-K-0724.

References

1 (a) D. Stoler, Phys. Rev. D **1**, 3217 (1970); Phys. Rev. D **4**, 1925 (1971); (b) for a review see D. Walls, Nature **306**, 141 (1983).

2 for a review of the recent results see (a) "Squeezed States of the Electromagnetic Field", ed. H. J. Kimble and D. Walls, special issue of J. Opt. Soc. Am. B **4**, (1987); (b) "Squeezed states", ed. R. Loudon and P. L. Knight, special issue of J. Mod. Opt. **34**, (1987).

3 R. J. Glauber, in "Quantum Optics and Electronics", ed. C. De Witt, A. Blandin and C. Cohen-Tannoudji, Gordon and Breach, New York 1965; Phys. Rev. **130**,2529 (1963); **131**, 2766 (1963).

4 E. H. Kennard, Z. Phys. **44**, 326 (1927); see also E. Schrödinger, Naturwiss. **14**, 664 (1926).

5 R. Graham, in ref. 2 (b), p.873.

6 S. Friberg and L. Mandel, Opt. Comm. **46**, 141 (1983); R. Loudon and T. J. Shephard, Optica Acta **31**, 1243 (1984); see also R. J. Glauber, in "Quantum Optics", ed. A. Kujawski and M. Lewenstein, Reidel, Dordrecht 1986 and references therein.

7 (a) L. Mandel, Phys. Rev. Lett. **49** 136 (1982); (b) W. Schleich and J. A. Wheeler, Nature **326**, 574 (1987).

8 B. R. Mollow and R. J. Glauber, Phys. Rev. **160**, 1076 (1967); **160**, 1097 (1967).

9 Z. Białynicka-Birula, I. Białynicki-Birula, in ref. 2, p. 1621.

10 I. Abram, Phys. Rev. A **35**, 4661 (1987).
11 L. Knöll, W. Vogel, D. -G. Welsch, Phys. Rev. A **36**, 3803 (1987).
12 E. M. Purcell, Phys. Rev. **69**, 681 (1946); D. Kleppner, Phys. Rev. Lett. **47**, 233 (1981); modifications of spontaneous decay in dielectric media are discussed by G. S. Agarwal in "Quantum Electrodynamics and Quantum Optics", ed. A. O. Barut, Plenum 1984.
13 E. Yablonovitch, Phys. Rev. Lett. **58**, 2059 (1987).

GRAVITATIONAL WAVE DETECTION AND QUANTUM OPTICS

G. Leuchs

Max-Planck-Institut für Quantenoptik und
Sektion Physik der Universität München
8046 Garching/FRG

INTRODUCTION

One of the most promising applications of squeezed states of light is to overcome the shot noise limit of strain measuring interferometers. This is especially important for the large scale laser interferometers currently being developed for gravitational wave detection.[1]

In the limit of weak gravitational fields as in our solar system the field equations of General Relativitiy predict gravitational waves quite analogous to the electro-magnetic counterpart. However, there is only one type of mass and consequently the strongest coupling between source or detector and field will be of the quadrupole type resulting in an oscillating ellipsoidal spatial strain pattern induced by a gravitational wave. Because the gravitational constant determining the coupling between field and source or detector is so small gravitational waves have so far not been observed directly and in addition it is impossible to detect artificially generated gravitational waves on earth. Possible sources are the gravitational collapse of the core of a supernova and compact binary stars spiralling towards each other finally also ending in a gravitational collapse. The final few oscillations in these violent processes occur at a frequency around 1 kHz given essentially by the speed of light and the initial dimensions of the highly relativistic masses. Such events should be just detectable with already existing detectors if they occured in our galaxy. Unfortunately the rate of these events is low, 3 to 10 per century.

If the smallest strain $\delta l/l$ measurable with gravitational wave detectors could be pushed from 10^{-18} (broad band at 1 kHz) down to 10^{-21}, the event rate would increase considerably. Since the detectors measure directly the amplitude and not the energy of the field, a thousand times better sensitivity means to look one thousand times further out into space. The estimated event rate goes up as the number of galaxies that can be reached, incidently also a factor of about one thousand in this case. With a detector sensitivity of $\delta l/l = 10^{-21}$ or better, one expects to see at least one supernova and may be 3 to 4 coalescences of binary stars per month.[2]

The first kind of gravitational wave detectors were resonant bars acting much like a tuning fork. These detectors pioneered by J.Weber are now operated at cryogenic temperatures (2 K) and have a sensitivity[3] of $\delta l/l \approx 10^{-18}$. A high sensitivity is reached by making δl small and l large. However, there is a limit to increasing l. The strain induced by the gravitational wave oscillates. After each half cycle the strain reverses sign and the effect averages out if the round trip time for phonons in the bar is too large. Thus the optimum length of the bar is of the order of one meter. Much larger detectors can be built if light is used to probe the strain. The gain in length is the ratio of the speed of light to the speed of sound. In addition, the laser interferometers have the advantage that the pulse shape of the gravitational wave can be deduced more easily than with the bars since the interferometer mirrors are virtually free.

THE 30 M LASER INTERFEROMETER AT GARCHING AND FUTURE PLANS

In Munich the development of laser interferometric gravitational wave detectors started in 1974, then at the Max-Planck-Institut für Physik. Today there is a Michelson interferometer at Garching with 30 m armlength. Additional mirrors near the beam splitter allow for multiple reflection of the splitted laser beams in each arm so that the effective armlength is about 3 km - still far from the value of 100 km which would correspond to the optimum of half a gravitational wave length round about 1.5 kHz. The current sensitivity of this prototype detector for time varying strains around 1 kHz in a bandwidth of 1 kHz is 3×10^{-18}, within a factor of 1.5 at the shot noise limit.[4]

Such a small strain corresponds to mirror motions with an amplitude of only 10 % of a proton diameter! At this level the instrument is sensitive to noise from various sources. The thermal excitation of mechanical resonances, seismic motion still penetrating the double pendulum mirror suspension system, effects of scattered light and laser frequency fluctuations are some of the effects that had to be identified and controlled. Strain sensitivities comparable to the one of the Garching prototype detector have been obtained by the groups at Glasgow University and at the California Institute of Technology.

These groups and others in Boston, Paris and Pisa aim for large kilometer sized interferometers where the effective arm length is matched to half a period of the gravitational wave. The goal of a strain sensitivity of $\delta l/l = 10^{-21}$ can be reached if in addition to increasing l also the smallest measurable length change δl_{min} is decreased by a factor of 100. The shot noise limited value for the length is

$$\delta l_{min} = \{\lambda/2\pi B\}\ [h\nu/\eta P\tau]^{1/2} \qquad (1)$$

where λ is the laser wavelength and B is the number of times the split light beams pass each interferometer arm before recombination at the beam splitter. The photon energy is $h\nu$, the incoming laser power P, η describes the overall transmission of the interferometer including the quantum efficiency of the detector, and τ is the sampling time intervall which has to be around 0.5 ms for a broad band measurement of signals at 1 kHz. Since B has already been fixed by optimizing the effective armlength, the only way to improve δl_{min} by a factor 100 is to increase the laser power by 10^4 according to Eq.1. This overall power enhancement can be achieved partly by increasing the incoming laser power and partly by recycling the light that leaves the interferometer through the second output port. This is especially effective if the interference at the first output port is controlled to be at a dark fringe. For too high laser powers the photon shot noise will no longer be the dominating noise source and light pressure fluctuations on the end mirrors will take over instead.

However, for the numbers discussed here one stays well inside the photon shot noise limited regime.

A different way to improve the interferometer sensitivity came up when Caves pointed out that the shot noise formula (Eq.1) is not an unsurmountable law of nature.[5]

SQUEEZED STATES FOR LASER INTERFEROMETER

Squeezed states of light do not change the ultimate sensitivity of an interferometer which is determined by the quantum mechanical uncertainties of the mirror positions. However, it is possible to improve the sensitivity in the regime where it is shot noise limited.[6]

It is important to note that the shot noise limit of the interferometer does not depend on fluctuations of the incoming laser beam.[7] It is rather determined by the noise imposed by the beam splitter when individual photons are either reflected or transmitted leading to binomial photon distributions in the output beams. In the field picture energy quantization leads to Heisenberg's uncertainty relation. Any attenuation of the light beam must go hand in hand with an enhancement of the fluctuations in the light field. Mathematically speaking a vacuum operator has to be added to the attenuated operator of the laser field mode in order to ensure unitarity of the field operator at the output port. In Caves' picture this vacuum operator describes nothing but the field mode entering the beam splitter through the second normally unused input port. When the zero point fluctuations at this second input port are squeezed the field fluctuations at the output port are modified. The shot noise limit is improved if those fluctuations are squeezed which are 90° out of phase with the laser field:

$$\delta l_{min} = \{\lambda/2\pi B\}\ [(h\nu/\eta P\tau)\ (1-\eta+\eta\varepsilon)]^{1/2} \qquad (2)$$

The factor ε describes the reduction of noise power due to squeezing. The formula shows that the effect of squeezing may be high, provided the detector quantum efficiency is close to unity.

Another serious limitation to the shot noise reduction arises because the minimum interference intensity at the output of the interferometer will never be zero. Consequently, any stabilization scheme[8] will have to probe the interference signal more or less far away from the minimum. This reduces the gain in sensitivity with the squeezed state technique. For a fringe visibility $V = (I_{max}-I_{min})/(I_{max}+I_{min})$ close to unity and for perfect squeezing ($\varepsilon = 0$) the factor $(1-\eta+\eta\varepsilon)$ has to be replaced by[9]

$$f(\eta,V) = \{3(1-V)/2V\}^{1/2} + 1 - \eta \qquad (3)$$

A nonzero fringe visibility does not only affect the sensitivity gain through squeezing but also the one through light recycling, the two methods that in their combined action make the theoretical strain sensitivity of an interferometer independent of the number of light reflections in each arm.[10]

CONCLUSION

Taking into account the practical limitations of the squeezed state technique a ten-fold reduction in noise power, corresponding to a three-fold increase in strain sensitivity of the interferometer seems within reach. This is based on available photodetectors ($\eta \approx 0.9$) and a fringe visibility of $V \approx 0.99$, demonstrated experimentally with the 30 m prototype interferometer.[4]

Such an increase in sensitivity may not seem much at first sight. However, an interferometer measures the amplitude of the gravitational wave and the amplitude drops off only inversely proportional to the distance from the source. As a result, one looks three times further out into space collecting signals from a 27-times larger volume and if one already starts from a strain sensitivity of 10^{-21} or less one may well assume that the universe is homogeneously filled with galaxies. From this discussion it becomes clear that even small changes in the sensitivity drastically change the estimated gravitational wave signal rate.

REFERENCES

1. A. D. Jeffries, P. R. Saulson, R. E. Spero, and M. E. Zucker, "Gravitational wave observatories", Scient. Am. 256, 50 (1987).
2. B. F. Schutz, ed., "Gravitational wave data analysis", Adam Hilger, in press.
3. P. F. Michelson, The low temperature gravitational wave detector at Stanford University, in: "Gravitational Radiation", Les Houches 1982, N. Deruelle and T. Piran, eds., p. 465, North Holland (1983).
4. D. Shoemaker, R. Schilling, L. Schnupp, W. Winkler, K. Maischberger, and A. Rüdiger, "Noise behaviour of the Garching 30 m prototype gravitational wave detector", MPI für Quantenoptik, report MPQ 130, Garching (1987).
5. C. M. Caves, "Quantum mechanical noise in an interferometer", Phys. Rev. D23, 1693 (1981).
6. See contributions by H. J. Kimble and by R. E. Slusher in this volume.
7. G. Leuchs, "Squeezing the Quantum Fluctuations of Light", Contemp. Phys., in press (1988).
8. J. Gea-Banacloche and G. Leuchs, "Squeezed states for interferometric gravitational wave detectors", J. Mod. Opt. 34, 793 (1987), see also the contribution of P. Grangier in this volume.
9. J. Gea-Banacloche and G. Leuchs, "Applying squeezed states to nonideal interferometers", J. Opt. Soc. Am. B4, 1667 (1987).
10. M. H. Muendel, G. Wagner, J. Gea-Banacloche, and G. Leuchs, Squeezed states of Light, in: "Gravitational wave data analysis", B.F. Schutz, ed., Adam Hilger in press.

SUBHARMONIC GENERATION AND SQUEEZING IN A DAMPED OSCILLATOR

Robert Graham

Fachbereich Physik
Universität GHS Essen
Essen, W. Germany

ABSTRACT

Various approximations and assumptions are recalled on which the conventional quantum optical approach to quantum fluctuations in parametric processes with dissipation is based (weak dissipation, i.e. weak couplinq of the basic oscillator to the environment so that lowest order perturbation theory applies, Markov approximation, 'high' temperatures $k_BT \gg \hbar\kappa$ where κ is the dissipation rate, assumption of negligible influence of the parametric driving on the oscillator-reservoir interaction). Some of these approximations may not always be satisfied and all of them, taken together, do not satisfy exactly a basic fundamental principle, the fluctuation-dissipation relation in thermodynamic equilibrium. To remedy this unsatisfactory situation we develop here a general theory of parametrically driven, linear, dissipative quantum oscillators which is exact within a specified framework, and which respects all fundamental physical principles. The conventional approach is contained as a limiting case. The theory is applied to the decay of a squeezed state of a quantum oscillator where we reproduce in a very easy and direct manner results obtained earlier in the literature by a more complicated method involving a triple functional integral. The theory is also applied to subharmonic generation. The exact result for the subharmonic fluctuation intensities still involves time-integrals over response functions, which have to be calculated by solving a Mathieu equation. An approximate evaluation, valid near the subharmonic instability of the Mathieu equation, is given.

1. INTRODUCTION: CONVENTIONAL APPROACH TO SUBHARMONIC SQUEEZING

In the present paper we shall be concerned with a theory of subharmonic generation. The conventional approach of quantum optics is most conveniently formulated in the Heisenberg picture where one has to consider equations of motion of the form

$$\dot{b} = - i\omega_0 b - \kappa b + g\, e^{-2i\omega_0 t}\, b^+ + F(t) \tag{1.1}$$

$$\dot{b}^+ = i\omega_0 b^+ - \kappa b^+ + g\, e^{2i\omega_0 t}\, b + F^+(t)$$

Equations of this form have been derived for parametric processes in ref. [1]. Here b and b^+ are the annihilation and creation operators of the subharmonic mode of frequency ω_0, κ is the amplitude decay rate of that mode. The parametric coupling term $gb^+\exp(-2i\, \omega_0\, t)$ is proportional to the nonlinear dynamic polarization at the frequency ω_0 created in some non- linear medium by the mixing of an external pump-field $\sim ge^{-2i\, \omega_0 t}$ at the frequency $2\, \omega_0$ and the internally generated field at the frequency ω_0.
The noise-operators F(t) and $F^+(t)$ in eq.(1.1) are necessary in order to preserve the commutation relation

$$[b(t), b^+(t)] = 1$$

in spite of the presence of dissipation. The basic theory from which the properties of F(t) and $F^+(t)$ follow in the absence of the parametric coupling (g=0) has been given by Senitzky [2], who assumed a weak interaction between the oscillator and the environment and found that F(t) is Gaussian, has vanishing expectation value, and the second order correlation functions

$$\langle [F(t), F^+(t')] \rangle = \kappa \left(\delta(t-t') + \frac{i}{\pi} \frac{P}{t-t'} \right) \tag{1.2}$$

$$\langle F^+(t)\, F(t') \rangle = \frac{\kappa}{e^{\beta\, \hbar\, \omega_0} - 1} \left(\delta(t-t') - \frac{i}{\pi} \frac{P}{t-t'} \right) \tag{1.3}$$

In eq.(1.3) $\beta = 1/k_B T$ is the inverse absolute temperature in energy units. Eq.(1.3) holds only if the exponential $e^{\beta \hbar \omega}$ changes negligibly in a frequency interval $\Delta\omega \sim \kappa$ around ω_0, i.e. if $k_B\, T >> \hbar\kappa$. It is customary, in applications [3], to recall that eqs.(1.2-3) are only needed in a frequency domain close to the basic oscillator frequency where eqs.(1.2-3) are simplified to

$$\langle [F(t), F^+(t')] \rangle \cong 2\, \kappa\, \delta(t-t')$$

$$\langle F^+(t)\, F(t') \rangle \cong \frac{2\, \kappa}{e^{\beta\, \hbar\, \omega_0} - 1}\, \delta(t-t') \tag{1.4}$$

The δ-functions in eq.(1.4) turn the quantum process (1.1) into a Markovian process. It is, furthermore, customary to assume in applications that the addition of the parametric coupling term in eq.(1.1) does not lead to any changes in the form of the fluctuating force F(t). To summarize: any theory of subharmonic generation based on or equivalent to eq.(1.1), (1.4) (e.g. master equations) is restricted to (i) weak dissipation, i.e. weak coupling of the subharmonic oscillation to the environment, (ii) sufficiently high temperature $k_B T \gg \hbar\kappa$, (iii) the Markov approximation and (iv) the uncontrolled assumption that the parametric driving term in eq. (1.1) does not change the required form for the forces F.

The great advantage bought at the rather heavy prize of all these assumptions is extreme simplicity. In fact, eqs. (1.1) can immediately be solved and one obtains

$$b(t) = e^{-i(\omega_0+\kappa)t}\,\frac{1}{2}\left\{(b(0) + b^+(0))e^{gt} + (b(0) - b^+(0))e^{-gt}\right\}$$

$$+\, e^{-i\omega_0 t}\int_0^t d\tau\; e^{-(\kappa-g)(t-\tau)}\,\frac{1}{2}\,(e^{i\omega_0\tau}F(\tau) + e^{-i\omega_0\tau}F^+(\tau))$$

$$+\, e^{-i\omega_0 t}\int_0^t d\tau\; e^{-(\kappa+g)(t-\tau)}\,\frac{1}{2}\,(e^{i\omega_0\tau}F(\tau) - e^{-i\omega_0\tau}F^+(\tau)) \qquad (1.5)$$

and the hermitian conjugate for $b^+(t)$. An inspection of eq. (1.5) shows that the solution contains an exponentially growing term if the threshold condition for subharmonic generation is satisfied

$$g > \kappa \qquad (1.6)$$

The exponential growth inevitably leads to the breakdown of the linear form of the parametric driving term in eq. (1.1), but this is not our concern here, and it only means that we have to restrict our interest to a sufficiently short times interval $t < (g-\kappa)^{-1}$ after the parametric interaction has been turned on. Further inspection of eq. (1.5) reveals that the solution exhibits squeezing - of course a well known fact -, namely the linear combination

$$x(t) = \frac{1}{2i}\,(b(t)\,e^{i\omega_0\tau} - b^+(t)\,e^{-i\omega_0\tau}) \qquad (1.7)$$

decays exponentially, and so do its fluctuations. The exact form of the linear superposition of $be^{i\omega_0 t}$ and its adjoint in eq. (1.7), depends on our choice of phase for the oscillating term on the right hand side of eq. (1.1). The squeezing becomes manifest if we use eq. (1.5) to compute the time-dependent expectation value of the square of (1.7), to obtain

$$\langle x^2(t)\rangle = \frac{\kappa}{4(\kappa+g)}\,\coth\frac{\beta\hbar\omega_0}{2}\,\left\{1+\frac{g}{\kappa}e^{-2(\kappa+g)t}\right\} \qquad (1.8)$$

In eq. (1.8) it is assumed that the oscillator is in thermal equilibrium at t=0. Squeezing below the noise level of the vacuum fluctuations is achieved if the value of $\langle x^2\rangle$ drops below its value in the vacuum $\langle x^2\rangle_{vac} = 1/4$ which is possible if

$$\kappa(\kappa+g)^{-1}\coth\frac{1}{2}\beta\hbar\omega_0 < 1 \qquad (1.9)$$

After the transients in eq. (1.8) have died out (assuming that this may occur within a time-interval $\simeq (\kappa+g)^{-1}$ sufficiently short compared to the time-scale $(g-\kappa)^{-1}$ after which the linear approach breaks down) the fluctuation intensity of the squeezed component x(t) of the fluctuations is spectrally distributed according to

$$\langle x^2(\omega)\rangle = \int_{+\infty}^{-\infty} d(t-t')\, e^{i\omega(t-t')} \langle x(t)\, x(t')\rangle \tag{1.10}$$

and we find from eq. (1.5)

$$\langle x^2(\omega)\rangle = \langle x^2 \rangle \frac{2\,(\kappa+g)}{\omega^2+(\kappa+g)^2} \tag{1.11}$$

The review given in eqs. (1.5) - (1.11) shows the extreme simplicity resulting from the approximations which lead to eq. (1.1). However, these approximations need not be valid in all circumstances. E.g. one may have spectrally distributed losses $\kappa(\omega)$ which lead to retardation effects, or the coupling of the oscillator to the environment may not be weak, a case one encounters in experiments without mirrors, or the temperature may not be sufficiently high to satisfy $k_B T >> \hbar\kappa$. Furthermore, even without parametric coupling, the approximations made in eq.(1.1) lead to a violation of the fluctuation dissipation relation in termal equilibrium [4], which requires that

$$C_+(\omega) = \int_{-\infty}^{+\infty} d(t-t')\, e^{i\omega(t-t')} \langle b^+(t)\, b(t')\rangle$$

$$C_-(\omega) = \int_{-\infty}^{+\infty} d(t-t')\, e^{i\omega(t-t')} \langle b(t)\, b^+(t')\rangle \tag{1.12}$$

are related according to

$$C_-(\omega) = e^{\beta\hbar\omega}\, C_+(-\omega) \tag{1.13}$$

while the evaluation of eq. (1.12) with the help of eq. (1.5) for g=0, and shifting the initial time from 0 to $-\infty$, yields

$$C_+(-\omega) = \frac{1}{e^{\beta\hbar\omega_0}-1}\, \frac{2\kappa}{\kappa^2+(\omega-\omega_0)^2}$$

$$C_-(\omega) = e^{\beta\hbar\omega_o}\, C_+(-\omega) \tag{1.14}$$

which satisfies eq. (1.13) only for $\omega = \omega_0$ but not otherwise. While this is consistent within the assumption $k_B T >> \hbar\kappa$ implicit in eq. (1.1), it is not entirely satisfactory that quantum optics, already in its fundamentals, does not satisfy basic physical principles exactly. Finally, the question whether the parametric driving makes a modification of the fluctuation forces necessary is so far completely open and, obviously, should be resolved. For all these reasons we undertake it in the following sections to develop a theory which is exact within certain general assumptions and free from all the short-comings explained above. The price we have to pay when doing this is a considerable loss of simplicity. However, the simple theory given above emerges as a limiting case and the more general theory presented below therefore may be valuable when one or several of the approximations of eq. (1.1) fail.

2. EXACT QUANTUM LANGEVIN-EQUATION OF PARAMETRICALLY DRIVEN LINEAR OSCILLATORS

In the following we review the results of a recent paper [5] on quantum Langevin equations of the form

$$\ddot{q}(t) = -(\omega_0^2 + f(t))\, q(t) - \int_{-\infty}^{t} d\tau\, \gamma(t-\tau)\, \dot{q}(\tau) + \frac{1}{m} F_q(\tau) \tag{2.1}$$

where q and the associated momentum $p = m\dot{q}$ are quantized, $[p,q] = -i\hbar$, where the term $f(t).q$ represents the (linear) parametric driving, where $\gamma(t-\tau)$ is a retarded dissipation-kernel and $F_q(t)$ is an unknown operator Langevin force. Our approach is based on and exact within the following assumptions: We assume that the operator force $F_q(t)$ and the dissipation-kernel $\gamma(t)$ are produced by and an exclusive property of a heat reservoir in thermal equilibrium to which the parametrically driven oscillator is coupled. It is assumed that the coupling does not modify any of the properties of the heat reservoir. This assumption does not imply that the coupling must be weak, i.e. the properties of the oscillator may be strongly modified by the coupling, it only implies that the reservoir must be very big. Furthermore, it is assumed that the expectation value of $F_q(t)$ vanishes at all times and that $F_q(t)$ is Gaussian - an assumption which seems reasonable in view of the central limit theorem. As we shall see these assumptions determine all properties of $F_q(t)$ completely. The fact that $F_q(t)$ is a property of a heat reservoir in thermal equilibrium implies that its two independent second order correlation functions

$$\begin{aligned} S_F(t-t') &= \frac{1}{2} \langle F_q(t)\, F_q(t') + F_q(t')\, F_q(t) \rangle \\ \chi_F(t-t') &= \Theta(t-t')\, \frac{i}{\hbar} \langle [F_q(t), F_q(t')] \rangle \end{aligned} \tag{2.2}$$

can be functions of (t-t') only and must be related by the fluctuation-dissipation theorem which states that

$$S_F(t-t') = \hbar \int \frac{d\omega}{2\pi} \chi_F^{||}(\omega).\coth \frac{1}{2} \beta\hbar\omega.\cos \omega t$$

$$\chi_F^{||}(\omega) = \int_0^\infty dt \ \chi_F(t) \sin \omega t \tag{2.3}$$

Thus, the knowledge of $\chi_F(t-t')$ determines $S_F(t-t')$ oompletely. By the Gaussian property, all higher order correlation functions of $F_q(t)$ can be expressed by the correlation functions (2.2). In order to determine $\chi_F(t-t')$ let us write down the formal solution of eq. (2.1) in the form

$$q(t) = \frac{1}{m} \int_{-\infty}^{t} d\tau \ \chi(t,\tau) F_q(\tau) \tag{2.4}$$

Here $\chi(t,\tau)$ is the response function of eq. (2.1) which is completely fixed by the conditions

$$\ddot{\chi}(t,\tau) + (\omega^2_0 + f(t)) \chi(t,\tau) + \int_\tau^t ds \ \gamma(t-s) \dot{\chi}(s,\tau) = 0$$

$$\chi(t,\tau) = 0 \qquad (t \leq \tau) \tag{2.5}$$

$$\dot{\chi}(\tau,\tau) = 1$$

Here and in the following we use the notation $\dot{\chi}(t,\tau) = d\chi(t,\tau)/dt$.The response function $\chi(t,\tau)$ is completely fixed by eq. (2.5) as a function of both of its time arguments. On the other hand, linear response theory [6] applied to the parametrically driven oscillator coupled to its reservoir shows that rigorously $\chi(t,\tau)$ must also be given by

$$\chi(t,\tau) = \frac{im}{\hbar} \langle [q(t), q(\tau)] \rangle \, \Theta(t,\tau) \tag{2.6}$$

where expectation values might be omitted on the right hand side, because the commutator in a linear system must be a c-number. We note that eq. (2.6) implies also that $[p(\tau),q(\tau)] = -i\hbar$ via the condition $\dot{\chi}(\tau,\tau) = 1$ in eq. (2.5), while $\chi(\tau,\tau)= 0$ is automatically satisfied by eq. (2.6). The commutator of F_q can now be obtained as follows. We use eq. (2.6) and eq. (2.1) to obtain a second differential equation satisfied by $\chi(t,\tau)$ of the form

$$\ddot{\chi}(t,\tau) + (\omega^2_0 + f(t)) \chi(t,\tau) + \int_\tau^t ds \ \gamma(t-s) \dot{\chi}(s,\tau) = R(t,\tau) \tag{2.7}$$

with

$$R(t,\tau) = \int_{-\infty}^{\tau} ds \left(\frac{d}{ds}\chi(\tau,s)\right)\gamma(t-s) + \frac{i}{\hbar}\langle[F_q(t), q(\tau)]\rangle \tag{2.8}$$

A partial integration of the first term and using eq. (2.4) in the second term transforms $R(t,\tau)$ into

$$R(t,\tau) = \int_{-\infty}^{\tau} ds \;\; \chi(\tau,s)\left(\dot{\gamma}(t-s) + \frac{i}{m\hbar}\langle[F_q(t), F_q(s)]\rangle\right) \tag{2.9}$$

Clearly, $R(t,\tau)$ must vanish if (2.7) and (2.5) are to define the same function. Therefore,

$$\langle[F_q(t), F_q(\tau)]\rangle = i\hbar m\,(\dot{\gamma}(t-\tau)\,\Theta(t-\tau) - \dot{\gamma}(\tau-t)\;\,\Theta(\tau-t)) \tag{2.10}$$

and

$$\begin{aligned} \chi_F{}^{\|}(\omega) &= \frac{m\omega}{2}\,(\gamma(\omega) + \gamma(-\omega)) \\ S_F(\omega) &= \frac{m\hbar\omega}{2}\,(\gamma(\omega) + \gamma(-\omega))\coth\frac{1}{2}\beta\hbar\omega \end{aligned} \tag{2.11}$$

where

$$\begin{aligned} S_F(\omega) &= \int_{-\infty}^{+\infty} d(t-t')\;\; e^{i\omega(t-t')}\; S_F(t-t') \\ \gamma(\omega) &= \int_{0}^{\infty} dt\;\; e^{i\omega t}\;\gamma(t) \end{aligned} \tag{2.12}$$

Within our assumptions we have thereby determined exactly all properties of the operator force $F_q(t)$, independent of the assumptions of weak coupling, Markov property or temperature. We have also shown in eq.(2.9) that the parametric driving does not appear explicitly in the condition R=0 in eq. (2.9), i.e. the operator force $F_q(t)$ is not modified by the parametric driving. Fortunately, this puts on firm ground the uncontrolled assumption made in the conventional approach. Finally, putting f(t) = 0 [6] in eq. (2.5) one easily evaluates the response function as

$$\chi(t,\tau) = \chi_0(t,\tau) = \int_{-\infty}^{+\infty} \frac{d\omega}{2\pi} \, e^{-i\omega(t-\tau)} \, \chi_0(\omega)$$

$$\chi_0(\omega) = (\omega^2_0 - \omega^2 - i\omega\,\gamma(\omega))^{-1} \tag{2.13}$$

and one then easily checks that <q(t)q(t')> in thermal equilibrium exactly satisfies the fluctuation dissipation theorem

$$C_q(\omega) = e^{\beta\hbar\omega} \; C_q(-\omega) \tag{2.14}$$

where

$$C_q(\omega) = \int_{-\infty}^{+\infty} d(t-t') \; e^{i\omega(t-t')} \; \langle q(t)\, q(t')\rangle$$

$$= \frac{2}{m} \hbar\, \omega\gamma'(\omega) \, [(\omega^2 - \omega^2_0 - \omega\gamma''(\omega))^2 + \omega^2 \gamma'(\omega)^2]^{-1} \, (1-e^{-\beta\hbar\omega})^{-1} \tag{2.15}$$

with

$$\gamma'(\omega) = \frac{1}{2} \, (\gamma(\omega) + \gamma(-\omega))$$

$$\gamma''(\omega) = \frac{1}{2i} \, (\gamma(\omega) - \gamma(-\omega)) \tag{2.16}$$

We note that asymptotically for $(\gamma(\omega_0)/\omega_0) \to 0$ and $\beta\hbar\gamma'(\omega_0) << 1$ and neglecting $\gamma''(\omega_0)$, $C_q(\omega)$ may be approximated by a function with only two poles in the complex ω-plane

$$C_q(\omega) \simeq \frac{\gamma'(\omega'_0)}{(\omega-\omega'_0)^2 + \frac{1}{4}\gamma'^2(\omega'_0)} \cdot \frac{\hbar}{2m\,\omega'_0} \cdot \frac{e^{\beta\hbar\omega'_0}}{e^{\beta\hbar\omega'_0}-1}$$

$$+ \frac{\gamma'(\omega'_o)}{(\omega+\omega'_o)^2 + \frac{1}{4}\gamma'^2(\omega'_0)} \cdot \frac{\hbar}{2m\,\omega'_0} \cdot \frac{1}{e^{\beta\hbar\omega'_0}-1} \tag{2.17}$$

with $\omega'_0 = \sqrt{\omega^2_0 - \gamma'^2(\omega'_0)/4} \simeq \omega_0$, which is equivalent to the result (1.14) obtained from eqs. (1.1), (1.4) for g=0 if we replace ω_0 by ω'_0 in eq. (1.1), (1.14) and put

$$\kappa = \frac{1}{2}\,\gamma'(\omega'_o)$$

$$b = \sqrt{\frac{m\,\omega'_0}{2\,\hbar}}\,\left(q + \frac{i}{\omega'_0}\,\dot{q}\right) \tag{2.18}$$

The conventional approximate approach is therefore recovered from our exact approach in the appropriate limit.

3. FIRST EXAMPLE: DECAY OF A SQUEEZED STATE

As an easy application of the formalism we have developed let us analyze how a squeezed state, prepared suddenly at time t=0, decays under the influence of dissipation. This problem has recently been posed and solved by Schramm and Grabert [7] using an extension of the Feynman Vernon path integral approach. While their approach, like ours, is exact and able to avoid all the approximations discussed in section 1, it requires the evaluation of rather complicated functional integrals. E.g. in order to solve the problem discussed here, a triple functional integral had to be evaluated by Schramm and Grabert. The necessary calculation was lengthy and not presented explicitly in the quoted reference. Let us now see how the problem may be solved explicitly in a rather short and easy manner by the approach of section 2.

Let the coordinate and momentum operator of the oscillator at time t=0 be given by

$$q_0(0) = \frac{1}{m}\int_{-\infty}^{o} d\tau\;\chi_o(-\tau)\,F_q(\tau)$$

$$p_o(0) = \int_{-\infty}^{o} d\tau\;\dot{\chi}_0(-\tau)\,F_q(\tau) \tag{3.1}$$

where $\chi_0(t)$ is given by eq.(2.13) and eq.(3.1) expresses the fact that the oscillator is always in contact with the reservoir. Let now the coordinate of the oscillator suddenly be squeezed at time t=0 by some suitable (and very fast) nonlinear mechanism. The coordinate and momentum operators after the squeezing are then given by the unitary transformation

$$q(0) = e^{-z}\,q_0(0); \qquad p(0) = e^{z}\,p_0(0) \tag{3.2}$$

where the c-number e^{-z} expresses the degree of the squeezing. It must be noted here that the squeezing is assumed to be fast even on the time-scale of the frequency of the oscillator, which, unfortunately, makes this example seem somewhat academical in an optical context. However, applications in different frequency domains are conceivable where this assumption may be less unrealistic. In any case, the solution of eq.(2.1) for f(t) = 0 with the initial condition (3.2) can be written down immediately.

$$q(t) = \dot{\chi}_0(t) e^{-z}\, q_0(0) + \frac{1}{m}\, \chi_0(t) e^{z}\, p_0(0) + \frac{1}{m} \int_0^t d\tau\;\; \chi_0(t-\tau)\, F_q(\tau) \tag{3.3}$$

and, introducing the free oscillator solution

$$q_0(t) = \int_{-\infty}^{t} d\tau\;\; \chi_0(t-\tau)\, F_q(\tau) \tag{3.4}$$

(3.3) may be recast in the form

$$q(t) = q_0(t) + \dot{\chi}_0(t)\, (e^{-z}-1)\, q_o(0) + \chi_o(t)\, (e^{z}-1)\, \dot{q}_0(0) \tag{3.5}$$

From this expression one easily calculates all expectacion values of $q(t)$ and $p(t) = m\dot{q}(t)$ in terms of the function $\chi_0(t)$ and $S_0(t) = 1/2\, \langle q_0(t)q_0(o)+q_0(o)q_0(t)\rangle$. E.g. for $\langle q^2(t)\rangle$ one obtains

$$\langle q^2(t)\rangle = \langle q_0^2\rangle\, (1+ \dot{\chi}_0^2(t)\, (1-e^{-z})^2) + \langle \dot{q}_0^2\rangle\, \chi_0^2(t)\, (e^{z}-1)^2$$

$$+\;\; 2\, s_0(t)\, \dot{\chi}_0(t)\, (e^{-z}-1) - 2\, \dot{s}_0(t)\, \chi_0(t)\, (e^{z}-1) \tag{3.6}$$

which is the result first obtained in [7], and further illustrated there with some numerical examples. Some points made in ref. [7] are worth mentioning here. The first is that an exact theory of a damped harmonic oscillation like the one presented here (see also [7,8]) predicts squeezing of the coordinate fluctuations of the oscillator already in thermal equilibrium [8]. This is easily seen by computing $\langle q^2{}_o\rangle$ in thermal equilibrium using eq. (2.15) in

$$\langle q_0^2\rangle = \int_{-\infty}^{+\infty} \frac{d\omega}{2\pi}\;\; C_q(\omega) \tag{3.7}$$

to obtain via contour integration

$$\langle q_0^2\rangle = \frac{1}{\beta m} \sum_{n=-\infty}^{+\infty} \chi_0\, (i\, \frac{2\pi}{\beta\hbar}\, |n|) \tag{3.8}$$

Since $\gamma(i\,|\omega|) \geq 0$, it is seen from eqs.(3.8), (2.13) that damping always decreases the value of $\langle q^2{}_o \rangle$ in thermal equilibrium. If $\gamma << \omega_0$ the sum in eq.(3.8) is effectively cut off at a value of $|n|_{max} \sim \beta\hbar\omega_0$ which leads to the result $\langle q^2{}_o \rangle \simeq \frac{\hbar}{2m\omega'_0} \coth \frac{1}{2}\beta\hbar\omega'_0$ which also follows from eq.(2.17) and excludes squeezing. However, if the dissipation rate γ is comparable with the frequency ω'_0 then $\langle q^2{}_o \rangle$ is smaller and squeezed below the vacuum limit if $k_B T << \hbar\gamma$. This kind of squeezing by dissipation therefore only occurs in a region with strong damping $\gamma \sim \omega'_0$ and low temperature $k_B T << \hbar\gamma$ where the conventional approach of section 1 does not apply. A second point concerning the result (3.6) is the fact that correlations between the initial state of the oscillator and the reservoir (i.e. the operator force F_q) are, in general, non-negligible and produce contributions of the same order and even the same form as the autocorrelation terms of the oscillator coordinate and the operator force. This point is treated incorrectly in all theories which make the rather common assumption that the reservoir and the oscillator are uncorrelated initially. The only limit in which this assumption becomes true is the Markov limit discussed in section 1, where the δ-functions in eq. (1.4) wipe out correlations between the initial state and the reservoir.

4. SECOND EXAMPLE: SUBHARMONIC GENERATION

In order to discuss subharmonic generation in our new framework we return to eq. (2.1) whose solution may be written in the form

$$q(t) = \frac{1}{m} \int_0^t d\tau \;\; \chi_1(t,\tau) \;\; F_q(\tau) + \chi_1(t,0)\, \dot{q}(0) + \chi_2(t,0)\, q(0) \tag{4.1}$$

where we now distinguish two response functions $\chi_i(t,\tau)$ $(i = 1,2)$ which both satisfy the differential equation (2.5) with the initial conditions

$$\begin{aligned} &\chi_1(\tau,\tau) = 0\,, \qquad && \dot{\chi}_1(\tau,\tau) = 1\,, \\ &\chi_2(\tau,\tau) = 1\,, \qquad && \dot{\chi}_2(\tau,\tau) = 0\,, \end{aligned} \tag{4.2}$$

Thus, $\chi_1(t,\tau)$ is the response function $\chi(t,\tau)$ of section 2 which also satisfies eq. (2.6). As eq. (2.5) cannot be solved analytically, the response functions $\chi_i(t,\tau)$ must be computed either numerically or in some reasonable approximation. An approximate calculation is given below. Once the response functions are known, eq. (4.1) can be used to calculate all expectation values of q(t) and $p(t)=m\dot{q}(t)$. E.g. the expectation value $<q^2(t)>$ is obtained from

$$\langle q^2(t)\rangle = \frac{2}{m} \int_0^t d\tau \; \{\chi_1(t,\tau)\, \chi_1(t,0)\; \langle F_q(\tau)\, \dot{q}_0(0)\rangle_s$$

$$+\chi_1(t,\tau)\, \chi_2(t,0)\; \langle F_q(\tau)\, q_0(0)\rangle_s\}$$

$$+\frac{1}{m^2} \int_0^t d\tau \int_0^t d\tau' \; \chi_1(t,\tau)\, \chi_1(t,\tau')\, S_F(\tau-\tau') \tag{4.3}$$

$$+\chi_1^2(t,0)\, \langle \dot{q}_0^2\rangle + \chi_2^2(t,0)\, \langle q_0^2\rangle$$

where $<...>_S$ denotes the symmetrized expectation value. Besides the response functions $\chi_i(t,\tau)$ and the correlation function $S_F(\tau-\tau')$ (eqs. (2.2),.(2.11), (2.12)) eq. (4.3) contains the thermal expectation value $<q^2_o>$, given by eq. (3.8), and $<\dot{q}^2_0>$, obtained from

$$\langle \dot{q}_0^2\rangle = \int \frac{d\omega}{2\pi}\, \omega^2\, C_q(\omega) \tag{4.4}$$

by contour integration [8]

$$\langle \dot{q}_0^2\rangle = \frac{1}{m\beta} \sum_{n=-\infty}^{+\infty} (\omega_0^2 + \frac{2\pi|n|}{\beta\hbar}\, \gamma(\frac{i2\pi|n|}{\beta\hbar}))\, \chi_0(i\, \frac{2\pi}{\beta\hbar}\, |n|) \tag{4.5}$$

Furthermore, eq. (4.3) contains also the two symmetrized expectation values $<F_q(\tau)\dot{q}_o(o)>_s$ and $<F_q(\tau)q_o(o)>_s$ which describe the correlations of the initial state at t=0 and the reservoir. These expectation values are defined by

$$\langle F_q(\tau)\, q_0(0)\rangle_s = \int_{-\infty}^{0} d\tau'\; \chi_0(-\tau')\, S_F(\tau-\tau')$$

$$\langle F_q(\tau)\, \dot{q}_0(0)\rangle_s = \int_{-\infty}^{0} d\tau'\; \dot{\chi}_0(-\tau')\, S_F(\tau-\tau') \tag{4.6}$$

These integrals may be evaluated by contour integration in the complex ω-plane, making use of the explicit forms of the Fourier transforms of $\chi_0(\tau)$ and $S_F(\tau)$. The explicit form of $\gamma(\omega)$ in the complex ω-plane enters into these integrals, and we shall therefore assume that $\gamma(\omega)$ has simple poles at $\omega=\omega_i$ ($i = 1,...,p$) with $(\omega_i-\omega_i^*)/i < 0$ and residua Γ_i. The integrals (4.6) are then evaluated as

$$\langle F_q(\tau)\, q_o(0)\rangle_s = -\frac{m}{2\beta} \sum_{n=-\infty}^{+\infty} e^{(-2\pi/\beta\hbar)\,|n|\tau}\, \frac{2\pi|n|}{\beta\hbar}\, \gamma\!\left(\frac{i2\pi|n|}{\beta\hbar}\right) \chi_o\!\left(i\,\frac{2\pi}{\beta\hbar}\,|n|\right)$$

$$- i \sum_{i=1}^{p} e^{-i\omega_i\tau}\, \frac{\hbar\,\omega_i\, m\, \Gamma_i}{2}\,.\, \coth\frac{1}{2}\,\beta\hbar\,\omega_i\, \chi_o(-\omega_i) \tag{4.7}$$

and

$$\langle F_q(\tau)\, \dot{q}_o(0)\rangle_s = -\frac{m}{2\beta} \sum_{n=-\infty}^{+\infty} e^{(-2\pi/\beta\hbar)\,|n|\tau} \left(\frac{2\pi}{\beta\hbar}n\right)^2 \gamma\!\left(\frac{i2\pi|n|}{\beta\hbar}\right) \chi_o\!\left(i\,\frac{2\pi}{\beta\hbar}\,|n|\right)$$

$$+ i \sum_{i=1}^{p} e^{-i\omega_i\tau}\, \frac{\hbar\,\omega_i^2\, m\, \Gamma_i}{2}\,.\, \coth\frac{1}{2}\,\beta\hbar\,\omega_i\,.\,\chi_o(-\omega_i) \tag{4.8}$$

Finally, we discuss how approximate analytical expressions for the yet unknown response functions $\chi_i(t,\tau)$ can be obtained. Their equation of motion (2.5) is most directly attacked by using the Laplace-Fourier Transform

$$\chi_i(\omega,\tau) = \int_0^{\infty} d(t-\tau)\, e^{i\omega(t-\tau)}\, \chi_i(t,\tau) \tag{4.9}$$

Specifying the parametric driving force f(t) as

$$f(t) = 2\,\Omega\, g \cos \Omega\, t$$

and using the initial conditions (4.2) we obtain

$$\chi_o^{-1}(\omega)\, \chi_i(\omega,\tau) + \Omega\, g\, e^{i\Omega\tau}\, \chi_i(\omega+\Omega,\tau) + \Omega\, g\, e^{-i\Omega\tau}\, \chi_i(\omega-\Omega,\tau) = I_i(\omega) \tag{4.10}$$

with

$$I_1(\omega) = 1\,; \qquad I_2(\omega) = -\, i\omega + \gamma(\omega) \tag{4.11}$$

The corresponding equations for $\chi_i(\omega \pm \Omega,\tau)$ show that these quantities are coupled back to $\chi_i(\omega,\tau)$ but also coupled to $\chi_i(\omega \pm 2\Omega,\tau)$ and so forth. A simple but reasonable approximation consists in replacing in these terms $\chi_{1,2}(\omega \pm 2\Omega,\tau)$ by

$\chi_0(\omega\pm2\Omega)$, $-i(\omega \pm 2\Omega)\, \chi_0(\omega \pm 2\Omega)$, respectively, which closes the equations. In this way we focus on the subharmonic instability of the Mathieu equation, while neglecting all other parametric instabilities, the most important of which occurs at the third harmonic of the oscillator frequency. In this approximation the three coupled equations for $\chi_i(\omega,\tau)$, $\chi_i(\omega\pm\Omega,\tau)$ can be solved analytically and we obtain

$$\chi_1(\omega,\tau) = \frac{1}{N(\omega)}[\chi_0^{-1}(\omega+\Omega)\, \chi_0^{-1}(\omega-\Omega) - g\Omega e^{-i\Omega\tau}\, \chi_0^{-1}(\omega+\Omega)$$

$$- g\Omega e^{i\Omega\tau}\, \chi_0^{-1}(\omega - \Omega) + g^2\Omega^2 \chi_0^{-1}(\omega+\Omega)\, \chi_0(\omega-2\Omega) e^{-2i\Omega\tau}$$

$$+ g^2\Omega^2 \chi_0^{-1}(\omega-\Omega)\, \chi_0(\omega+2\Omega) e^{2i\Omega\tau}] \qquad (4.12a)$$

$$\chi_2(\omega,\tau) = \frac{1}{N(\omega)}\, [\chi_0^{-1}(\omega+\Omega)\, \chi_0^{-1}(\omega-\Omega)(-i\omega+\gamma(\omega))$$

$$- (-i(\omega-\Omega) + \gamma(\omega-\Omega))\, \chi_0^{-1}(\omega+\Omega) g\Omega e^{-i\Omega\tau}$$

$$- (-i(\omega+\Omega) + \gamma(\omega+\Omega))\, \chi_0^{-1}(\omega-\Omega) g\Omega e^{i\Omega\tau}$$

$$-i(\omega-2i\Omega)\, \chi_0^{-1}(\omega+\Omega)\, \chi_0(\omega-2\Omega) g^2\Omega^2 e^{-2i\Omega\tau}$$

$$-i(\omega+2i\Omega)\, \chi_0^{-1}(\omega-\Omega)\, \chi_0(\omega+2\Omega) g^2\Omega^2 e^{2i\Omega\tau}] \qquad (4.12b)$$

with

$$N(\omega) = \chi_0^{-1}(\omega)\, \chi_0^{-1}(\omega+\Omega)\, \chi_0^{-1}(\omega-\Omega) - g^2\Omega^2 (\chi_0^{-1}(\omega+\Omega) + \chi_0^{-1}(\omega-\Omega)) \qquad (4.13)$$

Our explicit results for $\chi_1(\omega,\tau)$ and $\chi_2(\omega,\tau)$ show that the oscillator amplitude at frequency ω receives contributions from thermal and quantum noise at frequencies close to ω, and also $\omega \pm \Omega$, $\omega \pm 2\Omega$, the latter two contributions being due to the parametric driving. Such contributions are partially neglected within the simple treatment of section 1.

The roots of $N(\omega) = 0$ determine the modes of the system, and whether these correspond to squeezed or amplified components. E.g. for $\Omega=\pm\, 2\omega'_0$, $\gamma(\omega) = \gamma$ independent of ω and g, $\gamma << \omega'_0$, there are six roots, which are given, approximately, by

$$\omega_{1,2} \simeq \pm \ \omega_0' + ig - i\frac{\gamma}{2}$$
$$\omega_{3,4} \simeq \pm \ \omega_0' - ig - i\frac{\gamma}{2} \qquad (4.14)$$
$$\omega_{5,6} \simeq \pm \ 3\omega_0' - i\frac{\gamma}{2}$$

The root near $\pm 3\ \omega'_0$ describes parametric fluctuations near the third harmonic of the oscillator frequency. which are always damped within our approximation, because we have neglected feed-back effects which lead to gain near $3\ \omega'_0$ for sufficiently strong external pumping (g large). The other two pairs of roots correspond, respectively, to the amplified and the squeezed components of the oscillation at frequency ω'_0. Neglecting the poles at the third harmonic, using otherwise the poles as given by eq. (4.14), and evaluating their residua to zero order in g/ω'_0 we obtain

$$\chi_1(t,\tau) = \left\{ \frac{e^{i\omega_0'(t-\tau)}}{4i\ \omega_0'} [(1-i\ e^{2i\omega_0'\tau})\ e^{(g-\gamma/2)(t-\tau)} + (1+i\ e^{2i\omega_0'\tau})\ e^{-(g+\gamma/2)(t-\tau)}] + c.c. \right\} \qquad (4.15)$$

and

$$\chi_2(t,\tau) = \left\{ \frac{e^{i\omega_0'(t-\tau)}}{4} [(1+i\ e^{2i\omega_0'\tau})\ e^{(g-\gamma/2)(t-\tau)} \right.$$
$$\left. + (1-i\ e^{2i\omega_0'\tau})\ e^{-(g+\gamma/2)(t-\tau)}] + c.c. \right\} \qquad (4.16)$$

The initial conditions $\chi_1(\tau,\tau) = 0$, $\chi_2(\tau,\tau) = 1$ are satisfied exactly by these expressions, but, due to the approximations made in the step from eqs. (4.12), (4.13). to eqs. (4.15), (4.16), the further conditions $\dot{\chi}_1(\tau,\tau) = 1$ and $\dot{\chi}_2(\tau,\tau) = 0$ are violated in order $0(g/\omega'_0)$. Eqs. (4.15) and (4.16) can now be used, together with the results for $<q^2_o>$, $<\dot{q}^2_o>$, $S_F(\tau-\tau')$, $<F_q(\tau)\ \dot{q}_o(o)>_S$, $<F_q(\tau)\ q_o(o)>_S$ in equations like (4.3) to compute the desired expectation values.

In an experimental analysis for squeezing, the oscillator amplitude is analyzed by heterodyning techniques [9]. This means that one measures expectation values and correlation functions of that quadrature phase q_1 of the signal which is in phase with an external local oscillator at frequency ω'_0 and phase ϕ

$$q_1(\phi,t) = q^{(+)}(t)\ e^{-i\phi+i\omega_0't} + q^{(-)}(t)\ e^{i\phi-i\omega_0't} \qquad (4.17)$$

Here $q^{(+)}(t)$ and $q^{(-)}(t)$ are the positive and negative frequency parts of $q(t)$, respectively. To construct $q^{(-)}(t)$ we present the functions on the right hand side of eq. (4.1) by their Fourier transforms and find after some rearrangements for $t \geq 0$

$$q(t) = q^{(+)}(t) + q^{(-)}(t)$$

$$= \int_{-\infty}^{+\infty} \frac{d\omega}{2\pi} e^{-i\omega t} \Big[\int_0^{\infty} d\tau \; e^{i\omega\tau} \chi_1(\omega,\tau) \; F_q(\tau)$$

$$+ \chi_1(\omega,o) \; \dot{q}(o) + \chi_2(\omega,o) \; q(o) \Big] \tag{4.18}$$

from which the positive and negative frequency parts may be read off. Hence $q_1(\phi,t)$ is known in terms of the results we have given. In order to discuss squeezing one has to form the expectation value $<q_1(\phi,t)^2>$, analyze its dependence on ϕ, and compare its size with the vacuum value

$$\langle q_1(\phi,t)^2 \rangle = \frac{\hbar}{2 \; m \; \omega_0'} \tag{4.19}$$

it would have for a corresponding harmonic oscillator. The explicit evaluation of the expectation value $<q_1(\phi,t)^2>$ is not performed here. It is best done numerically, using the results obtained above. In the limits γ, $g << \omega'_0$, $k_B T >> \hbar\gamma$, $\hbar g$ an analytical evaluation can be performed using eqs. (4.16), (4.17) and the result (1.8) is recovered with ω_0 replaced by $\omega'_0$and $x(t) = \sqrt{m\omega'_0/2\hbar}\, q_1(\frac{\pi}{2},t)$. However, if any of these approximations breaks down one may observe Non-Markovian effects due to thermal correlations (if $k_B T >> \hbar\gamma$), additional noise near frequency ω'_0 due to parametric conversion of noise at frequency $\omega'_0 \pm \Omega$, $\omega'_0 \pm 2\Omega$ (if g is not negligible against ω'_0), or squeezing due to dissipation (if γ is not negligible against ω'_0). Such effects are most likely to show up in applications outside optics, e.g. for Josephson parametric amplifiers, for which squeezing of thermal microwave radiation has recently been reported [10].

ACKNOWLEDGEMENT

Financial support of this work by the Deutsche Forschungsgemeinschaft through the Sonderforschungsbereich 237 is gratefully acknowledged.

REFERENCES

1. R. Graham, H. Haken, Z. Physik 210, 276 (1968).
2. I.R. Senitzky, Phys. Rev. 119,670 (1960); 124, 642 (1961).
3. H. Haken, “Laser Theory”, Encyclopedia of Physics, Vol. XXV/2c, ed. L. Genzel, Springer, Berlin 1969.

4. P. Talkner, Ann. Phys. (N.Y.) 167, 390 (1986).
5. R.Graham, "Quantum Langevin Equation of Parametrically Driven Linear Oscillators", to appear in Europhysics Letters.
6. R. Kubo, Rep. Progr. Phys. 29,.255 (1966).
7. P. Schramm, H. Grabert, Phys. Rev. A34,.4515 (1986); cf. also G.J. Milburn, D.F. Walls, Am. J. Phys. 51, 1134 (1983).
8. H. Grabert, U. Weiss, P. Talkner, Z. Physik B55, 87 (1984); cf. also F.Haake, R.Reibold, Phys.Rev.A32, 2462 (1985).
9. H.P. Yuen, J.H. Shapiro, IEEE Trans. Inform. Theory IT-26, 78 (1980); H.P. Yuen, V.W.S. Chan, 0pt. Lett. 8, 177 (1983).
10. B. Yurke, "Squeezinq Thermal Microwave Radiation", these Proceedings.

MECHANISMS FOR THE GENERATION OF NONCLASSICAL LIGHT

G.S. Agarwal

School of Physics, University of Hyderabad

Hyderabad - 500 134, India

In this lecture I review some of my recent proposals[1-5] for the production of the nonclassical states of the radiation field. In the first part I demonstrate how the fields at the Rabi side bands and the subharmonics[2,3] of the Rabi side bands which are produced in a number of nonlinear mixing processes possess very strong quantum characteristics. I present theories[2,3] that describe the generation of the Rabi side bands and their subharmonics starting from vacuum fluctuations. These theories yield the distribution function of the radiation field so generated. I then present the unusual photon number distributions[6,7] associated with such fields. In the second part of my lecture I consider a scheme[1,4,5,8] for producing steady state fields with strong quantum features. Clearly steady state is possible by controlling the growth of fields in a parametric process. This could happen in a dense medium where several nonlinear processes are simultaneously possible. I thus consider a two photon medium in which a strong competition among four wave mixing, two photon absorption, amplified spontaneous emission and the reabsorption of the generated photons leads to steady state fields with remarkable quantum properties such as squeezing, sub Poissonian statistics, violations of the Cauchy-Schwarz inequality. I also demonstrate the generation of a new class of coherent states called pair coherent states which are fundamental in understanding the behavior of the radiation fields in situations where the photons are destroyed or created in pairs.

I. QUANTUM GENERATION OF THE RABI SIDE BANDS IN OPTICAL TRANSMISSION

Some years back Harter et al[9] studied the transmission of light tuned close to the D2 line of sodium. They reported the appearance of the side bands.

They confirmed that the frequencies of the side bands correspond to the generalized Rabi frequency for a field interacting with a two-level optical transition. Thus the side bands appeared at the frequencies

$$\omega_{\pm} = \omega \pm \Omega', \quad \Omega' = \frac{\Delta}{|\Delta|} \sqrt{\Delta^2 + \Omega^2}, \quad \Delta = \omega - \omega_0, \tag{1.1}$$

where ω is the pump frequency, ω_0 is the frequency of the optical transition and $\Omega = \frac{2\vec{d}\cdot\vec{\varepsilon}}{\hbar}$. It should be noted that no input fields were present at $\omega_{\pm}$. The appearance of these side bands can be understood in terms of a four wave mixing process for example the fields at ω and $\omega + \Omega'$ can lead to the generation of the fields at $\omega - \Omega'$ through the four wave mixing susceptibility $\chi^{(3)}(\omega, \omega, -(\omega + \Omega'))$. Since initially no field is applied at either of the frequencies $\omega_{\pm}$, such a generation is a quantum mechanical process in which the noise photons at $\omega_{\pm}$ participate and these grow to give rise to the macroscopic fields. We thus examine how these side bands are produced. We then report on the quantum statistics of the generated fields at the Rabi side bands.

The quantum theory[3] of the side band generation in optical transmission can be formulated as follows - . We write the total field as a sum of the pump field $\varepsilon e^{i\vec{k}.\vec{r}-i\omega t}$ and the generated fields. We assume the pump to be strong so that the pump is treated semi-classically and we ignore it's depletion. The generated fields $\vec{E}$ are assumed to be weak and we write these in the form

$$\vec{E}(\bar{r},t) = \beta_s \hat{\varepsilon}_s a e^{i\vec{k}_s.\vec{r}-i\omega_s t} + \beta_c \hat{\varepsilon}_c b e^{i\vec{k}_c.\vec{r}-i\omega_c t} + \text{H.c.} \quad . \tag{1.2}$$

For phase matching the wave vectors must satisfy

$$\vec{k}_s + \vec{k}_c = 2\vec{k} \ . \tag{1.3}$$

In (1.2) a, b are the annihilation operators, $\hat{\varepsilon}$'s are polarization vectors and $\beta_\alpha = -i\left(\frac{2\pi\hbar\omega_\alpha}{V}\right)^{1/2}$. Note that we can choose ω_s to be $\omega + \Omega'$. We further assume that the response of the medium is fast and that there are no correlations among different atoms. Starting from the density matrix equation for the coupled atoms-field system and using the above stated assumptions we can derive the equation for the reduced density matrix ρ_F for the field E alone. We also especialize to the case of a two level optical transition and thus the polarization vectors essentially become unimportant. Calculations[3,10] show that ρ_F satisfies the master equation

$$\frac{\partial \rho_F}{\partial t} = \frac{\alpha c}{4T_2} \left(\hat{Q}^{++}(-i\nu_c)[a^+, [b^+, \rho_F]] + \hat{C}^{++}(-i\nu_c)[a^+, \{b^+, \rho_F\}] + \hat{Q}^{++}(-i\nu_s) \right.$$

$$\left. \times [b^+, [a^+, \rho_F]] + \hat{C}^{++}(-i\nu_s)[b^+, \{a^+, \rho_F\}] \right) e^{-it(2\omega - \omega_s - \omega_c)}$$

$$- \frac{\alpha c}{4T_2} \{ \hat{Q}^{+-}(i\nu_s)[a^+, [a, \rho_F]] + \hat{C}^{+-}(i\nu_s)[a^+, \{a, \rho_F\}] + \hat{Q}^{+-}(i\nu_c)[b^+, [b, \rho_F]]$$

$$+ \hat{C}^{+-}(i\nu_c)[b^+, \{b, \rho_F\}]\} + \text{H.c.}, \quad \nu_{s,c} = \omega - \omega_{s,c} \quad , \tag{1.4}$$

where T_2 is the transverse relaxation time in the medium and $\alpha = 4\pi n \omega |d|^2 T_2/\hbar c$ is the unsaturated line center absorption coefficient, n is the density of atoms. For simplicity we have assumed a common α for all the fields. The parameters $Q^{\pm\pm}$, $C^{\pm\pm}$ depend on the medium and the intensity of the applied pump. These can be related to the two time correlation functions of the medium. For the two level system $\hat{Q}$'s(z) etc., are the Laplace transforms of Q's(τ) defined by

$$\left.\begin{matrix} Q^{+\pm}(\tau) \\ C^{+\pm}(\tau) \end{matrix}\right\} = \lim_{t\to\infty} \langle [S^{(-)}(t+\tau) - \langle S^-(t+\tau)\rangle , S^{\mp}(t) - \langle S^{\mp}(t)\rangle]_{\pm} \rangle \quad . \tag{1.5}$$

Thus Q's are given by the anticommutator and C's by the commutator. The correlation functions can be calculated from optical Bloch equations and the quantum regression theorem. The correlation functions C's have simple interpretation[10] in terms of the nonlinear susceptibilities of the medium. The correlation functions Q's have no counterpart in semiclassical theories and these represent the quantum fluctuations in the medium. It should be borne in mind that since we are dealing with a system that is far from thermal equilibrium C's and Q's are not related via the fluctuation dissipation theorem.

In order to solve Eq. (1.4) we transform it into a differential equation for the Wigner function $\Phi(z_a, z_b)$ defined by

$$\Phi(z_a, z_b) = \frac{1}{\pi^4} \text{Tr}\, \rho_F \iint d^2\alpha\, d^2\beta \exp\{ -[\alpha(z_a^* - a^+) + \beta(z_b^* - b^+) - \text{H.c.}]\} \quad . \tag{1.6}$$

Calculations show that

$$\frac{\partial \Phi}{\partial t} = \frac{\alpha c}{4T_2} [(\hat{Q}^{++}(-i\nu_c) + \hat{Q}^{++}(-i\nu_s)) \frac{\partial^2 \Phi}{\partial z_a \partial z_b} - \hat{C}^{++}(-i\nu_c) \frac{\partial}{\partial z_a}(2z_b^* \Phi)$$

$$- \hat{C}^{++}(-i\nu_s) \frac{\partial}{\partial z_b}(2z_a^* \Phi)] e^{-it(2\omega - \omega_s - \omega_c)} - \frac{\alpha c}{4T_2} [\hat{Q}^{+-}(i\nu_s) \frac{\partial^2 \Phi}{\partial z_a \partial z_a^*}$$

$$+ \hat{Q}^{+-}(i\nu_c) \frac{\partial^2 \Phi}{\partial z_b \partial z_b^*} + \hat{C}^{+-}(i\nu_s) \frac{\partial}{\partial z_a}(2z_a \Phi) + \hat{C}^{+-}(i\nu_c) \frac{\partial}{\partial z_b}(2z_b \Phi)] + \text{c.c.} \quad . \tag{1.7}$$

It should be borne in mind that for the steady state propagatation problems t should be replaced by z/c. The equation (1.7) for the Wigner function has the form of a linearized Fokker-Planck equation. The drift terms in (1.7) are determined by the atomic correlations C's i.e., by the nonlinear susceptibilities. The diffusion terms depend on Q's and thus the diffusion of the Wigner function is determined by the quantum fluctuations in the medium. The solution of (1.7) yields the symmetrized expection values for example

$$< \frac{a^+a + aa^+}{2} > = \iint d^2z_a \, d^2z_b \, |z_a|^2 \, \Phi \, (z_a, z_b) \quad . \qquad (1.8)$$

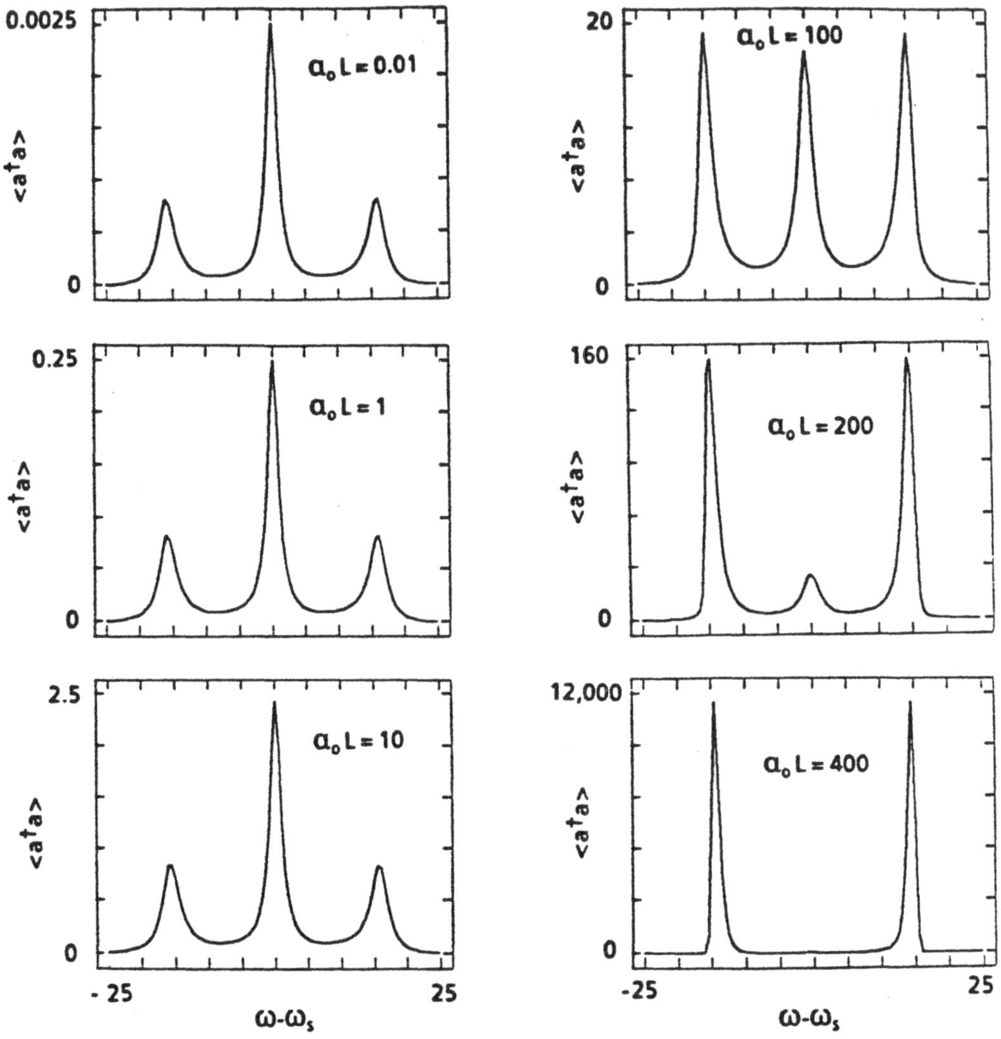

Fig. 1. The mean number of photons [Ref. 3] plotted as a function of $\omega - \omega_s$ for several value of $\alpha_0 L$ for the case of radiatively broadened optical transition. Other parameters are chosen as $\Delta = 0$, $\Omega \, T_2 = 16$.

If initially the fields a and be are in the vacuum state which is indeed the case for the experiment of Harter et al, then

$$\Phi\ (z_a,z_b,\ 0)\ =\ \frac{4}{\pi}\ \exp\{\ \ -2|z_a|^2 - 2|z_b|^2\ \}\quad . \tag{1.9}$$

It can then be shown that the Wigner function at time t is also Gaussian i.e., it has the form

$$\Phi(z_a,\ z_b,\ t)\ =\ \exp\ \{\ \text{Quadratic form in } (z_a,z_b)\ \}\quad . \tag{1.10}$$

Using the solution of (1.7) we have calculated the generation of the Rabi side bands in optical transmission. The Fig. 1 gives the growth of the side bands for various values of the parameter $\alpha_0 l$. For the case of a thin medium $\alpha_0 l = 0.01 << 1$, the emission line shape is identical to the well known Mollow spectrum. For increasing path-length the line shapes change considerably. In particular the number of photons per mode increases rapidly especially for signal modes near the Rabi side band frequencies. The Rabi side bands become more and more prominent whereas the central component becomes unimportant. The side bands also display considerable gain narrowing. The results for the collisionally broadened medium are similar except for the collisional redistribution effects.

We have also examined the quantum statistics of the radiation generated at the two side bands. We find that the two side modes are strongly correlated with each other. To see this we calculate the correlation function

$$C_{ab}\ =\ <a^+a\,b^+b> - <a^+a><b^+b>\quad . \tag{1.11}$$

Using the Gaussian form of Φ, one can show that

$$C_{ab}\ =\ |<ab>|^2\ . \tag{1.12}$$

In Fig. 2 we show the behavior of C_{ab} for a typical case. Clearly the two side bands are strongly correlated. These correlations persist even when the mean number of photons per mode is small. Since the photons are produced in pairs once a photon is detected, the probability of detecting a second photon in the mode b is quite large.

We have also studied in detail the phase sensitive properties of the radiation at the side bands. In particular we have found that the radiation has considerable squeezing for a certain range of parameters. In squeezing studies one looks at the properties of a field which is a linear superposition of the fields a and b. We thus introduce the maximum squeezing parameter

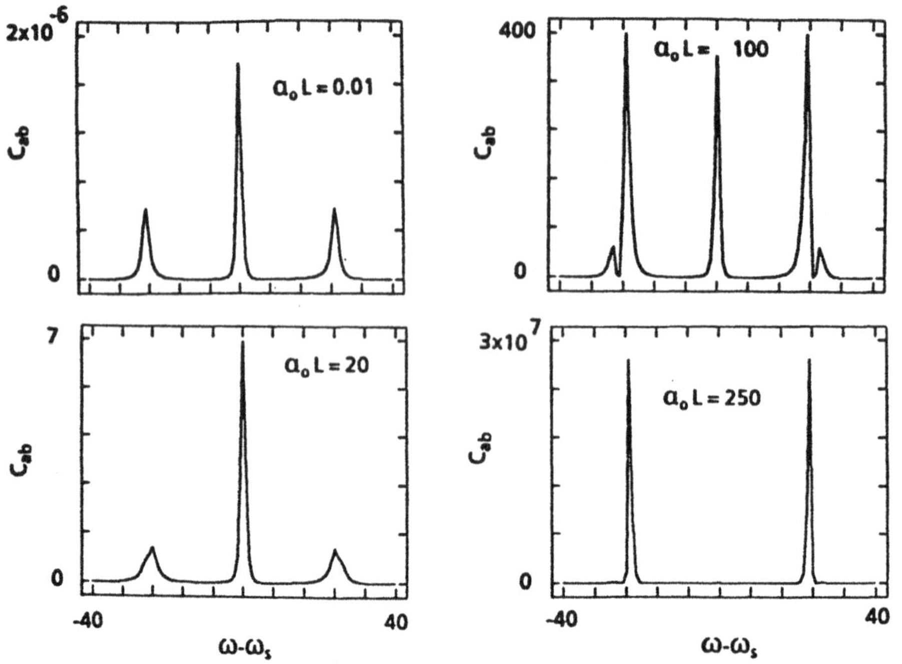

Fig. 2. The correlation C_{ab} [Ref. 3] as a function of $\omega - \omega_s$ for $(T_2/T_1) = 2$, $\Delta T_2 = -8$, $\Omega' T_2 = 25$ and for different values of $\alpha_o L$.

$$S_{max} = \frac{1}{4}[< a^+ a > + < b^+ b > - 2 |< ab >|] , \quad -1 \le S_{max} \le 0. \quad (1.13)$$

In Fig. 3 we show the squeezing characteristics. We find broad regions of squeezing centered about the Rabi side band frequency. Appreciable levels of squeezing occur for a range of the propagation lengths.

Other statistical features of the mode, that is a linear combination of a and b, can be obtained from (1.10). The Wigner function for this new mode A can be written as

$$\Phi (z) = \frac{1}{\pi\sqrt{\tau^2 - 4|\mu|^2}} \exp [- \frac{\mu z^2 + \mu^* z^{*2} + \tau |z|^2}{\pi\sqrt{\tau^2 - 4|\mu|^2}}] \quad , \tau > \mu + \mu^* . \quad (1.14)$$

The parameters μ and τ depend on the characteristics of the atomic medium. These yield the fluctuations in A i.e.,

$$< A^2 > = -2\mu^* , < \frac{A^+ A + AA^+}{2} > = \tau \quad . \quad (1.15)$$

Using (1.14) one can prove the following important property[10,11]. If the mode A is squeezed to second order, then it is also squeezed to all even orders.

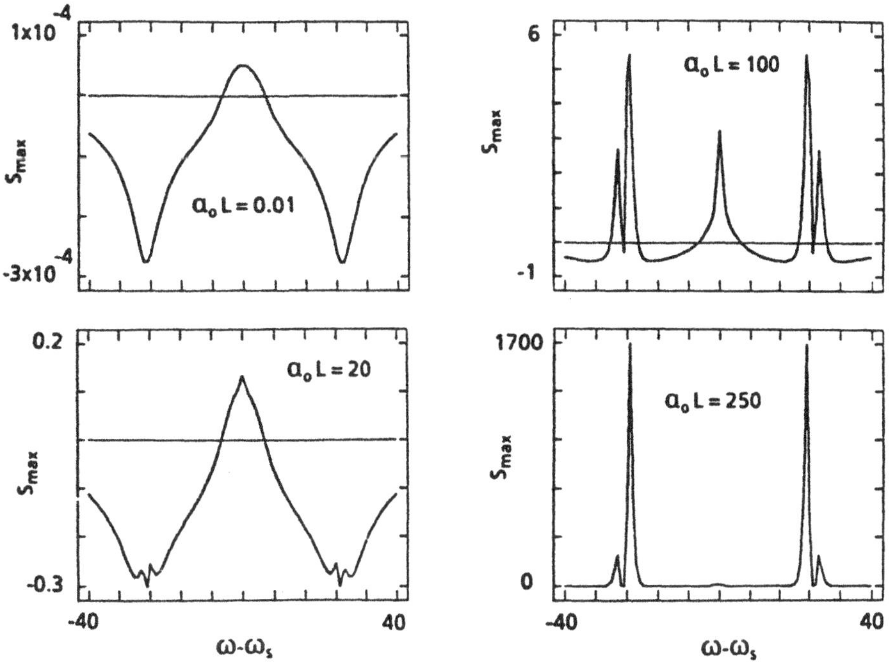

Fig. 3. Maximum squeezing parameter [Ref. 3] plotted as a function of the detuning $\omega - \omega_s$ for the same parameters as in Fig. 2.

This property essentially follows from the moment theorem for the Gaussian processes and the fact that for squeezing purposes one needs to know the moments of say $(A + A^+)$ i.e., one needs to know the symmetrized moments.

II. PHOTON NUMBER DISTRIBUTIONS

The form (1.14) is quite attractive[12] as it includes the cases of various types of the fields for example it describes : (i) Incoherent field for $\mu = 0$, which goes over to the vacuum field if we set $\tau = \frac{1}{2}$. (ii) squeezed vacuum for

$$\mu = \frac{1}{4} \sinh x \, e^{-i\theta}, \tau = \frac{1}{2} \cosh x \quad . \tag{2.1}$$

Thus squeezed vacuum corresponds to $\tau^2 - 4|\mu|^2 = \frac{1}{4}$. The distribution (1.14) is much more general as it includes losses in the medium. Generally we can write

$$\tau = \frac{Q}{2} \cosh x, \quad \mu = \frac{Q}{4} \sinh x \, e^{-i\theta} \quad , \tag{2.2}$$

with parameters Q and x depending on the nature of the four wave mixing process. We will now calculate the photon number distributions for the field described by the Gaussian Wigner function. We would like to emphasize the rather universal nature of the Gaussian Wigner function as both trilinear and quartic interactions in the parameteric approximation i.e., when the effective Hamiltonian can be approximated by a bilinear interaction, would lead to a Gaussian Wigner function if initially the modes a and b are in vacuum state. Note further that in most problems in quantum optics, the quantum fluctuations are described by a linearized Fokker- Planck equation for an appropriate quasi-probability distribution. Thus in effect the fluctuations are described by (1.14). Thus the number distributions that we calculate can also be used for other systems in quantum optics. All we need to do is to use the values of Q and x for the system under consideration. It can also be proved that the distribution (1.14) can be regarded as that for a field which is produced by the coherent superposition of a thermal field and a field in the squeezed vacuum.

It can be shown that the component $\frac{A+A^{+}}{2}$ is squeezed $(\theta = 0)$ if

$$Q\, e^{-X} \leq 1 \quad . \tag{2.3}$$

The photon number distribution can be calculated in terms of the phase space integral

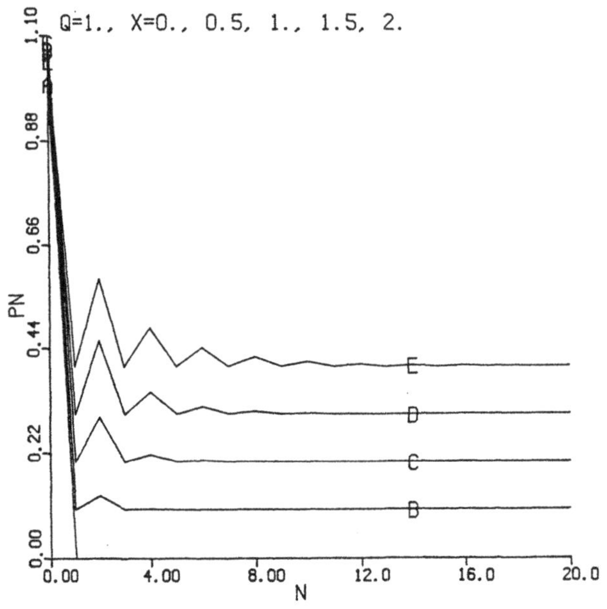

Fig. 4. Photon number distribution PN [Ref. 6] for a field characterized by a Gaussian Wigner function with Q = 1. The value of x is increasing from bottom to the top curve.

$$p(n) \;=\; \int d^2z\, \Phi(z)\, \frac{2}{n!}\, (-1)^n\, L_n(4|z|^2)\, e^{-2|z|^2} \;, \tag{2.4}$$

where L_n is the Laguerre polynomial. The result can be written in terms of the hypergeometric function

$$p(n) \;=\; \frac{(-1)^n}{Q} \sum_{k=0}^{n} (-1)^k \binom{n}{k} \frac{1}{\sqrt{k+1}}\, F\!\left(\frac{1}{2}\,,\, k+1,\, 1,\, -\frac{\sinh x}{\sqrt{Q}}\right) \;, \tag{2.5}$$

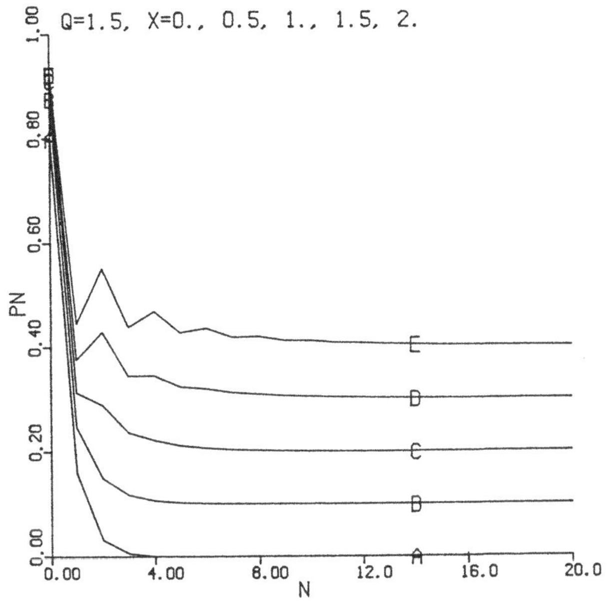

Fig. 5. Same as in Fig. 4 but now Q = 1.5 (after Ref. 6).

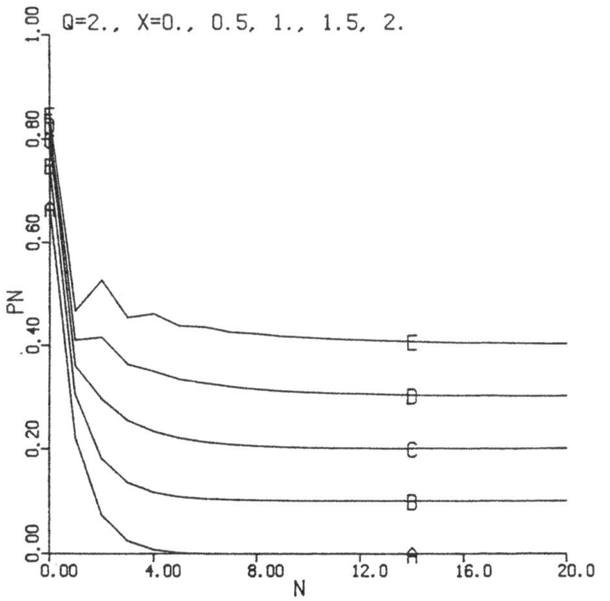

Fig. 6. Same as in Fig. 4 but now Q = 2 (after Ref. 6).

$$v = \frac{1}{2}\left(1 + \frac{e^{-x}}{Q}\right) \quad . \tag{2.6}$$

Note that p(n) reduces to the well known Bose-Einstein distribution for $x = 0$. We show in Figs. 4-6 the photon number distributions[6] for different values of the parameters Q and x. These distributions show oscillatory character in the region when the field starts exhibiting squeeqing i.e., when $Qe^{-x} < 1$. For $Q = 1$ (usual squeezed vacuum) Schleich and Wheeler[7] have interpreted these oscillations as due to the interference phenomena in phase space. Such an interference can also be understood in terms of the formula (2.4). We also note that the entropy $S(\alpha \mathrm{Tr}\, \rho \ln \rho)$ associated with the distribution (1.14) can also be computed using the result of Ref. 13. The result is

$$S \;\alpha\; (Q + \tfrac{1}{2}) \ln(Q + \tfrac{1}{2}) - (Q - \tfrac{1}{2}) \ln(Q - \tfrac{1}{2}) \tag{2.7}$$

and thus the entropy is independent of the squeezing parameter x.

III QUANTUM PROPERTIES OF THE RADIATION IN THE NEIGHBORHOOD OF THE SUBHARMONICS OF THE RABI SIDEBANDS

We next consider six wave mixing in presence of a strong pump of frequency ω. Let ω_2 be the frequency of the probe. The fifth order nonlinearity of the medium can produce a radiation at the frequency $3\omega - 2\omega_2$. This radiation

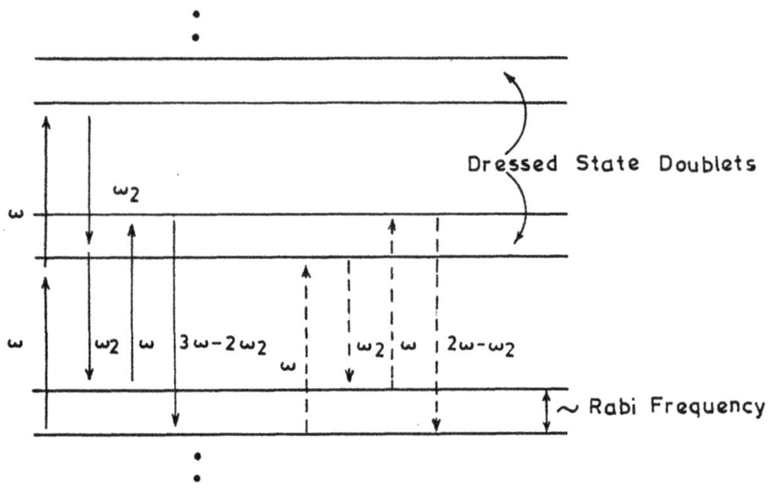

Fig. 7. Schematic illustration of the appearance of the Rabi side band and it's subharmonic in a six wave mixing process. Each doublet represents the dressed states for the atom-field system for a fixed number of pump photons. The separation between the two states of the doublet is approximately equal to the Rabi frequency.

will be phase matched in the direction $3\vec{k} - 2\vec{k}_2$. The nonlinear susceptibility responsible for a six wave mixing process is denoted by $\chi^{(5)}(\omega, \omega, \omega, -\omega_2, -\omega_2)$. If the pump is weak, then $\chi^{(5)}$ has resonances at

$$\omega_2 - \omega = 0, \pm\Delta, \pm\frac{\Delta}{2}, \Delta = \omega_0 - \omega . \tag{3.1}$$

However if the pump is strong enough to saturate the two level optical transition, then the above $\chi^{(5)}$ is to be renormalized to account for the saturation effects. Thus the renormalized $\chi^{(5)}$ depends on all the powers of the pump field and has resonances[14] at

$$\omega_2 - \omega = 0, \pm\sqrt{\Delta^2 + \Omega^2}, \pm\frac{1}{2}\sqrt{\Delta^2 + \Omega^2} . \tag{3.2}$$

Thus in six wave mixing not only Rabi side bands appear but submultiple or subharmonic of the Rabi side bands also appear. The existence of the subharmonic of the Rabi side bands in six wave mixing can be understood from the Fig. 7 which shows the possible transitions between the dressed states of the atom and pump system. The solid lines show how the subharmonic resonance arises in six wave mixing since

$$\omega + \omega - \omega_2 - \omega_2 = \text{Rabi frequency} . \tag{3.3}$$

The dashed lines show the existence of the usual Rabi side bands.

The question that we now investigate is- what are the quantum statistical properties of the radiation generated in the region of the subharmonics of the Rabi side bands. For this purpose we write an effective Hamiltonian. The semiclassical form of the effective Hamiltonian is clearly

$$H = -P(3\omega - 2\omega_2) \cdot \varepsilon^*(3\omega - 2\omega_2) + \text{c.c.} , \tag{3.4}$$

where $P(3\omega - 2\omega_2)$ is the induced polarization at $3\omega - 2\omega_2$. Let us now imagine the following- we apply fields at the frequencies ω and $3\omega - 2\omega_2 \approx \omega - \Omega$, then a field at the frequency $\omega_2 = \omega + \frac{\Omega}{2}$ will be generated. We treat this generated field quantum mechanically. Thus the effective Hamiltonian (3.4) can be written in the form[2]

$$H = \beta a^{+2} + \text{H.c.} . \tag{3.5}$$

The parameter β is discussed below. Note that the effective Hamiltonian (3.5) has the form of the down conversion Hamiltonian. In the present problem three pump photons have produced two photons of frequency $\omega + \frac{\Omega}{2}$ and one photon of frequency $\omega - \Omega$. The field at $\omega + \frac{\Omega}{2}$ has squeezing characteristics given by

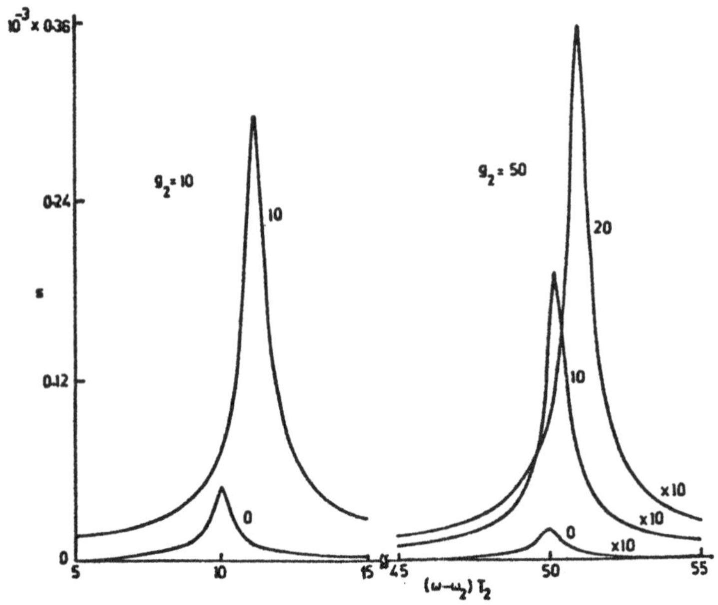

Fig. 8. The parameter s [Ref. 2] which determines the amount of squeezing in the radiation generated in the region of the subharmonic Rabi resonance. Curves are labelled by the values of the detuning and the Rabi frequency $g_2 = \Omega/2$.

$$\pm \frac{1}{4} \langle (a(t)\, e^{i\theta} \pm a^{+}(t)\, e^{-i\theta})^2 \rangle = \frac{1}{4} e^{\pm 4|\beta| t} , \quad -i\beta = |\beta|\, e^{-2i\theta} . \tag{3.6}$$

Thus the amount of squeezing depends on $|\beta| t$ which in turn depends on the renormalized fifth order susceptibility and absorption length. It can be shown that

$$4\beta t = (2\alpha L)\,(\text{Rabi frequency of the field at } \omega - \Omega)\, s , \tag{3.7}$$

where

$$s = \frac{i}{2T_2} \Omega^3 \left[1 + \frac{\Omega^2 T_1 T_2}{(1 + \Delta^2 T_2^2)}\right]^{-1} P^{-1}[i(\omega_2 - \omega)] P^{-1}[2i(\omega_2 - \omega)][1 + i\Delta T_2]^{-1} \times \{2 + 3iT_2(\omega_2 - \omega)\}\{2 + iT_2(\omega_2 - \omega)\}, \tag{3.8}$$

and where P(z) is the polynomial that gives the characteristic roots of the optical Bloch equations

$$P(z) = (z + \frac{1}{T_1})\left[(z + \frac{1}{T_2})^2 + \Delta^2\right] + \Omega^2 (z + \frac{1}{T_2}) . \tag{3.9}$$

In Fig. 8 we show the dependence of the parameter $|\beta| t$ on the parameters of the medium. This figure shows that the six wave mixing could be an important mechanism for producing squeezed light. We have also shown that the medium losses are rather unimportant in the region of the subharmonics of the Rabi side bands. It may be noted that one way to produce such subharmonics would be to repeat the experiment of Harter et al but now a field is also externally applied at one of the side band frequencies.

IV A MECHANISM TO PRODUCE STEADY STATE SQUEEZING AND OTHER QUANTUM PROPERTIES

We next investigate a mechanism that can produce quantum characteristics in the steady state. In the problems we discussed so far no steady state was possibile as in any parametric process the fields grow. In order to get a steady state there must be a mechanism to control this growth. In a dense nonlinear medium, other nonlinear optical processes start occuring which might stabilize such a growth. We thus consider a very interesting case[8] of a two photon medium where competing nonlinear processes result in steady state fields which possess remarkable squeezing and other quantum properties. Malcuit et al first demonstrated the existence of such competing nonlinear processes in a two photon medium. In Fig. 9 we show schematically a two photon medium and the various optical processes. The medium is interacting with a strong pump tuned to the two photon resonance 3s-3d. We assume that the pump is not too strong so that two photon saturation effects can be ignored. Because of the four wave mixing characterized by the susceptibility

$$\chi_{FWM} = \frac{N|d_{ba}|^2 |d_{cb}|^2}{\hbar^3 \Delta_1 \Delta_3 (\Delta_2 - i\Gamma_{ca})} , \qquad (4.1)$$

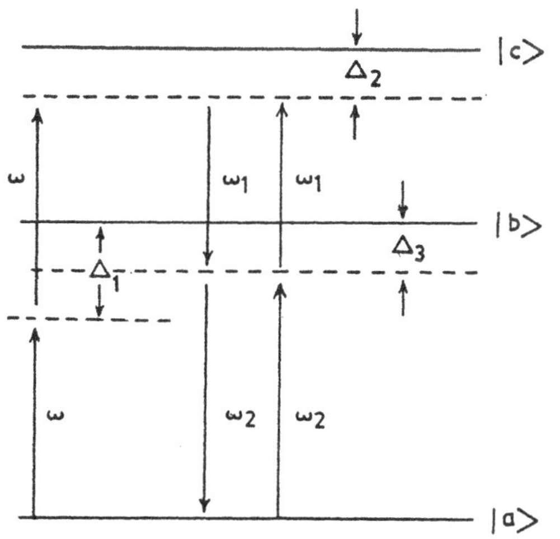

Fig. 9. Schematic scheme of the two photon medium. ω is the pump field. The fields ω_1 and ω_2 are generated by four wave mixing. These can be reabsorbed by two photon absorption. Δ's represent various detunings.

the coherent fields of frequencies ω_1 and ω_2 such that $\omega_1 + \omega_2 = 2\omega$ are generated. In (4.1) Γ_{ca} is the two photon line width. If the medium is dense enough then the reabsorption of ω_1 and ω_2 via a two photon process described by the nonlinearity

$$\chi^{TPA}_{\omega_1,\omega_2} = \frac{N|d_{ba}|^2|d_{cb}|^2}{\hbar^3\Delta_3^2(\Delta_2 - i\Gamma_{ca})} \tag{4.2}$$

becomes significant. In addition one has the possibility of incoherent spontaneous emission from the state $|c\rangle$. We have demonstrated[1,4] that the quantum generation of the fields ω_1 and ω_2 can be described by the master equation

$$\frac{\partial\rho}{\partial t} = -i\,[Ga^+b^+ + G^*ab,\rho] - i\eta\,\frac{\kappa}{2}[b^+ba^+a,\rho] - \frac{\kappa}{2}[a^+b^+ab\rho - 2ab\rho\, a^+b^+ + \rho a^+b^+ab]. \tag{4.3}$$

Here G describes four wave mixing

$$G = 2\pi\sqrt{\omega_1\omega_2}\,\varepsilon_p^2\,\chi_{FWM}\,, \tag{4.4}$$

and κ gives the two photon absorption

$$\kappa = \frac{8\pi^2\hbar\omega_2\omega_1}{V}\,\mathrm{Im}\,\chi_{TPA} \tag{4.5}$$

The parameter η gives the two photon dispersion $\eta = -\Delta_2/\Gamma_{ca}$. A solution of (4.3) will give the quantum statistics of the generated modes. Both time dependent and steady state features are discussed in Refs. 1,4,5. Here we discuss some salient features.

We first note that for short time (or short propagation legnths), the two photon absorption is negligible and the solution of (4.3) is the well known squeezed vacuum state

$$\rho = |-iGt\rangle_{TP}\;{}_{TP}\langle -iGt|\,,$$

$$|\psi\rangle_{TP} = \sum_{n=0}^{\infty}(\mathrm{sech}|\psi|)\left(\frac{\psi}{|\psi|}\tanh|\psi|\right)^n|n,n\rangle\,. \tag{4.6}$$

The solution for long times turns out to be

$$\zeta = |\varphi\rangle_p\;{}_p\langle\varphi|\,, \tag{4.7}$$

where $|\varphi\rangle_p$ is a pair coherent state defined by

$$|\varphi\rangle_p = N_0\sum_m\frac{\varphi^m}{m!}|m,m\rangle\,,\quad 1 = |N_0|^2\sum_m\frac{|\varphi|^{2m}}{(m!)^2}\,, \tag{4.8}$$

and where N_0 is the normalization constant. Thus the steady state quantum features of the generated fields can be studied in terms of the pair coherent state $|\varphi\rangle_P$.

The pair coherent state $|\zeta, q\rangle$ can be defined by the eigenvalue equation[5,15]

$$ab|\zeta, q\rangle = \zeta\, |\zeta, q\rangle ,$$

$$(a^+a - b^+b)\, |\zeta, q\rangle = q|\zeta, q\rangle \quad , |\varphi\rangle_P = |\varphi, 0\rangle . \tag{4.9}$$

Note that the pair coherent state is not the same as the two photon coherent state $|\psi\rangle_{TP}$. The pair coherent state is the eigenstate of the operator which simultaneously annihilates the photons in the modes a and b. The second of equations (4.9) ensures that the relative number of photons in the two modes remains fixed. It is clear that the states $|\zeta, q\rangle$ should be quite important in any optical process where photons are either created or destroyed in pairs.

The Glauber-Sudarshan function $P(z_a, z_b)$ defined by

$$|\zeta, q\rangle\langle\zeta, q| = \iint d^2z_a\, d^2z_b\, P(z_a, z_b)\, |z_a, z_b\rangle\langle z_a, z_b| , \tag{4.10}$$

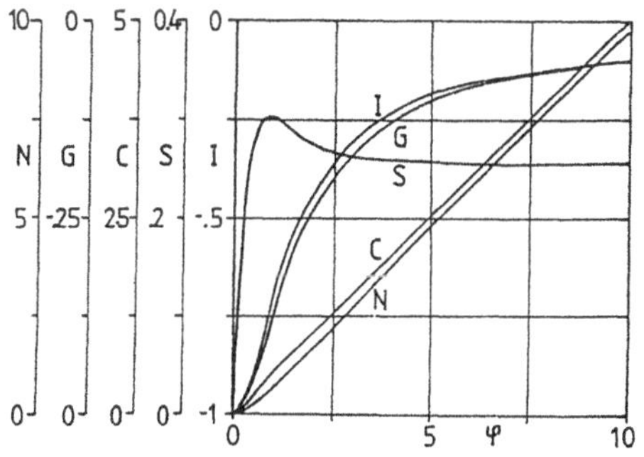

Fig. 10. The quantum statistical properties, N,C, $G = (\langle b^{+2}b^2\rangle - \langle b^+b\rangle^2)/\langle b^+ b\rangle^2$,S and I as a function of $|\zeta|$. We set $q = \eta = 0$. The scales for various curves are as shown [after Ref. 4].

does not exist which suggests that the generated fields must have striking quantum features many of which are displayed in Fig. 10. For example the distribution p_n of photons in the b mode is sub-Poissonian

$$p_n = |<n+q, n|\zeta, q>|^2 = N_q \frac{|\xi|^{2n}}{n!\,(n+q)!} \quad . \tag{4.11}$$

This can be seen from the explicit expression for the number fluctuations

$$C_b = <(b^+b)^2> - <b^+b>^2 = |\zeta|^2 - N_b^2 - qN_b, \quad N_b = <b^+b> \tag{4.12}$$

and the inequality

$$|<AB>|^2 \leq <AA^+><B^+B> \quad . \tag{4.13}$$

Calculation shows that

$$C_b \leq N_b \quad . \tag{4.14}$$

In contrast the two photon coherent state leads to super-Poissonian distribution if $<a> = 0$. Strictly quantum features also lead to the violations of the classical Cauchy-Schwarz (C-S) inequality[16] for the distribution functions. For example if P function exists and has the properties of a classical distribution, then

$$I = \frac{(<b^{+2}\, b^2><a^{+2}\, a^2>)^{1/2}}{|<a^+a\, b^+b>|} - 1 \tag{4.15}$$

should be positive. Thus negative values of I imply quantum characteristics. Calculations show that for $q = 0$

$$I = -\frac{N_b}{|\zeta|^2} < 0 \quad . \tag{4.16}$$

Thus the steady state fields generated by the two photon medium exhibit strong quantum properties. We have shown that the modulation index σ in a two photon interference experiment[17] characterized by the fourth order field correlation functions is a direct measure of the violations of the C-S inequality. The index σ turns out to be

$$\sigma = (2 + I)^{-1} \quad . \tag{4.17}$$

Thus more than 50% modulation implies a violation of the C-S inequality. For parameter $|\zeta| \sim 1$, the modulation could be 100% as is seen from Fig. 10.

The state (4.7) has also very strong squeezing. For this purpose the relevant operator is

$$d = (b e^{-i\psi/2} + a^{+} e^{i\psi/2} + H.c.) / 2\sqrt{2} \quad . \tag{4.18}$$

It can be shown that

$$S = 2 < {}^{o}_{o} d^{2} {}^{o}_{o} > = \frac{1}{2}(\zeta e^{-i\psi} + c.c.) + \frac{q}{2} + N_b \tag{4.19}$$

$$\xrightarrow{\text{large } \zeta} -\frac{1}{4} + |\zeta| \, (1 + \cos(\psi - \theta)) = -\frac{1}{4} \text{ if } \psi - \theta = \pi \, . \tag{4.20}$$

Thus the fields generated in a two photon medium can exhibit 50% squeezing. In Fig. 10 we show the quantum nature of the generated fields for a range of the parameter values $|\zeta|$. Note that ζ is determined by the ratio of the four wave mixing and two photon absorption. A better appreciation of the phase sensitive noise can be obtained by examining the distribution P(x) of d. Such a distribution can be defined by

$$P(x) = \frac{1}{2\pi} \int_{-\infty}^{+\infty} e^{-ihx} \, dh < e^{ihd} > \quad . \tag{4.21}$$

The result of the calculation is shown in Fig.11 where it is also compared with the distribution $P_c(x)$

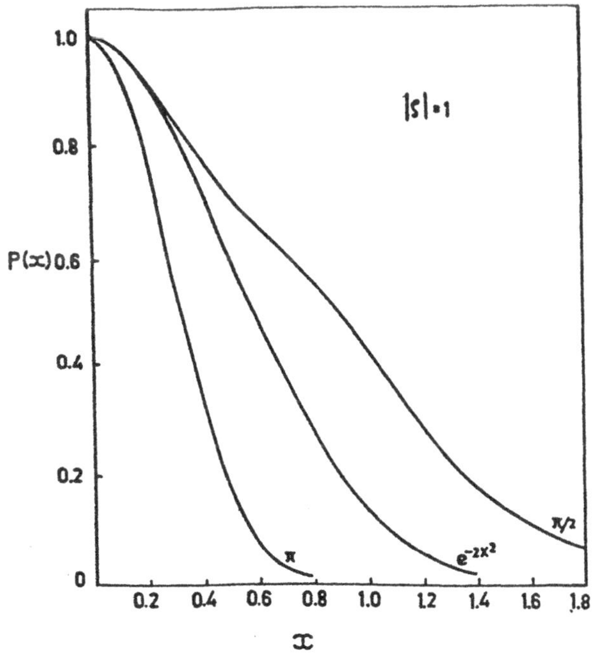

Fig. 11. The distribution P(x) of the cosine component of the field detected in an heterodyne experiment for $|\zeta| = 1$, q = 0 and $\psi - \theta = \pi$ and $\frac{\pi}{2}$. The corresponding distribution for the vacuum field is also shown [after Ref. 5].

$$P_c(x) = \frac{2}{\sqrt{2\pi}} \exp\{-2x^2\} \quad (4.22)$$

for the vacuum field. The distribution is squeezed or antisqueezed depending on the parameter $\psi - \theta$. Finally we mention that the time dependent solutions of (4.3) also exhibit very interesting quantum properties.

Part of this work has been supported by a grant from the Department of Science and Technology, Government of India.

REFERENCES*

1. G.S. Agarwal, Phys. Rev. Lett. 57, 827 (1986).
2. G.S. Agarwal, J. Opt. Soc. Am. B5, (1988), in press.
3. G.S. Agarwal and R.W. Boyd, sub to Phys. Rev. A.
4. G.S. Agarwal and F. Rattay, Phys. Rev. A37, (1988), in press.
5. G.S. Agarwal, J. Opt. Soc. Am. B5, (1988), in press.
6. G.S. Agarwal and G. Adam, sub to Phys. Rev. A.
7. W. Schleich and J.A. Wheeler, Nature 326, 574 (1987).
8. M.S. Malcuit, D.J. Gauthier and R.W. Boyd, Phys. Rev. Lett. 55, 1086 (1985).
9. D.J. Harter, P. Narum, M.G. Raymer and R.W. Boyd, Phys. Rev. Lett. 46, 1192 (1981).
10. G.S. Agarwal, Phys. Rev. A34, 4055 (1986).
11. C.K. Hong and L. Mandel, Phys. Rev. A32, 974 (1985).
12. G.S. Agarwal, J. Modern Optics. 34, 909 (1987).
13. G.S. Agarwal, Phys. Rev. A3, 828 (1971).
14. G.S. Agarwal, and N. Nayak, Phys. Rev. A33, 391 (1986).
15. D. Bhaumik, K. Bhaumik and B. Dutta-Roy, J. Phys. A9, 1507 (1976).
16. M.D. Reid and D.F. Walls, Phys. Rev. A34, 1260 (1986).
17. R. Ghosh and L. Mandel, Phys. Rev. Lett. 59, 1903 (1987).

---

*This list is far from complete. The references to the work of other authors can be found in my publications quoted below and I have quoted only those publications which are absolutely must for reading the present lecture.

INTERFEROMETRIC MEASUREMENTS BEYOND THE SHOT-NOISE LIMIT

P. Grangier*, R.E. Slusher, B. Yurke, and A. La Porta

AT & T Bell Laboratories, Murray Hill

NJ 07974, USA

I - INTRODUCTION

The ultimate performance in the measurement of very small optical phase shifts is currently of great interest, due to the realization of extremely sensitive interferometric devices, such as gravitational wave antennas[1], laser gyroscopes[2], or interferometric electro-optic sensors for measuring small voltage or charge densities in semiconductors[3,4]. These devices can already operate at the so-called "shot-noise limit" (SNL); i.e., the noise in the measurement of a phase shift Φ is due to the quantum fluctuations of the light which are injected in the interferometer. This measurement noise for an optimally designed interferometer is:

$$(\Delta\Phi)_{SNL} = \frac{1}{\sqrt{\eta N}} \tag{1}$$

where N is the number of photons from the input beam, entering the interferometer during the chosen integration time, and η a detection efficiency factor measuring the ratio between the number of detected photoelectrons and the number of input photons. However, the shot noise limit is a practical rather than fundamental limit, and it can be overcome using squeezed light, as discussed by many theoretical papers[5-11] and as demonstrated in recent experiments[12,13]. Let us emphasize that overcoming the SNL is interesting only to the extent that both the laser power and the integration time are limited by some practical considerations, yielding a value of $(\Delta\Phi)_{SNL}$ larger than the expected magnitude of the phenomena under study. For instance, both gravitational wave antennas[1] and electro-optic measurements of weak electric fields in biological systems (see § III.3) require integration times no longer than 1 ms, while the laser powers are limited respectively to tens of watts or tens of milliwatts. Another important consideration for practical purposes is the frequency range of interest. For instance, both above examples are expected to yield signals in the kHz range; in this range, the fluctuations of the input light are no more quantum mechanical in origin, but are due to technical sources of noise (acoustical noise, 1/f

* Permanent Address: Institut d'Optique, BP 43, F 91496 Orsay, Cédex, France

noise of electronics, etc...). It is nevertheless possible to achieve quantum-limited sensitivity in that range, is some modulation techniques are used in order to transfer the phase information to a higher frequency range, where the technical noise is negligible. As shown by theoretical analysis[10,11], squeezed light enhancement can be used together with these modulation techniques if one uses broadband or multi-frequency squeezed light.

In this paper we describe two sub-shot-noise interferometry experiments, involving a polarization interferometer and a $KTiOPO_4$ squeezer[13]. The signal-to-noise improvement is first obtained at 400 kHz, which is above the technical noise range, then at 10 kHz, using an amplitude modulation technique. Let us emphasize that this last result is new with respect to Refs. 12 and 13, and has important implications for practical applications of squeezed light.

II - EXPERIMENT

II.1 - Polarization interferometer

Figure 1 shows the polarization interferometer used in our experiments. The first polarizer (P1) is a polarization dependent beam splitter. The coherent laser light and the squeezed light have orthogonal polarizations, so that this beam splitter directs these two beams along the input direction of the interferometer. A half-wave plate rotates these polarizations by 45° and the second polarizer (P2) acts as the beam splitter for a balanced homodyne detector at the output of the interferometer. The coherent laser light serves as the local oscillator (LO) for this homodyne detector. When there is no polarization rotation between the polarizers, the differenced photocurrent i_D from the two balanced photodetectors fluctuates at the SNL. Any polarization rotation in the interferometer (due, for instance, to Faraday effect in the rotator PR in Fig. 1) will be detected as a change in the photocurrent i_D. This polarization interferometer is actually equivalent to a Mach-Zehnder interferometer[12], where the spatially separated "arms" have been changed to copropagating left- and right-handed circular polarizations between P1 and P2 in Fig. 1. When no squeezed light enters the interferometer, the measurement noise for a polarization rotation Φ (at frequencies above technical noise) is the SNL for phase measurement as given by Eq. (1). This measurement noise can also be expressed in terms of the power P_0 of the coherent beam in units of photon/sec and the bandwidth B of the detector electronics, as:

$$(\Delta\Phi)_{SNL} = \sqrt{\frac{2B}{\eta\, P_0}} \tag{2}$$

where η is the quantum efficiency of the detectors. Typical values for these quantities in our experiment are B = 30 kHz, $P_0 = 5 \times 10^{15} sec^{-1}$ corresponding to an optical power of 1 mW and $\eta = 0.9$. For these values

$$(\Delta\Phi)_{SNL} \simeq 3.5 \times 10^{-6} \text{ rad} \tag{3}$$

Now consider the effect of squeezed light entering the dark port of the interferometer through the polarizing beam splitter. One can easily show[14] that the measurement noise becomes:

$$(\Delta\Phi)_{squeezed} = \sqrt{\frac{2B}{\eta\, P_0}}\sqrt{R} \tag{4}$$

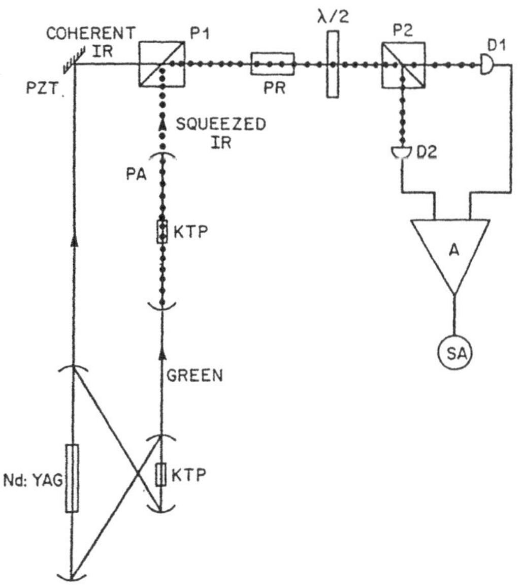

Figure 1. Experimental setup. Coherent IR light from the CW Nd:YAG laser (left) and squeezed light from the parametric amplifier PA (right) are injected into the two input ports of the polarization interferometer (top) formed by polarizing beam splitters P1 and P2. A rotation of the polarization axes due to a polarization rotator PR is measured by an imbalance in detector currents D1 and D2 on a spectrum analyszer SA.

where R is the degree of squeezing achieved for the input light. R is defined as the minimum ratio of noise powers obtained for a squeezed field entering the dark port of the input beam splitter relative to the vacuum field entering this port. The effective R value is increased by any losses in the optical system. These loss limitations in the present experiment are not due to the interferometer itself, but to imperfect detectors efficiencies and to crystal losses in the optical cavity generating the squeezed light.

II.2 - Generation of squeezed light

Squeezed light is generated using optical parametric amplification[15].We use as a primary source a CW ring Nd:YAG laser, operating at 1.064 µm in single transverse and longitudinal modes. This laser is not frequency stabilized; its free-running frequency jitter is small enough to allow direct locking to the optical cavity in which the squeezed light is generated. A frequency doubling $KTiOPO_4$ (KTP) crystal in the laser cavity provides light at 0.532 µm, used to pump the parametric amplifier. The squeezing cavity is locked to this green pump light using an FM modulation technique[16]. The squeezing cavity is resonant for both the green and infrared light, with finesses for the empty cavity which are respectively 160 and 70. In this cavity we use two KTP crystals. This pairing of the nonlinear elements allows for compensation of the walk-off which occurs for non-90° phase-matching[17] in a single nonlinear KTP element. The observed threshold for parametric oscillation in this cavity range between 150 mW and 200 mW, depending on the quality of the KTP crystals and on the degree of focusing (the cavity is close to a concentric configuration). The phase-matching configuration in KTP is type II [17]; i.e., the signal and idler parametric decay fields have

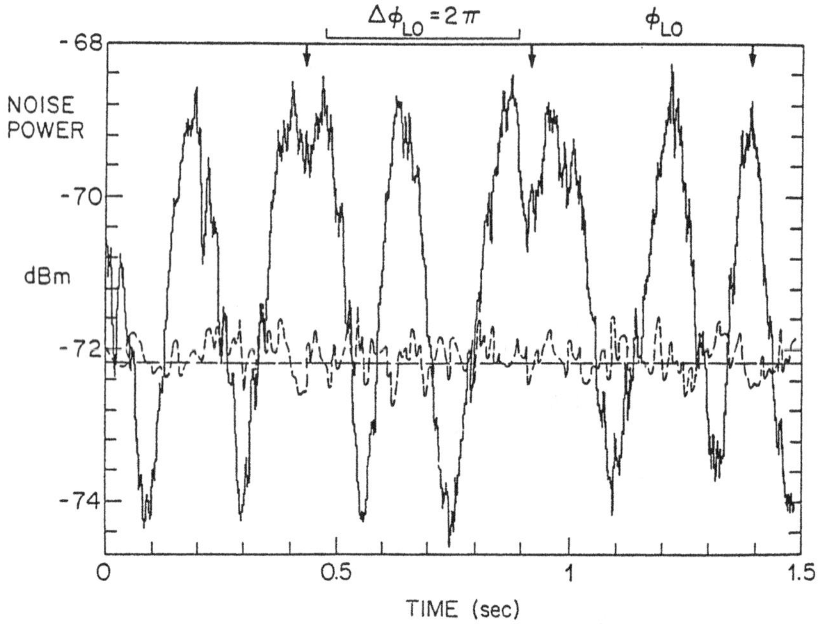

Figure 2. Noise power at the output of the balanced detector, as a function of the local oscillator phase. The dashed and solid lines correspond respectively to the squeeze cavity blocked and unblocked. The arrows on the Φ_{LO} axis indicate the reversals of the PZT motion. The center frequency is 400 kHz, the RF bandwidth is 30 kHz, and the video bandwidth is 30 Hz.

orthogonal polarizations. In this case[14], three different polarization modes must resonate together in the cavity: the ordinary and extraordinary polarizations for the infrared light, and the extraordinary polarization for the green pump light. This is achieved by adjusting both the temperature and orientation of the crystals, so that the single pass optical path shift between ordinary and extraordinary polarizations be equal to an integral number of wavelengths. When the resonant conditions are achieved, one can define two orthogonal polarizations modes, which are oriented 45° from the principal axes of the KTP crystals. These orthogonally polarized modes both have squeezed fluctuations[14,18].

II.3 - Noise reduction above the technical noise frequency range

Figure 2 shows the output of the homodyne detector when the local oscillator phase is varied with no Faraday rotation, for a center frequency of 400 kHz and a RF bandwidth 30 kHz. The maxima of phase-dependent amplification and deamplification from the vacuum noise level are respectively + 3.4 dB and - 2.0 dB. By deconvolving the photodetector amplifier noise, these values increase to + 3.9 dB and - 2.6 dB. Taking into account the 2% intracavity and 20% extracavity losses, this corresponds to a "lossless" squeezing value of ± 5 dB. The extracavity losses include heterodyne efficiency (95%) and detector quantum efficiencies (90%). In this regime the observed squeezing is actually limited by the parametric gain. The parametric gain is difficult to stabilize at higher gain values due to a thermal instability associated with green absorption in the KTP crystals.

The noise reduction in the interferometer is illustrated in Fig. 3. An oscillating current at 400 kHz is applied to the Faraday rotator,

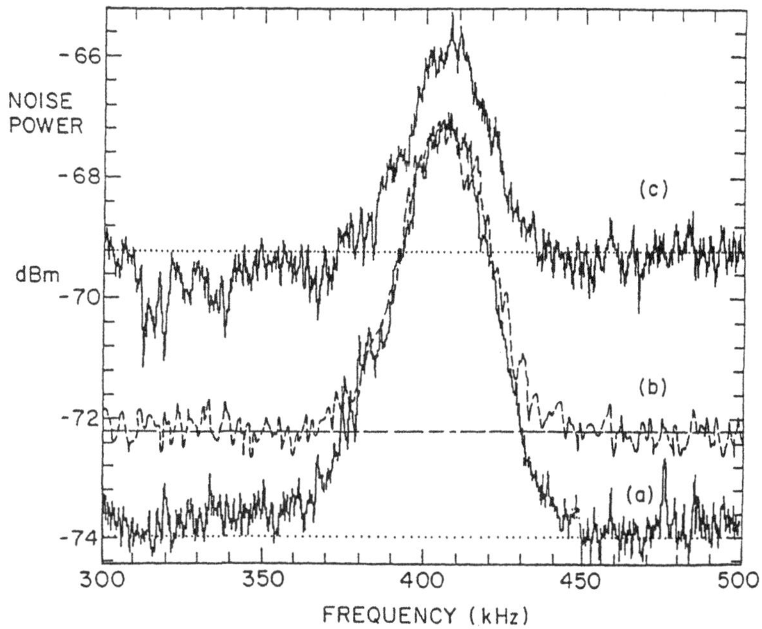

Figure 3. Polarization rotation signal-to-noise measurements. Current at a frequency of 400 kHz is applied on the Faraday rotator to produce an oscillating polarization angle. The spectrum analyzer is frequency scanned across the resulting signal (30 kHz RF bandwidth). Trace (a) is with LO phase adjusted for maximum squeezed noise reduction, trace (b) is obtained with vacuum input at the dark port, and trace (c) with LO phase adjusted for maximum noise increase. The horizontal lines indicate the noise levels with no polarization rotation signal. Fluctuations in the base line of traces (a) and (c) are due to slow fluctuations in the LO phase. The duration of the scan is 1.5 s.

with an amplitude comparable to the SNL. The frequency of the spectrum analyzer is then scanned between 300 and 500 kHz. Without squeezed light, the signal-to-noise ratio is 5.2 dB (trace (b)). With squeezed light entering the interferometer, and LO phase at its optimum value for noise reduction, the signal-to-noise ratio improves to 7 dB; that is, a 1.8 dB improvement beyond the SNL is obtained. During shorter time periods than shown in Fig. 3, the noise decreases by 2 dB below the vacuum noise level (as shown in Fig. 2). On the other hand, a 90° LO phase shift from trace (a) increases the noise by about 3 dB (trace (c)) as expected for the antisqueezed quadrature.

To be more quantitative, we have studied the signal-to-noise ratio as a function of the spectrum analyzer radio-frequency bandwidth. The results are plotted in Fig. 4, for both vacuum and squeezed light inputs, with the signal amplitude and LO power held constant. These curves clearly show that a given S/N ratio can be obtained with the integration time reduced by a factor of 1.6, if squeezed light is used. Equivalently, one could choose to keep the bandwidth constant. A given S/N ratio is then obtained with an input coherent light power 1.6 times smaller.

With respect to possible applications, an important step further is to extend the noise reduction into the low frequency domain, i.e.,

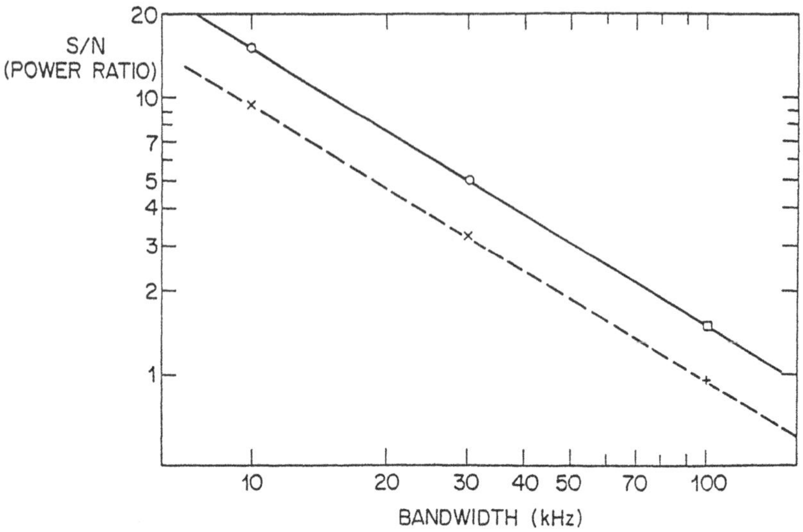

Figure 4. Signal-to-noise power ratios as a function of the RF bandwidth. The dashed line corresponds to a vacuum input and the solid line to a squeezed light input. These lines correspond to the expected 1/B dependence. The open circle and crossed data points have been measured at 400 kHz, and the open square and plus sign data points at 850 kHz. These results illustrate the improvements in response time of the interferometer due to the use of squeezed light.

within the frequency range where the amplitude (technical) noise of the laser limits the interferometer sensitivity. This can be done by using modulation techniques, in order to transfer the phase information at a frequency range where the technical noise is negligible. Both amplitude [10] and phase[11] modulations have been proposed in the literature. In the next paragraph we describe the implementation of the amplitude modulation technique.

III - SUB-SHOT-NOISE MEASUREMENTS AT LOW FREQUENCIES

III.1 - Theoretical overview of the amplitude modulation technique

This scheme has been introduced by B. Yurke, P. Grangier and R.E. Slusher [10] as a nondegenerate generalization of the first proposal by Caves[5]. As previously shown, coherent light is injected into one port of the interferometer, while squeezed light is fed into the other input port. Photodetectors are placed at each of the two output ports of the interferometer, and the readout signal is obtained by taking the difference of the photocurrents delivered by these two detectors. The phase difference between the two arms of the interferometer (i.e. the polarization rotation in our case), is adjusted to get the same currents on both detectors. That is, one works at the "side" of a fringe. The main feature here is that the coherent light has two frequency components, $\omega_0 + \omega_s/2$ and $\omega_0 - \omega_s/2$, and one looks at the beat frequency ω_s laying beyond the region dominated by technical noise. The signal frequency component at ω_s is then detected via radio-frequency mixing with a local oscillator at frequency ω_s.

A wideband analysis of the interferometer[10] shows that the squeezed light must involve the four frequencies $\omega_0 \pm \omega_s/2$ and $\omega \pm 3\ \omega_s/2$. As previously, we use a parametric amplifier for squeezed light generation. For this device, the squeezing spectrum extends well over the squeeze cavity bandwidth, and the optimum phase for noise reduction does not depend on frequency[18]. Squeezing can therefore be obtained simultaneously for both pairs of noise sidebands, provided ω_s is small enough compared to the cavity bandwidth Γ_c. In our experimental case, $\Gamma_c/2\pi$ = 7 MHz, $\omega_s/2\pi$ = 0.4 MHz, and the reduction in measured squeezing for the outer sidebands ($\pm 3\ \omega_s/2$) relative to the inner sidebands ($\pm \omega_s/2$) is less than 0.2 dB. The broadband squeezed light emitted by the parametric amplifier is therefore well adapted for implementing the amplitude modulation scheme.

III.2 - Experiment

As the previous one, this experiment uses a polarization interferometer. The signal to be measured is due to a Faraday rotator inside the interferometer (25 mm of silica in a small coil). We will first describe the two-frequency local oscillator and the noise performance without squeezed states, then the squeezed light enhancement.

In order to get a two-frequency LO, two acousto-optics modulators are used. The first one, driven by a pure 80 MHz radio frequency, shifts up the LO frequency by 80 MHz. The second one is driven by the same RF, amplitude modulated with a carrier suppression greater than 40 dB, and down shifts the LO frequency by 80 ± 0.4 MHz. One thus gets two optical modes shifted up and down from the original frequency, by the 400 kHz amplitude modulation frequency. Figure 5 (solid line) shows the detected photocurrent when one detector is blocked. The main beat note at 800 kHz is more than 30 dB above other frequency components. Figure 5 (dashed line) shows the output of the balanced detector. The common mode rejection at 800 kHz is greater than 60 dB.

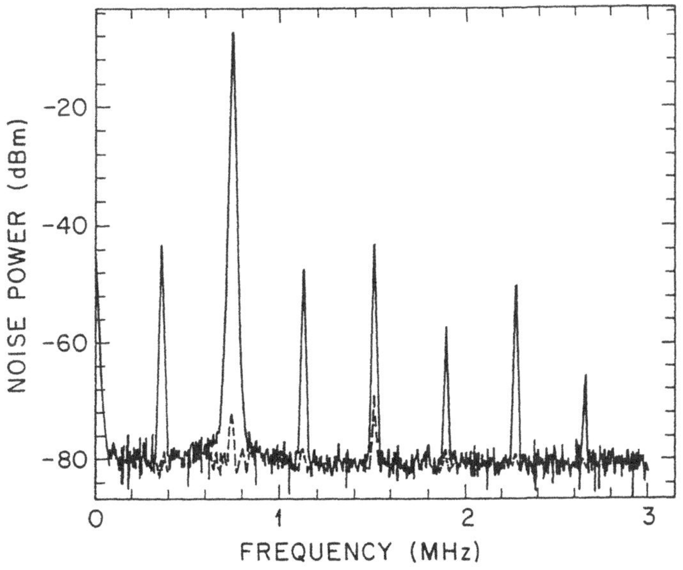

Figure 5. Signal at the output of the balanced detector. The solid line is obtained when one detector is blocked, and the dashed line shows that the 800 kHz modulation is balanced out when both detectors are used.

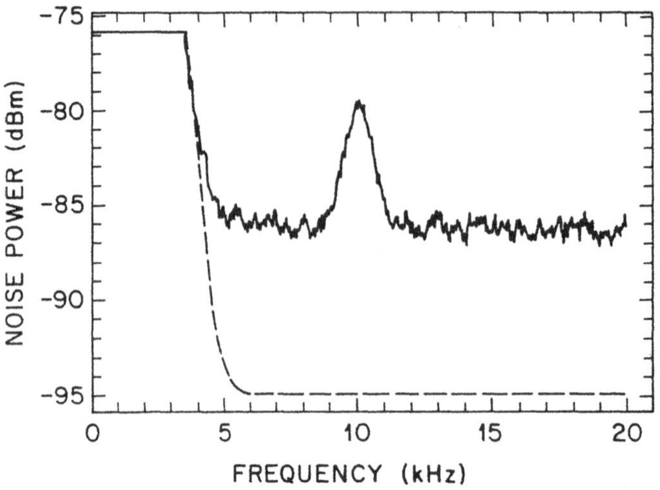

Figure 6. Low frequency noise spectrum of the polarization interferometer, using the amplitude modulation scheme. The radio-frequency bandwidth is 1 kHz, the video bandwidth is 3 Hz and the shot-noise level is - 86 dBm. The peak at 10 kHz is a calibration signal, and the dashed line indicates the amplifier noise level.

When no modulation is used, our polarization interferometer is shot-noise limited above 60 kHz. With the two-mode LO, this values is improved by more than one order of magnitude, down to about 5 kHz. Figure 6 shows a typical low frequency noise spectrum at the output of the RF mixer, with the RF phase adjusted to optimize the sensitivity. The peak at 10 kHz is due to a Faraday rotation with rms amplitude 1 μrad. The shot noise floor is 6.5 dB below that peak, corresponding to the SNL

$$(\Delta\Phi)_{SNL} \simeq 0.45\ 10^{-6}\ \text{rad}$$

for 1 kHz bandwidth, 2 mW injected power and detection quantum efficiency equal to 90%.

In order to demonstrate the noise reduction, the spectrum analyser frequency was set at 10 kHz, and the noise floor was recorded as a function of the LO and squeezed light relative phase. As shown on Fig. 7, the noise floor drops by 1.6 dB below the shot-noise level for optimum adjustement of the phase. The decrease in noise reduction from 2.0 dB at 400 kHz to 1.6 dB at 10 kHz can be almost entirely attributed to phase jitter, since the video bandwidth had to be decreased from 30 Hz to 3 Hz. As demonstrated theoretically[8], this problem would be eliminated by locking the relative phase of the LO and squeezed light for optimum noise reduction.

III.3 - Discussion

The amplitude modulation scheme has two main features. First, it has to be used with low coherent light power in the interferometer, since all the input light must be detected. Second, it is relatively insensitive to defects of the interferometer, and it can make full use of a large degree of squeezing. Interesting examples of applications are the polarization interferometers which are currently used to detect charge

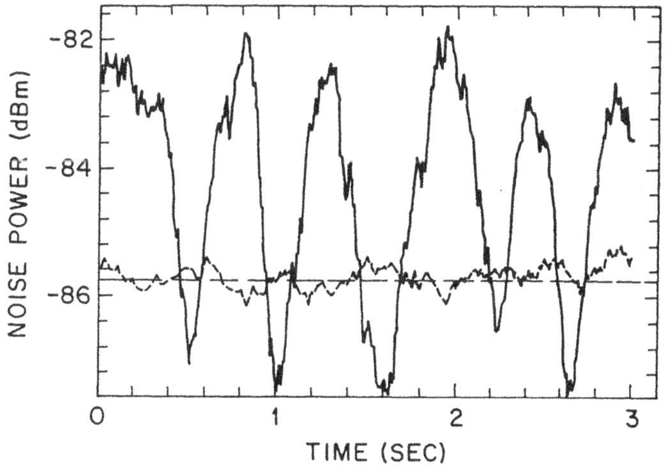

Figure 7. Noise power at the output of the balanced detector, detected through the RF mixer, as a function of the local oscillator phase. The dashed and solid lines correspond respectively to the squeeze cavity blocked and unblocked. The center frequency is 10 kHz, the RF bandwidth is 1 kHz, and the video bandwidth is 3 Hz.

or voltage in semiconductors, using induced birefringence[3]. These devices already operate at the shot noise limit[4]. Polarization interferometer can also be used in biophysics to detect the change of the birefringence of an axon induced by a nerve pulse[19]. Here the integration time must be shorter than the nerve pulse (~ 2 ms), and the light power has to be low enough to avoid destroying the axon. A simple estimation then indicates that the expected effect is very close to the shot noise limit. Use of squeezed states should allow to avoid averaging over many pulses, which has been necessary so far to get a usable signal[19].

IV - CONCLUSIONS

IV.1 - Quantum noise reduction at 400 kHz

In this experiment, we have demonstrated an interferometer that measures polarization rotation with a measurement noise 2.0 dB below the shot-noise limit. This enhanced precision is accomplished by generating squeezed light, which is injected into the dark port of the interferometer. In these experiments, KTP has been used for the first time to generate the squeezed light. This crystal is readily available and has a very high damage threshold and moderately low linear absorption, which make it attractive for generating quantum states of light. The pair correlated photons from the squeezed cavity are orthogonally polarized and thus easily separable with a polarizing beam splitter. This property is useful in correlation experiments with quantum light[20,21].

The enhanced sensitivity of this polarization interferometer should be especially useful for very rapid phase measurements where shot noise dominates because of the small number of photons in the measurement time interval. Interesting examples are the shot-noise limited polarization interferometers currently used to measure voltage or charge density variations in semiconductors[3,4] or in biological systems[19]. Polarization interferometry can also be used with pulsed squeezed light[22], allowing measurement of time variable phase shifts in the picosecond range[23].

IV.2 - Quantum noise reduction at 10 kHz

In this experiment, we have shown that broadband squeezed light emitted by an optical parametric amplifier (subthreshold OPO) allows us to implement an amplitude modulation scheme which involves squeezed noise sidebands at various frequencies. Though the sensitivity improvement is still modest, it has been extended for the first time to the low-frequency range, which is usually dominated by technical noise. Use of squeezed light for actual measurements purposes is therefore no more out of reach, and it can be envisioned that results of measurements which could not have been done without squeezed light will appear soon.

REFERENCES

1. K.S. Thorne, Rev. Mod. Phys. 52, 285 (1980). See also Physics Today 39, 17 (Feb. 1986).
2. G.A. Sanders, N.G. Prentiss and S. Ezekiel, Optics Lett., 6, 569 (1981).
3. J.A. Valdmanis, G. Mourou and C.W. Gabel, Appl. Phys. Lett. 41, 211 (1982).
4. H.K. Heinrich, D.M. Bloom and B.R. Hemenway, Appl. Phys. Let. 48, 1066 (1986).
5. C.M. Caves, Phys. Rev. D23, 1693 (1981).
6. R.S. Bondurant and H.J. Shapiro, Phys. Rev. D30, 2548 (1984).
7. A. Heidmann, S. Reynaud and C. Cohen-Tannoudji, Opt. Commun. 52, 235 (1984).
8. B. Yurke, S.L. Mc Call and J.R. Klauder, Phys. Rev. A33, 4033 (1986).
9. B. Yurke and E.A. Whittaker, Opt. Lett. 12, 236 (1987).
10. B. Yurke, P. Grangier and R.E. Slusher, JOSA B.4, 1677 (1987).
11. J. Gea-Banacloche and G. Leuchs, JOSA B.4, 1667 (1987).
12. Min Xiao, Ling-An Wu and H.J. Kimble, Phys. Rev. Lett. 59, 278 (1987).
13. P. Grangier, R.E. Slusher, B. Yurke and A. La Porta, Phys. Rev. Lett. 59, 2153 (1987).
14. P. Grangier and B. Yurke, to be published.
15. Ling-An Wu, H.J. Kimble, J.L. Hall and H. Wu, Phys. Rev. Lett. 57, 2520 (1986).
16. R.W. Drever, J.L. Hall, F.V. Kowalski, J. Hough, G.N. Ford, A.J. Nunsley and H. Ward, Appl. Phys. B31, 97 (1983).
17. J.Q. Yao and T.S. Fahlen, J. Appl. Phys. 55, 65 (1984).
18. N.J. Collett and D.F. Walls, Phys. Rev. A32, 2887 (1985).
19. L.B. Cohen, B. Hille and R.D. Keynes, J. Physial 211, 495 (1970).
20. B. Yurke and D. Stoler, Phys. Rev. A36, 1955 (1987).
21. A. Heidmann, R.J. Horowicz, S. Reynaud, E. Giacobino and C. Fabre, Phys. Rev. Lett. 59, 2555 (1987).
22. R.E. Slusher, P. Grangier, A. La Porta, B. Yurke and M.J. Potasek, Phys. Rev. Lett. 59, 2566(1987).
23. See R.E. Slusher Contribution in this volume.

REALIZATION OF MEASUREMENT AND THE STANDARD QUANTUM LIMIT

Masanao Ozawa

Department of Mathematics
College of General Education
Nagoya University
Nagoya 464, Japan

1 Introduction

What measurement is there? It is a difficult question but the importance of this question has increased much in connection with the effort to detect gravitational radiation. For monitoring the position of a free mass such as the gravitational-wave interferometer [1], it is usually supposed that the sensitivity is limited by the so called standard quantum limit (SQL) [2,3]. In the recent controversy [4]–[8], started with Yuen's proposal [4] of a measurement which beats the SQL, the meaning of the SQL has been much clarified. In order to settle this controversy, rigorous treatment of the question on what measurement there is seems the key point.

Usually, the first approach to such an ontological question is very mathematical. Fortunately, in the last two decades, mathematical theory for describing wide possibilities of quantum measurement was developed in the field of mathematical physics [9]–[18]. Indeed, the question on what measurement there is was given a solution by the present author [15] under a physically reasonable mathematical formulation. Unfortunately, this result and the mathematics for this result have not been familiar with physicists who need that result.

The purpose of this paper is two folds. The first purpose is to present the results from the mathematical theory of quantum measurement in a form accessible for physicists. As a consequence of this theory, I shall settle the controversy of the SQL in two ways; by a general consideration giving new vistas concerning such a nonstandard measurement and by giving a model of measuring interaction which breaks the SQL [19]. The second purpose is, of course, to present this result.

2 Standard Quantum Limit for Free-Mass Position

The uncertainty principle is a physicist's wisdom which gives correct answers to many quantum mechanical problems without so much cumbersome analysis of the problem. An application of this wisdom to analysis of the performance of interferometric gravitational-wave detector leads to the limit for sensitivity of monitoring the free-mass position, which has long been a topic of controversy within the quantum optics and general relativity community. This limit — referred to as the ***standard quantum limit*** (SQL) for monitoring the position of a free mass — is usually stated as follows: In the repeated measurement of its position x of a free mass m with the time τ between two measurements, the result of the second measurement cannot be predicted with uncertainty smaller than $(\hbar\tau/m)^{1/2}$.

2.1 Yuen's proposal of breaching SQL

In the standard argument [2,3] deriving this limit, the uncertainty principle

$$\Delta x(0)\Delta p(0) \geq \hbar/2 \tag{1}$$

is applied to the position uncertainty $\Delta x(0)$ and the momentum uncertainty $\Delta p(0)$ at the time $t = 0$ just after the first measurement so that by the time τ of the second measurement the squared uncertainty (variance) of x increases to

$$\Delta x(\tau)^2 = \Delta x(0)^2 + \Delta p(0)^2\tau^2/m^2 \geq 2\Delta x(0)\Delta p(0)\tau/m \geq \hbar\tau/m. \tag{2}$$

The SQL is thus obtained as

$$\Delta x(\tau) \geq (\hbar\tau/m)^{1/2}, \tag{3}$$

and it is usually explained as a straightforward consequence of the uncertainty principle (1).

Yuen [4] pointed out a serious flaw in the standard argument. Since the evolution of a free mass is given by

$$\hat{x}(t) = \hat{x}(0) + \hat{p}(0)t/m, \tag{4}$$

the variance of x at time τ is given by

$$\Delta x(\tau)^2 = \Delta x(0)^2 + \Delta p(0)^2\tau^2/m^2 + \langle\Delta\hat{x}(0)\Delta\hat{p}(0) + \Delta\hat{p}(0)\Delta\hat{x}(0)\rangle\tau/m \tag{5}$$

where $\Delta\hat{x} = \hat{x} - \langle\hat{x}\rangle$ and $\Delta x^2 = \langle\Delta\hat{x}^2\rangle$, etc. Thus the standard argument implicitly assumes that the last term — we shall call it the *correlation term* — in Eq. (5) is non-negative. Yuen's assertion [4] is that some measurements of x leave the free mass in a state with the negative correlation term.

In probability theory, it is an elementary fact that the variance of the sum of two random variables is the sum of their variances plus the correlation term which is twice their covariances. The covariance may be negative and it vanishes if these random variables are independent. In quantum mechanics, if the state at $t = 0$ is a minimum-uncertainty one then the correlation term vanishes. However, there are states with negative correlation terms. The ratio of the covariance to the product of the standard deviations is called the *correlation coefficient* in probability theory. The negative correlation expresses the tendency that the larger than the mean one variable, the less than the mean the other. From this, one can expect that in such a state the momentum works as attracting the free mass around the mean position and that the free evolution narrows the wave packet of the mass. Gaussian states [26] with this property are analyzed as follows [4].

Let a be the following operator in the Hilbert space of the mass states:

$$\hat{a} = (m\omega/2\hbar)^{1/2}\hat{x} + 1/(2\hbar m\omega)^{1/2}i\hat{p}, \quad [\hat{a}, \hat{a}^\dagger] = 1, \tag{6}$$

where ω is an arbitrary parameter with unit $\sec^{-1}$. The *twisted coherent state* (TCS) $|\mu\nu\alpha\omega\rangle$ is the eigenstate of $\mu\hat{a} + \nu\hat{a}^\dagger$ with eigenvalue $\mu\alpha + \nu\alpha^*$:

$$(\mu\hat{a} + \nu\hat{a}^\dagger)|\mu\nu\alpha\omega\rangle = (\mu\alpha + \nu\alpha^*)|\mu\nu\alpha\omega\rangle, \quad |\mu|^2 - |\nu|^2 = 1. \tag{7}$$

The free-mass Hamiltonian is expressed by

$$\hat{H} = \hat{p}^2/2m = (\hbar\omega/2)(\hat{a}^\dagger\hat{a} + \frac{1}{2} - \frac{1}{2}\hat{a}^2 - \frac{1}{2}\hat{a}^{\dagger 2}). \tag{8}$$

Within the choice of a constant phase the wave function $\langle x|\mu\nu\alpha\omega\rangle$, where $\hat{x}|x\rangle = x|x\rangle$, is given by

$$\langle x|\mu\nu\alpha\omega\rangle = \left(\frac{m\omega}{\pi\hbar|\mu-\nu|^2}\right)^{1/4} \exp\left(-\frac{m\omega}{2\hbar}\frac{1+2\xi i}{|\mu-\nu|^2}(x-x_0)^2 + ip_0(x-x_0)\right) \tag{9}$$

where

$$\xi = \mathrm{Im}(\mu^*\nu), \quad \alpha = (m\omega/2\hbar)^{1/2}x_0 + 1/(2\hbar m\omega)^{1/2}ip_0, \quad x_0, p_0 \quad \text{real}. \tag{10}$$

When $\xi = 0$, the wave function (9) is the usual minimum-uncertainty state. In the context of oscillators, the squeezed states are the wave functions of the form (9) with $\mu \neq 0$. The first two moments of $|\mu\nu\alpha\omega\rangle$ are

$$\langle x\rangle = x_0, \quad \langle p\rangle = p_0, \tag{11}$$

$$\Delta x^2 = \hbar|\mu-\nu|^2/2m\omega, \quad \Delta p^2 = \hbar m\omega|\mu+\nu|^2/2, \tag{12}$$

$$\langle\Delta\hat{x}\Delta\hat{p} + \Delta\hat{p}\Delta\hat{x}\rangle = -2\hbar\xi, \quad \Delta x\Delta p = \hbar(1+\xi^2)^{1/2}/4, \tag{13}$$

$$\langle H\rangle = (\langle p\rangle^2 + \Delta p^2)/2m. \tag{14}$$

When $\xi > 0$ the x-dependent phase in (9) leads to a narrowing of the $\Delta x(t)$ from $\Delta x(0)$ during free evolution. Because of this behavior, Yuen called mass states (9) with $\xi > 0$ as *contractive states.*

The position fluctuation for a free-mass starting in an arbitrary TCS (9) is immediately obtained from Eqs. (5) and (12)–(13):

$$\Delta x(t)^2 = (\hbar/2m\omega)(|\mu-\nu|^2 - 4\xi\omega t + |\mu+\nu|^2(\omega t)^2) \tag{15}$$

$$= (1/4\xi)(\hbar\tau/m) + (2\hbar/m\omega)(|\mu+\nu|\omega/2)^2(t-\tau)^2, \tag{16}$$

where

$$\tau = 2\xi/(\omega|\mu+\nu|^2) = \xi\hbar m/\Delta p(0)^2. \tag{17}$$

For $\xi > 0$, the position fluctuation decreases during time $t = 0$ to $t = \tau$. The minimum fluctuation achieves $1/(2\xi^{1/2})$ times the SQL at the time $\tau = 2\xi/(\omega|\mu+\nu|^2)$. Thus the minimum fluctuation $\Delta x(\tau)$ is related only to the momentum uncertainty as follows:

$$\Delta x(\tau) = \hbar/2\Delta p(0) = \Delta x(0)/(1+4\xi^2)^{1/2}. \tag{18}$$

This shows that $\Delta x(\tau)$ can be arbitrarily small for arbitrarily large τ with sufficiently large ξ. It should be noted that the minimum uncertainty product is realized between the momentum uncertainty at $t = 0$ and the position uncertainty at $t = \tau$.

Thus the SQL formulated by Eq. (3) can be breached if there is a measurement of free-mass position which leaves the free-mass in a contractive state just after the measurement.

The reader may have the following question: Is the state after the measurement always an eigenstate of the position observable? If so, it never has any such contractive character. It is natural to say so in the textbook description of measurement. However, any real measuring apparatus cannot leave the free mass in any eigenstate. This statement has been repeatedly emphasized in the study of measurement of continuous observables (observables with continuous spectrum). In the study of measurement, there is a deep gap between discrete observables and continuous observables. Once von Neumann criticized Dirac's formulation in this point [20, pp.222–223]: "The division into quantized and unquantized quantities corresponds ... to the division into quantities R with an operator $\hat{R}$ that has a pure discrete spectrum, and into such quantities for which this is not the case. And it was for the former, and only for these, that we found a possibility of an absolutely precise measurement — while the latter could be observed only with arbitrarily good (but never absolute) precision. (In addition, it should be observed that the introduction of an eigenfunction which is 'improper,' i.e., which does not belong to Hilbert space ... gives a less good approach to reality than our treatment here. For such a method pretends the existence of such states in which quantities with continuous spectra take on certain values exactly, although this never occurs. Although such idealizations have often been advanced, we believe that it is necessary to discard them on these grounds, in addition to their mathematical untenability.)"

2.2 Caves's defense of SQL

After Yuen's proposal [4] of measurement with a contractive state, Caves published a further analysis [8] of the SQL, where he gave the following improved formulation of the SQL:

> Let a free mass m undergo unitary evolution during the time τ between two measurements of its position x, made with identical measuring apparatuses; the result of the second measurement cannot be predicted with uncertainty smaller than $(\hbar\tau/m)^{1/2}$, in average over the possible results of the first measurement.

Caves [8] showed that the SQL holds for a specific model of position measurement due to von Neumann [20, pp.442–445] and he also gave the following heuristic argument for the validity of the SQL. His point is the notion of the imperfect resolution σ of one's measuring apparatus. His argument runs as follows. *The first assumption* is that the variance $\Delta(\tau)^2$ of the measurement of x is the sum of σ^2 and the variance of x at the time of the measurement; i.e.,

$$\Delta(\tau)^2 = \sigma^2 + \Delta x(\tau)^2, \tag{19}$$

and this is the case when the measuring apparatus is coupled linearly to x. *The second assumption* is that just after the first measurement, the free mass has position uncertainty not greater than the resolution:

$$\Delta x(0) \leq \sigma. \tag{20}$$

Under these conditions, he derived the following estimate for the uncertainty $\Delta(\tau)$ of the second measurement at time τ:

$$\begin{aligned} \Delta(\tau)^2 &= \sigma^2 + \Delta x(\tau)^2 \geq \Delta x(0)^2 + \Delta x(\tau)^2 \geq 2\Delta x(0)\Delta x(\tau) \\ &\geq \hbar\tau/m. \end{aligned} \tag{21}$$

According to this argument, the SQL is a consequence from the uncertainty relation

$$\Delta x(0)\Delta x(\tau) \geq |\langle[\hat{x}(0), \hat{x}(\tau)]\rangle|/2 = \hbar\tau/2m, \tag{22}$$

under assumptions (19)–(20).

However, his definition of the resolution of a measurement is ambiguous and so a critical examination for his argument is necessary. In fact, he used three different definitions in his paper: 1) If the free mass is in a position eigenstate at the time of a measurement of x, then σ is defined to be the uncertainty in the result. 2) The measurement determines the position immediately after the measurement to be within roughly a distance σ of the measured value. 3) The square σ^2 of the resolution is the variance of the pointer-position just before the system-meter interaction. These three notions are essentially different, although they are the same for von Neumann's model.

The notion of resolution of measurement has been often mentioned in literature but up to now we have not yet reached any satisfactory definition for it. What does the readout value tell one the states of the free mass? There are two principal thoughts about this question. One thinks that the readout tells the position of the free mass just before the measurement, since the prior position is the cause of the effect of the measurement that is the readout value. Another one thinks that the readout tells the position of the free mass just after the measurement, since the measurement changes the position of the free mass so that the two effects of the measurement — the posterior position and the readout — may be in concordance. Which is true is hardly answered. The best way to attack the problem seems to recognize that there are two types of concept of resolution of measurement. From this reason, we make the following distinction: If the free mass is in a position eigenstate at the time of a measurement of x then the *precision* ε of the measurement is defined to be the uncertainty in the result and the *resolution* σ is defined to be the deviation of the position of the free mass just after the measurement from the readout just obtained. The mathematical definitions of these concepts including the case of superposition will be given and discussed thoroughly in the later sections.

We shall return to the problem of the validity of the SQL. By the improvement of the formulation of the SQL, the idea of measurement leaving the free mass in a contractive state can not readily vitiate the SQL. However, there is a possibility for circumventing the heuristic proof given by Caves [8], since his assumption $\sigma \geq \Delta x(0)$ is formulated in an ambiguous manner. Nonetheless, it is a heavy burden for one who wants to vitiate the SQL to make a realizable model of measurement which circumvents Caves's assumption. In the next section, we shall give general considerations of realization of measurement.

Before going further, I shall mention certain attempts of breaking the SQL. Recently, Ni [21] proposed a scheme of repeated position measurements for which one can predict the result of the next measurement with arbitrarily small errors. However, a close examination of this scheme leads to the conclusion that the improved formulation of the SQL due to Caves [8] is not broken by this repeated measurement scheme. This scheme uses a combination of the Arthurs-Kelley measurements [22] which measure the position and momentum simultaneously and approximately. Accordingly, one measurement of this scheme has four meters, two of which measure the free-mass position with a high resolution by one meter and with a low resolution by the other meter, and the other two of which measure the free-mass momentum with lower resolutions. The prediction of the result of the next position measurement is done using these four readouts. If one of these meter is left alone, the prediction cannot have the desired accuracy. This means that one meter reading with the highest position resolution serves the position measurement and the other three meter readings serve the preparation procedure for the next measurement. There are several similar models proposed with preparation procedures for the next measurement and these examples do not vitiate the improved SQL, since it disallows explicitly any tinkering between two identical position measurements. Indeed, in these proposed models (e.g., [21,23]), there is at least one auxiliary meter with a lower resolution, the reading of which prepares the state for the next position measurement really done by the other high resolution meter-reading. Thus any many-meter systems are excluded out of the scope of the improved SQL. The true problem is thus whether we can make a measurement with only one meter, the reading of which gives the precise position of the mass and simultaneously prepares the state for the next identical position measurement, for instance, in a contractive state.

3 Quantum Mechanics of Measurement

What can one say about quantum measurement from quantum mechanics? Von Neumann is acknowledged to be the first scientist who made a route to this problem. Although his original motivation was to show the consistency of the repeatability hypothesis (usually referred to as the projection postulate) with the axioms of quantum mechanics, his argument opened the way to analyze the quantum measurement within quantum mechanics. However, his method caused the controversy about the difficult problem of interpretation of quantum mechanics. In what follows, I shall attempt to present von Neumann's method with some elaborations which avoid difficulties of the problem of interpretation and discuss several consequences from the quantum mechanics of measurement.

3.1 Statistics of measurement

By the axioms of quantum mechanics we shall mean the axioms of nonrelativistic quantum mechanics without any superselection rules, which are reduced to (a) the definitions of observables and states as self-adjoint operators and state vectors of a Hilbert space, (b) the Schrödinger time-dependent equation and (c) the Born statistical formula for probability distributions of commuting observables. The projection postulate is excluded from our axioms and its status will be discussed below.

In order to discuss all possible quantum measurements, it is convenient to classify them into two classes. A measurement is of the *first kind* if it does not destroy the quantum mechanical description of the system to be measured so that we can determine in principle the state just after the measurement corresponding to the result of measurement. A measurement is of the *second kind* if it destroys the quantum mechanical description of the system. The whole process of a direct interaction with a macroscopic detector such as a photon counter is considered as a measurement of the second kind.

The statistics of a measurement of the first kind is specified by the following two elements: For the system state ψ just before the measurement, let $P(a|\psi)$ the probability density of obtaining the result a and let ψ_a be the system state just after the measurement with result a. Then the physical design and the indicated preparation of the apparatus determine $P(a|\psi)$ and the transition $\psi \to \psi_a$ for all possible ψ. These two elements will be called the *statistics* of a given measurement of the first kind; $P(a|\psi)$ will be called the *measurement probability* and $\psi \to \psi_a$ will be called the *state reduction*, further we shall call the state ψ just before the measurement as the *prior state* and the state ψ_a just after the measurement as the *posterior state*. (For notational convenience, ψ_a will be denoted sometimes by $\psi[a]$.) This specification of measurement statistics implies that if two given measurements are identical then the corresponding two statistics are identical. On the other hand, the statistics of a measurement of the second kind is specified only by its measurement probability, for such a measurement does not allow to describe the system state after the measurement.

3.2 Scheme of measurement

Once we accept the axioms of quantum mechanics, it is natural to accept in principle the following fact as a basis of our discussion.

Postulate 1 *For any observable A with its spectral decomposition*

$$\hat{A} = \int x \, d\hat{A}(x), \tag{23}$$

there is a measurement which may be of the second kind such that its measurement probability $P(a|\psi)$ satisfies the Born statistical formula

$$P(a|\psi)da = \langle\psi|d\hat{A}(a)|\psi\rangle, \tag{24}$$

for all prior state ψ.

We shall call any measuring apparatus satisfying Eq.(24) as a *detector* for an observable A. It should be noticed that Postulate 1 alone never implies existence of any measurements of the first kind. Our fundamental problem is thus *what measurement of the first kind is allowed within our axioms and postulates of quantum mechanics.* In order to solve this problem, we adopt the following scheme of measurement instituted by von Neumann.

Suppose that a quantum system S — called the *object system* — with the unknown system state ψ just before the measurement is to be measured by a measuring apparatus. For the observer — called the *first observer* — who applies quantum mechanics only to the object system, this measurement is described by the statistics of this measurement specified by the given measuring apparatus. Suppose that this measurement is of the first kind and we shall denote the statistics of this measurement by $P_I(a|\psi)$ and $\psi \to \psi_a$. Then by nondestructive nature of measurement of the first kind, there are other possibilities of application of quantum mechanics to this physical phenomena of the measurement. One possibility of such a quantum mechanical description of measurement arises if one separates the measuring apparatus into two parts. The first part — called the *probe system* — is a microscopic system which directly interacts with the object system and the second part is a detector which makes a second kind measurement of an observable — usually called the pointer position in somewhat misunderstanding manner — of the probe system.

For the observer — called the *second observer* — who applies quantum mechanics to the composite system of the object and the probe, this measurement is described as a combination of an object-probe interaction and a second kind measurement of the probe in the following manner. Let P be the probe system. By the arrangement of the measuring apparatus the following elements can be specified as controllable elements; the system state φ of the probe system just before the measurement, the time evolution $\hat{U}$ of the composite system $S+P$ during the interaction and the observable A of the probe system to be measured by the detector. Then just before the interaction the state of the composite system is $\psi \otimes \varphi$ and hence just after the interaction the composite system is in the state $\hat{U}(\psi \otimes \varphi)$. What can the second observer tell about the statistics of this measurement? The measurement probability $P_{II}(a|\psi)$ for the second observer is obviously the postulated probability distribution of the observable A in the state $\hat{U}(\psi \otimes \varphi)$:

$$P_{II}(a|\psi)\, da = \langle \hat{U}(\psi \otimes \varphi)|1 \otimes d\hat{A}(a)|\hat{U}(\psi \otimes \varphi)\rangle. \tag{25}$$

Thus from the consistency of the measurement probabilities of these two observers, we have

$$P_{I}(a|\psi)\, da = \langle \hat{U}(\psi \otimes \varphi)|1 \otimes d\hat{A}(a)|\hat{U}(\psi \otimes \varphi)\rangle. \tag{26}$$

In order to determine the state reduction, we may assume that immediately after the interaction the object system would be subjected to a detector of an arbitrary observable X of the object system. Since the probe system is also to be subjected to the detection of A immediately after the interaction in the second-observer description, quantum mechanics predicts the joint probability density $P_{II}(x, a|\psi)$ of obtaining the result $A = a$ and $X = x$ for the second observer by the relation

$$P_{II}(x, a|\psi)\, da\, dx = \langle \hat{U}(\psi \otimes \varphi)|d\hat{X}(x) \otimes d\hat{A}(a)|\hat{U}(\psi \otimes \varphi)\rangle. \tag{27}$$

For the first observer, the detection of the observable X occurs in the system state ψ_a given the result a of the first measurement and hence the probability density $P_I(x, a|\psi)$ of the same event is calculated by the first observer as follows:

$$P_I(x, a|\psi)\, da\, dx = \langle \psi_a|d\hat{X}(x)|\psi_a\rangle P_I(a|\psi)\, da. \tag{28}$$

From the consistency of the statistics of these two observers, we have

$$\langle \psi_a|d\hat{X}(x)|\psi_a\rangle = \frac{\langle \hat{U}(\psi \otimes \varphi)|d\hat{X}(x) \otimes d\hat{A}(a)|\hat{U}(\psi \otimes \varphi)\rangle}{\langle \hat{U}(\psi \otimes \varphi)|1 \otimes d\hat{A}(a)|\hat{U}(\psi \otimes \varphi)\rangle}. \tag{29}$$

The arbitrariness of X yields the following relation for any basis $\{|i\rangle\}$,

$$\langle i|\psi_a\rangle\langle \psi_a|j\rangle = \frac{\langle \hat{U}(\psi \otimes \varphi)|\left(|j\rangle\langle i| \otimes d\hat{A}(a)\right)|\hat{U}(\psi \otimes \varphi)\rangle}{\langle \hat{U}(\psi \otimes \varphi)|1 \otimes d\hat{A}(a)|\hat{U}(\psi \otimes \varphi)\rangle}, \tag{30}$$

and consequently,

$$|\psi_a\rangle\langle \psi_a| = \sum_{i,j} |i\rangle \frac{\langle \hat{U}(\psi \otimes \varphi)|\left(|j\rangle\langle i| \otimes d\hat{A}(a)\right)|\hat{U}(\psi \otimes \varphi)\rangle}{\langle \hat{U}(\psi \otimes \varphi)|1 \otimes dA(a)|\hat{U}(\psi \otimes \varphi)\rangle}\langle j|. \tag{31}$$

Thus we have shown that the second observer can calculate the statistics of this measurement from his knowledge about the measuring apparatus and predict what will happen to the first observer. For example, if the first observer observes that the system state ψ is reduced to the state ψ_a just after the measurement with the readout value a, the second observer can calculate just the same state reduction $\psi \to \psi_a$ by Eq.(31). The difference between those two observers is that the first observer knows only the relation between the readout a and the reduced state ψ_a but the second observer does know the statistical correlation between the object system and the probe system from which he can predict all statistics of this measurement. Some authors have made a misunderstanding at this point. They usually say that the state reduction is a consequence of the observation or of the knowing of the readout. If this would be the case, we would get an obvious contradiction between the first and the second observer; in fact, then the first observer would say that the state reduction occurs after the detection of the probe system contrary to the second observer's saying that it occurs already before the detection of the probe system just after the object-probe interaction. The point is that there is no causality relation between the readout a and the reduced state ψ_a but there is only a statistical correlation. Usually, statistical correlation between two events does not imply any causality relation. In fact, for another observer who observes the result of succeeding measurement of the object system first, the state reduction of the prove system occurs. We can only say that there is a statistical correlation between the results of successive measurements. Quantum mechanics shows that to get information from one system is to make a statistical correlation with another system by an interaction to be described by quantum mechanics. However, there is another problem — indeed, a different problem. When does the statistical interference between the object system and the probe system vanish? This is the content of Schrödinger's cat paradox. This is a deep problem. However, in our formalism, we can avoid the difficulty — as promised before — by boldly saying "During the second kind measurement of the probe system."

3.3 Operation measures and measurement statistics

From the preceding analysis of the process of quantum measurement, the problem as to what measurement there is can be reduced to the problem what statistics can be realized by the measurement scheme. An important step to the mathematical solution of the latter problem is to get a neat mathematical representation of the statistics of measurement. In what follows, we shall show that any plausible description of measurement statistics can be expressed by a single mathematical object called an operation measure. A thorough discussion of operations and effects may be found in [13,14], and of operation measures and effect measures in [9,10,15,16].

Let $\mathcal{H}$ be a Hilbert space of an object system. Suppose that a statistics of a measurement is given; this means that for any state vector ψ in $\mathcal{H}$ the measurement probability $P(a|\psi)$ and the state reduction $\psi \to \psi_a$ is presupposed. Our first task is to extend this statistical description to the case that the prior state is a mixture. Let $\hat{\rho}$ be a density operator which represents the prior state with its spectral decomposition

$$\hat{\rho} = \sum_i \lambda_i |\psi^i\rangle\langle\psi^i|. \tag{32}$$

In this case, the measurement probability — denoted by $P(a|\hat{\rho})$ — is given by

$$P(a|\hat{\rho}) = \sum_i \lambda_i P(a|\psi^i). \tag{33}$$

Then the posterior state $\hat{\rho}_a$ of the system for the readout a is obviously a mixture of all $|\psi_a^i\rangle\langle\psi_a^i|$'s , where ψ_a^i is the posterior state for the prior state ψ^i, such that their relative frequency is proportional to λ_i and $P(a|\psi^i)$. Thus we have

$$\hat{\rho}_a = \frac{\sum_i \lambda_i P(a|\psi^i)|\psi_a^i\rangle\langle\psi_a^i|}{\sum_i \lambda_i P(a|\psi^i)}. \tag{34}$$

Let Δ be an interval — or more generally a Borel set — in the real line and let S_Δ the subensemble of the object system selected by the condition that the readout a of this measurement lies in Δ. Then we can ask what is the state of S_Δ— denoted by $\hat{\rho}_\Delta$ —just after the measurement. It is well known that the state of this ensemble is not the superposition of all ψ_a's with a in Δ but a mixture of all $|\psi_a\rangle\langle\psi_a|$'s with a in Δ; their relative frequency is proportional to $P(a|\psi)$. Thus the state $\hat{\rho}_\Delta$ is

$$\hat{\rho}_\Delta = \frac{\int_\Delta da\, P(a|\psi)|\psi_a\rangle\langle\psi_a|}{\int_\Delta da\, P(a|\psi)}. \tag{35}$$

Further, for the case that the prior state is given by a mixture represented by a density operator (32), the state $\hat{\rho}_\Delta$ of the ensemble S_Δ just after the measurement is

$$\hat{\rho}_\Delta = \frac{\int_\Delta da\, P(a|\hat{\rho})\hat{\rho}_a}{\int_\Delta da\, P(a|\hat{\rho})}, \tag{36}$$

where $\hat{\rho}_a$ is given in Eq.(34).

These considerations lead to the following mathematical definition of the transformation $\mathcal{I}(\Delta)$ which maps a density operator $\hat{\rho}$ to a trace class operator $\mathcal{I}(\Delta)\hat{\rho}$ given by the relations

$$\mathcal{I}(\Delta)\hat{\rho} = \int_\Delta d\mathcal{I}(a)\hat{\rho}, \quad d\mathcal{I}(a)\hat{\rho} = \hat{\rho}_a P(a|\hat{\rho})\, da. \tag{37}$$

An obvious requirement for this transformation $\mathcal{I}(\Delta)$ is as follows: If $\hat{\rho}$ is the mixture of $\hat{\sigma}_1$ and $\hat{\sigma}_2$ then $\mathcal{I}(\Delta)\hat{\rho}$ is the mixture of $\mathcal{I}(\Delta)\hat{\sigma}_1$ and $\mathcal{I}(\Delta)\hat{\sigma}_2$. From this $\mathcal{I}(\Delta)$ can be extended to the mapping of all trace class operators. Then the following properties can be easily deduced from Eq.(37):

1. For any Borel set Δ, the mapping $\hat{\rho} \rightarrow \mathcal{I}(\Delta)\hat{\rho}$ is a linear transformation on the space of trace class operators which maps density operators to a positive trace class operator.

2. For any Borel set Δ and any density operator $\hat{\rho}$,

$$0 \le \mathrm{Tr}[\mathcal{I}(\Delta)\hat{\rho})] \le 1 \quad \text{and} \quad \mathrm{Tr}[\mathcal{I}(\mathbf{R})\hat{\rho}] = 1. \tag{38}$$

3. For any countable disjoint sequence $\{\Delta_i\}$ of Borel sets and density operator $\hat{\rho}$,

$$\sum_i \mathrm{Tr}[\mathcal{I}(\Delta_i)\hat{\rho}] = \mathrm{Tr}[\mathcal{I}(\bigcup_i \Delta_i)\hat{\rho}]. \tag{39}$$

A mapping $\hat{\rho} \rightarrow \mathcal{I}(\Delta)\hat{\rho}$ with properties 1–2 is called an *operation* and a mapping $\Delta \rightarrow \mathcal{I}(\Delta)$ with all properties 1–3 is called an *operation measure.* Thus we have shown that if we are given a plausible statistical description of measurement we can construct an operation measure. The mathematical importance of the operation measures is that it unifies two statistical data of measurement — the measurement probability and the state reduction — into a single mathematical object. In fact, the measurement probability can be recovered by the relation

$$P(a|\hat{\rho})da = \mathrm{Tr}[d\mathcal{I}(a)\hat{\rho}], \tag{40}$$

and the state reduction can be retained by the relation

$$\hat{\rho}_a = \frac{d\mathcal{I}(a)\hat{\rho}}{\mathrm{Tr}[d\mathcal{I}(a)\,\hat{\rho}]}. \tag{41}$$

The mathematical justification of the above differentiation is given in [16] for arbitrary operation measure — so that, without any presupposition of measurement statistics, any operation measure gives a measurement statistics. To put it simply, integration of any measurement statistics gives an operation measure and differentiation of any operation measure gives a measurement statistics.

Let F be a (random) variable the values of which shows the readout of a measurement. Then the probability distribution $P(F \in \Delta|\hat{\rho})$ that the value of F is in Δ given the prior state $\hat{\rho}$ is obtained from Eq.(40), i.e.,

$$P(F \in \Delta|\hat{\rho}) = \int_{\Delta} P(a|\hat{\rho})da. \tag{42}$$

In this case there is a unique positive operator valued measure $\Delta \rightarrow \hat{F}(\Delta)$ such that

$$0 \le \hat{F}(\Delta) \le 1, \quad \text{and} \quad \hat{F}(\mathbf{R}) = 1, \tag{43}$$

$$\mathrm{Tr}[\hat{F}(\Delta)\hat{\rho}] = P(F \in \Delta|\hat{\rho}). \tag{44}$$

We shall call this $\hat{F}$ as the *effect measure* of an operation measure $\mathcal{I}$. If an operation measure $\mathcal{I}$ represents an exact measurement of an observable $\hat{X} = \int x\, d\hat{X}(x)$, then the corresponding effect measure $\hat{F}$ satisfies

$$\begin{aligned} P(a|\hat{\rho})da &= \mathrm{Tr}[d\hat{X}(a)\hat{\rho}], \\ \mathrm{Tr}[\hat{F}(\Delta)\hat{\rho}] &= \int_{\Delta} \mathrm{Tr}[d\hat{X}(x)\hat{\rho}], \end{aligned} \tag{45}$$

and hence the effect measure of $\mathcal{I}$ coincides with the spectral measure of operator $\hat{X}$, i.e., $d\hat{F}(x) = d\hat{X}(x)$. In this sense, the notion of an effect measure generalizes the conventional presupposition that the measurement probability is represented by a spectral measure to the case of non-exact measurements.

3.4 Characterization of realizable measurement

In the preceding subsections, we have shown the following two facts:

1. Any measurement scheme consisting of an object-probe interaction and a probe detection determines the unique measurement statistics by means of the second-observer description.

2. Any plausible measurement statistics of the first-observer description gives an operation measure which unifies the measurement statistics in a single mathematical object.

However, it is not at all clear whether every operation measure is consistent with the second-observer description of the measurement — or what operation measures are consistent with the second-observer description.

This problem has the following rigorous mathematical formulation: Let $\mathcal{H}$ be a Hilbert space describing the object system and let $\mathcal{I}$ be an operation measure for the space $\mathcal{T}(\mathcal{H})$ of the trace class operators on $\mathcal{H}$. The problem is to determine when we can find another Hilbert space $\mathcal{K}$ describing the probe system with a self-adjoint operator $\hat{A}$ in $\mathcal{K}$ describing the observable actually detected, a state vector φ in $\mathcal{K}$ describing the preparation of the probe system and a unitary operator $\hat{U}$ on $\mathcal{H} \otimes \mathcal{K}$ describing time evolution of the object-probe composite system during the measurement interaction such that this second-observer description of measurement leads to the same measurement statistics as the first-observer description of the given operation measure $\mathcal{I}$. For the last part of this formulation, recall that the second-observer description of the measurement leads to the statistics given by Eqs.(25) and (31). On the other hand, the operation measure is given by Eq.(37). Thus the condition for these two to give the same statistics is the following relation:

$$\langle i|d\mathcal{I}(a)(|\psi\rangle\langle\psi|)|j\rangle = \langle\hat{U}(\psi\otimes\varphi)|\left(|j\rangle\langle i|\otimes d\hat{A}(a)\right)|\hat{U}(\psi\otimes\varphi)\rangle \tag{46}$$

for all ψ and a basis $\{|i\rangle\}$ in $\mathcal{H}$. We shall call any operation measure satisfying Eq.(46) for some $\hat{A}$, ϕ and $\hat{U}$ as *realizable.*

In order to present the solution of the problem which has been obtained in [15], we need one more mathematical concept concerning the positivity property of the operation. Let $\mathcal{I}$ be an operation measure. Then the transformation $\hat{\rho} \to \mathcal{I}(\Delta)\hat{\rho}$ on the trace class operators is positive, in the sense that for any density operator $\hat{\rho}$ the trace class operator $\mathcal{I}(\Delta)\hat{\rho}$ is a positive operator. It follows that for any vector φ, ψ we have

$$\langle\varphi|d\mathcal{I}(a)(|\psi\rangle\langle\psi|)|\varphi\rangle \geq 0. \tag{47}$$

An operation measure is called *completely positive* if it has the following stronger positivity property; for any finite sequences of vectors $\xi_1, \xi_2, \ldots, \xi_n$ and $\eta_1, \eta_2, \ldots, \eta_n$,

$$\sum_{i,j=1}^{n}\langle\xi_i|d\mathcal{I}(a)(|\eta_i\rangle\langle\eta_j|)|\xi_j\rangle \geq 0. \tag{48}$$

If an operation measure is realizable then by Eq.(46) we obtain

$$\begin{aligned}
&\sum_{i,j=1}^{n}\langle\xi_i|d\mathcal{I}(a)(|\eta_i\rangle\langle\eta_j|)|\xi_j\rangle \\
&\quad= \sum_{i,j=1}^{n}\langle\hat{U}(\eta_j\otimes\varphi)|\left(|\xi_j\rangle\langle\xi_i|\otimes d\hat{A}(a)\right)|\hat{U}(\eta_i\otimes\varphi)\rangle \\
&\quad= \|\sum_{i=1}^{n}\left(|\xi_i\rangle\langle\xi_i|\otimes d\hat{A}(a)\right)\hat{U}(\eta_i\otimes\varphi)\|^2 \\
&\quad\geq 0.
\end{aligned}$$

Thus every realizable operation measure is completely positive. The converse statement of this has been proved in [15] by mathematical construction of the Hilbert space $\mathcal{K}$ with unit vector φ, self-adjoint operator $\hat{A}$ on $\mathcal{K}$ and unitary operator $\hat{U}$ on $\mathcal{H}\otimes\mathcal{K}$ which satisfy Eq.(46) for a given completely positive operation measure $\mathcal{I}$ and thus we have

Theorem 1 *Every completely positive operation measure is realizable.*

For a particular type of measurement statistics, this problem is the one originally considered by von Neumann. In order to clarify the relation between the conventional approach and our general approach, we shall review his well-known result. Let $X =$

$\sum_i x_i|x_i\rangle\langle x_i|$ be an observable with a simple discrete spectrum $\{\ldots, x_{-1}, x_0, x_1, \ldots\}$ and unit eigenvectors $|x_i\rangle$. The statistics of the precise measurement of X is usually presupposed as follows:

$$\text{measurement probability:} \quad P(x_i|\psi) = |\langle x_i|\psi\rangle|^2, \tag{49}$$

$$\text{state reduction:} \quad \psi \to \psi[x_i] = |x_i\rangle. \tag{50}$$

For real numbers a outside the spectrum, we have $P(a|\psi) = 0$ and we can put $\psi[a]$ as arbitrary. This leads to the operation measure $\mathcal{I}$ such that

$$\mathcal{I}(\Delta)\hat{\rho} = \sum_{x_i \in \Delta} |x_i\rangle\langle x_i|\hat{\rho}|x_i\rangle\langle x_i|. \tag{51}$$

Von Neumann showed that this statistics is consistent with the second-observer description: One can construct the probe system from an arbitrary Hilbert space $\mathcal{K}$ with basis $\{\ldots, \varphi_{-1}, \varphi_0, \varphi_1, \ldots\}$. Let φ_0 be the probe system preparation and $\hat{U}$ the unitary operator on $\mathcal{H} \otimes \mathcal{K}$ such that

$$\hat{U}(|x_j\rangle \otimes |\varphi_i\rangle) = |x_j\rangle \otimes |\varphi_{i+j}\rangle,$$

describing the measurement interaction. Let $A = \sum_i x_i|\varphi_i\rangle\langle\varphi_i|$ be the probe observable. Suppose that the prior state of the object is $\psi = \sum_k c_k|x_k\rangle$. We have the following time evolution starting with $\psi \otimes \varphi_0$

$$\hat{U}|\psi \otimes \varphi_0\rangle = \sum_k c_k|x_k\rangle \otimes |\varphi_k\rangle. \tag{52}$$

Then with this setting, the joint probability distribution $P(X = x_j, A = x_i)$ obtaining the result $A = x_i$ and $X = x_j$ in the joint detection just after the object-probe interaction is

$$\begin{aligned} P(X = x_j, A = x_i) &= |\langle x_j| \otimes \langle\varphi_i|\hat{U}(\psi \otimes \varphi_0)\rangle|^2 \\ &= \sum_k |c_k|^2 |\langle x_j|x_k\rangle|^2 |\langle\varphi_i|\varphi_k\rangle|^2 \\ &= |\langle x_j|\psi\rangle|^2 \delta_{i,j}. \end{aligned} \tag{53}$$

Since, for the second-observer, the measurement probability $P_{II}(x_j|\psi)$ of this measurement is the probability $P(A = x_j)$ obtaining the result $A = x_j$, we have from Eq.(53),

$$\begin{aligned} P_{II}(x_j|\psi) &= P(A = x_j) = \sum_i P(X = x_i, A = x_j) \\ &= |\langle x_j|\psi\rangle|^2. \end{aligned} \tag{54}$$

Thus Eq.(49) holds for the second-observer.

Let $P(X = x_j|A = x_i)$ be the conditional probability of obtaining the result $X = x_j$ given $A = x_i$. Then by Eq.(53) we have

$$P(X = x_j|A = x_i) = \frac{P(X = x_j, A = x_i)}{P(A = x_j)} \tag{55}$$

$$= \delta_{i,j}. \tag{56}$$

This enables the second observer to make the following statistical inference: If the first observer were to make the detection of the observable X immediately after the fist measurement, then the results of these two measurement always coincides. This is possible only if, for the first observer, the first measurement changes the state as $\psi \rightarrow |x_j\rangle$ depending on the result $X = x_j$ of the first measurement. Thus we obtain Eq.(50) from the second-observer description. Obviously these reasoning is a particular case of general consideration presented in subsection 3.2 and we can thus obtain the operation measure (51) directly from Eqs.(31) and (52) by the following computation.

$$
\begin{aligned}
\mathcal{I}(\Delta)(|\psi\rangle\langle\psi|) &= \int_{\Delta} d\mathcal{I}(da)(|\psi\rangle\langle\psi|) \\
&= \sum_{x_k\in\Delta} |\psi[x_k]\rangle\langle\psi[x_k]|P(x_k|\psi) \\
&= \sum_{x_k\in\Delta}\sum_{i,j} |x_i\rangle\langle\hat{U}(\psi\otimes\varphi_0)|\,(|x_j\rangle\langle x_i|\otimes|\varphi_k\rangle\langle\varphi_k|)\,|\hat{U}(\psi\otimes\varphi_0)\rangle\langle x_j| \\
&= \sum_{x_k\in\Delta}\sum_{i,j} |x_i\rangle\langle\hat{U}(\psi\otimes\varphi_0)|x_i\rangle|\varphi_k\rangle\langle x_j|\langle\varphi_k|\hat{U}(\psi\otimes\varphi_0)\rangle\langle x_j| \\
&= \sum_{x_k\in\Delta}\sum_{i,j} |x_i\rangle\left(\sum_l c_l^*\langle x_l|x_i\rangle\langle\varphi_l|\varphi_k\rangle\right)\left(\sum_l c_l\langle x_j|x_l\rangle\langle\varphi_k|\varphi_l\rangle\right)\langle x_j| \\
&= \sum_{x_k\in\Delta}\left(\sum_{i,l} c_l^*|x_i\rangle\langle x_l|x_i\rangle\langle\varphi_l|\varphi_k\rangle\right)\left(\sum_{l,j} c_l\langle x_j|x_l\rangle\langle\varphi_k|\varphi_l\rangle\langle x_j|\right) \\
&= \sum_{x_k\in\Delta} |c_k|^2|x_k\rangle\langle x_k| \\
&= \sum_{x_k\in\Delta} |x_k\rangle\langle x_k|\psi\rangle\langle\psi|x_k\rangle\langle x_k|.
\end{aligned}
$$

Now I shall give some remarks about the conventional approach to the determination of the state reduction from the second-observer description. In the conventional argument, they apply the so-called projection postulate to the state $\hat{U}(\psi\otimes\varphi)$ (see Eq.(52)) just after the interaction and conclude that if the second observer get the result $A = x_i$ then the state of the object-probe composite system changes into $|x_i\rangle|\varphi_i\rangle$ and the state of the object changes into $|x_i\rangle$. Although this argument has an apparent advantage that it never uses the explicit statistical inference, it has the several definite weak points. First, this argument applies only when the measurement of the probe system is of the first kind and when it satisfies the projection postulate. However, any first kind measurement is subjected to the Schrödinger equation for the object-probe composite system and hence we can never realize the dynamics causing the projection postulate. Thus they need to assume a process of the second kind measurement at some point between the object and the real observer and try to describe dynamics causing the projection postulate in this process. However, this means the contradiction that they can use the projection postulate nowhere since after the second kind measurement the system state cannot be described by the standard quantum mechanics and hence the projection postulate cannot apply to it. Second, their state reduction occurs only after the detection of the probe system contrary to the fact that the measurement interaction finishes before the detection of the probe system. This implies that their argument puts an apparent limitation for the time interval of the successive measurements. When one can perform the second measurement of the same system at the earliest time? We can say that after the object-probe interaction but they must say that after the macroscopic interaction between the probe and the detector. Thus, from the first point they cannot describe any successive measurements of the one system. Last, their argument cannot be used for the measurement of observables with continuous spectrum such as the position observable, since we have no canonical state reduction postulate for continuous observables. In this case, only statistical inference can apply.

Historically speaking, von Neumann did not explicitly use the projection postulate in the second-observer description. Indeed he only mentioned the probability correlation between the object system and the probe system and wrote "If III [the second observer] were to measure (by process 1. [by the subsequent detection of observables]) the simultaneously measurable quantities A [the object observable], B [the probe observable] (in I [the object system] or II [the probe system] respectively, or both in I + II), then the pair of values a_n , b_n would have the probability 0 for $m \neq n$, and the probability w_n [= the measurement probability for a_n] for $m = n$. ... If this is established, then the measuring process so far as it occurs in II, is 'explained' theoretically,..." [20, p.440]. (The notes in the brackets above are due to the present author.)

4 Measurement Breaking SQL

In the preceding sections, we have discussed what is the problem of the SQL and what we can tell about quantum measurement from quantum mechanics. One conclusion is that every measurement statistics described by a completely positive operation measure is realizable. In this last section, I shall show that a measurement of the free-mass position which breaks the SQL is realizable.

4.1 Statistics of approximate measurement

Let $\mathcal{I}$ be a completely positive operation measure. Then $\mathcal{I}$ describes a statistics of a measurement. Now a problem arise — when can one consider $\mathcal{I}$ as a statistics of a measurement of a given observable? In what follows, we shall consider the problem as to when a given operation measure can be considered as a position measurement of a mass with one degree of freedom. Our analysis will lead to mathematical definitions of precision and resolution of measurement. This subsection will be concluded with the mathematical formulation of the SQL.

In the textbook description of position measurement, the statistics is so characterized as

$$\text{measurement probability:} \quad P(a|\psi) = |\psi(x)|^2, \tag{57}$$

$$\text{state reduction:} \quad \psi \to \psi_a = |a\rangle. \tag{58}$$

However, it is proved that there are no operation measures which satisfies these conditions both. In general, it is proved [15] that there are no weakly repeatable operation measures for non-discrete observables, where operation measure $\mathcal{I}$ is called *weakly repeatable* if

$$\mathrm{Tr}[\mathcal{I}(B \cap C)\hat{\rho}] = \mathrm{Tr}[\mathcal{I}(B)\mathcal{I}(C)\hat{\rho}], \tag{59}$$

for all density operators $\hat{\rho}$ and Borel sets B, C — Eqs.(57) and (58) lead to this condition. Thus every realizable position measurement is an approximate measurement. In order to clarify the meaning of *approximate* measurement, we shall introduce two error criteria for the preciseness of measurement.

In Subsection 2.2, we have introduced the following distinction: If the free mass is in a position eigenstate at the time of a measurement of x then the *precision* ε of the measurement is defined to be the uncertainty in the result and the ***resolution*** σ is defined to be the deviation of the position of the free mass just after the measurement from the readout just obtained. Now, we shall give precise definitions for the case of superposition.

A difficult step in defining the precision is to extract the noise factor from the readout distribution — this may be the reason why the above distinction has hardly ever discussed in literature. Let $\mathcal{I}$ be an operation measure and $\hat{F}$ be its effect measure. Consider the requirements for $\mathcal{I}$ to describe some approximate measurement of the position observable. Our first requirement is that $\hat{F}$ is *compatible* with the position observable, i.e.,

$$[\hat{x}, \hat{F}(\Delta)] = 0, \quad \text{for all Borel sets } \Delta. \tag{60}$$

This condition may be justified by the compatibility of the information obtained from this measurement with the original information of the position. Under this condition, there is a kernel function $G(a, x)$ such that

$$d\hat{F}(a) = da \int dx\, G(a, x)|x\rangle\langle x|. \tag{61}$$

Even if the measuring apparatus measures the position observable approximately, the readout distribution $P(a|\psi)$ is expected to be related to the position distribution $|\psi(x)|^2$ — from Eq.(61), this relation is expressed in the following form

$$P(a|\psi) = \int dx\, G(a, x)|\psi(x)|^2. \tag{62}$$

Note that $G(a, x)$ is independent of a particular wave function $\psi(x)$ and thus it expresses the noise in the readout. Obviously, $P(a|\psi) = |\psi(a)|^2$ for all ψ (i.e. the noiseless case) if and only if $G(a, x) = \delta(a - x)$. From Eq.(9) of [8], in the case of von Neumann's model, $G(a, x) = |\Psi(a - x)|^2$ where Ψ is the prepared state of the probe (See also [25], where $\Psi(a - x)$ is called the resolution amplitude.) Roughly speaking, $G(a, x)$ is the (normalized) conditional probability density of the readout a, given that the free-mass is in the position x at the time of measurement; hence the precision $\varepsilon(x)$ of this case should be

$$\varepsilon(x)^2 = \int da\, (a - x)^2 G(a, x). \tag{63}$$

Thus if for the prior state ψ of the mass, the *precision* $\varepsilon(\psi)$ of the measurement is given by

$$\varepsilon(\psi)^2 = \int dx\, \varepsilon(x)^2 |\psi(x)|^2. \tag{64}$$

By the similar reasoning, for the mass state ψ at the time of measurement, the *resolution* $\sigma(\psi)$ of the measurement is given by

$$\sigma(\psi)^2 = \int da\, \sigma(a)^2 P(a|\psi), \tag{65}$$

where

$$\sigma(a)^2 = \int dx\, (a - x)^2 |\psi_a(x)|^2. \tag{66}$$

The second requirement is that the noise is ***unbiased*** in the sense that the mean value of the readout is identical with the mean position in the prior state, i.e.,

$$\int dx\, x P(x|\psi) = \langle\psi|\hat{x}|\psi\rangle, \tag{67}$$

for all possible ψ.

Let $\Delta(\psi)$ be the uncertainty of the readout for the prior state ψ of the mass and $\Delta x(\varphi)$ the uncertainty of the mass position at any state φ. Then in general we can prove the following

Theorem 2 *Let $\mathcal{I}$ be an operation measure for one-dimensional system. Under conditions of compatibility and unbiasedness expressed by Eqs.(60) and (67), the following relations hold:*

$$\varepsilon(\psi)^2 = \Delta(\psi)^2 - \Delta x(\psi)^2, \tag{68}$$

$$\sigma(\psi)^2 = \int da\, P(a|\psi)\Delta x(\psi_a)^2 + \int da\, P(a|\psi)(a - \langle\psi_a|\hat{x}|\psi_a\rangle)^2. \tag{69}$$

Let ε_{max} be the maximum of $\varepsilon(\psi)$ ranging over all ψ. Then we have $\varepsilon_{max} = 0$ if and only if $P(x|\psi) = |\psi(x)|^2$ for all ψ.

The above theorem shows that the conditions of compatibility and unbiasedness are plausible conditions for characterizing approximate position measurements and for those measurements further specification concerning approximation can be done through the precision ε and the resolution σ. Thus in the following, we shall say that an operation measure $\mathcal{I}$ is of *approximate position measurement* if it satisfies the compatibility condition (60) and the unbiasedness condition (67).

Now we shall give a mathematical formulation of the SQL. Let $\mathcal{I}$ be an operation measure for a one-dimensional system. Suppose that a free mass m undergo unitary evolution during the time τ between two identical measurement described by the operation measure $\mathcal{I}$. Let $\hat{U}_\tau$ be the unitary operator of the time evolution. Suppose that the free mass is in a state ψ just before the first measurement. Then just after the measurement (at $t = 0$) the free mass is in the posterior state ψ_a with probability density $P(a|\psi)$. From this readout value a, the observer make a prediction $h(a)$ for the readout of the second measurement at $t = \tau$. Then the squared uncertainty of this prediction is

$$\Delta(\tau, \psi, a)^2 = \int dx\, (x - h(a))^2 P(x|\hat{U}_\tau\psi_a). \tag{70}$$

In literature, the following mean-value-prediction strategy is adopted for determination of $h(a)$:

$$h(a) = \langle\psi_a|\hat{x}(\tau)|\psi_a\rangle, \tag{71}$$

where

$$\hat{x}(\tau) = \hat{U}_\tau^\dagger x(0)\hat{U}_\tau = \hat{x} + \hat{p}\tau/m. \tag{72}$$

The predictive uncertainty $\Delta(\tau, \psi)$ of this repeated measurement with prior state ψ and time duration τ is defined as the squared average of $\Delta(\tau, \psi, a)$ over all readouts of the first measurement,

$$\Delta(\tau, \psi)^2 = \int da\, \Delta(\tau, \psi, a)^2 P(a|\psi). \tag{73}$$

The SQL asserts the relation

$$\Delta(\tau, \psi)^2 \geq \hbar\tau/m, \quad \text{for all prior state } \psi. \tag{74}$$

If the operation measure $\mathcal{I}$ is of approximate position measurement, we have from Eq.(68)

$$\begin{aligned}\Delta(\tau,\psi,a)^2 &= \Delta(\hat{U}_\tau\psi_a)^2,\\ &= \varepsilon(\hat{U}_\tau\psi_a)^2+\Delta x(\tau)(\psi_a)^2\end{aligned}\tag{75}$$

and hence

$$\begin{aligned}\Delta(\tau,\psi)^2 &= \int da\,P(a|\psi)\left(\varepsilon(\hat{U}_\tau\psi_a)^2+\Delta x(\tau)(\psi_a)^2\right),\\ &= [\varepsilon(\hat{U}_\tau\psi_a)^2]+[\Delta x(\tau)(\psi_a)^2],\end{aligned}\tag{76}$$

where the brackets means the average due to $P(a|\psi)$.

Now from refinement of Caves's argument in 2.2 (see Eq.(21)), we have the following sufficient condition for the SQL.

Theorem 3 *Let $\mathcal{I}$ be an operation measure of approximate position measurement with precision ε. If for any prior state ψ the relation*

$$[\Delta x(\psi_a)^2]\le[\varepsilon(\hat{U}_\tau\psi_a)^2]\tag{77}$$

holds, then the SQL holds for this measurement, i.e,

$$\Delta(\tau,\psi)^2\ge\hbar\tau/m,\quad \textit{for all prior state }\psi.\tag{78}$$

In fact, we have the following estimate using the uncertainty relation (22),

$$\begin{aligned}\Delta(\tau,\psi)^2 &= [\varepsilon(\hat{U}_\tau\psi_a)^2]+[\Delta x(\tau)(\psi_a)^2]\\ &\ge [\Delta x(0)(\psi_a)^2]+[\Delta x(\tau)(\psi_a)^2]\\ &\ge [2\Delta x(0)(\psi_a)\Delta x(\tau)(\psi_a)]\\ &\ge \hbar\tau/m.\end{aligned}$$

The above theorem shows that the concept of precision is relevant for derivation of the SQL among other candidates for error of measurement.

If the following simple relation

$$\sigma(\psi)^2\le[\varepsilon(\hat{U}_\tau\psi_a)^2].\tag{79}$$

between the precision and the resolution holds then assumption (77) of Theorem 3 follows from Eq.(69); and hence condition (79) implies the SQL. Thus it can be said that one's intuition which leads to the SQL is supported by the following statements; (1) every approximate measurement satisfies the compatibility condition and the unbiasedness conditions for the noise and for the state reduction, and (2) the resolution is no greater than the precision. However, it is not clear at all that every realizable measurement satisfies the last statement even if the first statement is admitted.

4.2 Von Neumann's model of approximate measurement

In [8], Caves showed that von Neumann's model of approximate position measurement satisfies the assumption of Theorem 3 and so the SQL holds for this model. Since the object-probe coupling of von Neumann's model is a simple linear coupling and it illustrates some proposed models of quantum nondemolition measurement, we shall review this result in our framework below; see [20,8] for original treatment.

This model of measurement is presented by the second-observer description. The probe system is a one dimensional system with coordinate Q and momentum P as well as the object system (the free mass) with coordinate x and momentum p. The object-probe coupling is turned on from $t = -\tilde{\tau}$ to $t = 0$ ($0 < \tilde{\tau} \ll \tau$), it is described by an interaction Hamiltonian $K\hat{x}\hat{P}$ where K is a coupling constant and it is assumed to be so strong that the free Hamiltonians of the mass and the probe can be neglected; so we choose units such that $K\tilde{\tau} = 1$. Then if $\Psi_0(x, Q) = f(x, Q)$, the solution of the Schrödinger equation is

$$\Psi_t(x, Q) = f(x, Q - Ktx). \tag{80}$$

At $t = -\tilde{\tau}$, just before the coupling is turned on, the unknown free-mass wave function is $\psi(x)$, and the probe is prepared in a state with wave function $\Phi(Q)$; for simplicity we assume that $\langle\Phi|\hat{Q}|\Phi\rangle = \langle\Phi|\hat{P}|\Phi\rangle = 0$. The total wave function is $\Psi_0(x, Q) = \psi(x)\Phi(Q)$. At the end of the interaction ($t = 0$) the total wave function becomes

$$\Psi(x, Q) = \hat{U}\Psi_0(x, Q) = \psi(x)\Phi(Q - x). \tag{81}$$

In order to obtain the measurement statistics, recall that the result of this measurement — the inferred value of x — is the value $\overline{Q}$ called the "readout" obtained by the detection of Q of the probe system turned on at $t = 0$ with the subsequent stage called the "detector" in the overall macroscopic measuring apparatus. Thus the measurement probability obtaining the result $\overline{Q}$ is given by the Born statistical formula

$$P(\overline{Q}) = \int dx\, |\Psi(x, \overline{Q})|^2 = \int dx\, |\psi(x)|^2 |\Phi(\overline{Q} - x)|^2. \tag{82}$$

The posterior wave function $\psi(x|\overline{Q})$ of the free mass (at $t = 0$) is obtained (up to normalization) by evaluating $\psi(x, Q)$ at $Q = \overline{Q}$:

$$\begin{aligned} \psi(x|\overline{Q}) &= \frac{\Psi(x, \overline{Q})}{P(\overline{Q})^{1/2}} \\ &= \frac{\psi(x)\Phi(\overline{Q} - x)}{P(\overline{Q})^{1/2}} \end{aligned} \tag{83}$$

Note that the posterior wave function (83) can be obtained directly from application of Eq.(31) up to normalization.

To write down the operation measure $\mathcal{I}$ of this measurement, notice that if the prior state of the free mass is a mixture $\hat{\rho}$ then the posterior state $\hat{\rho}_{\overline{Q}}$ satisfies the relations

$$d\mathcal{I}(\overline{Q})\hat{\rho} = \hat{\rho}_{\overline{Q}} P(\overline{Q}|\hat{\rho})\, d\overline{Q} = \Psi(\overline{Q}1 - \hat{x})\hat{\rho}\Psi(\overline{Q}1 - \hat{x})^\dagger\, d\overline{Q}. \tag{84}$$

Thus we have the operation measure $\mathcal{I}$ of this measurement:

$$\mathcal{I}(\Delta)\hat{\rho} = \int_\Delta d\mathcal{I}(\overline{Q})\hat{\rho} = \int_\Delta d\overline{Q}\, \Psi(\overline{Q}1 - \hat{x})\hat{\rho}\Psi(\overline{Q}1 - \hat{x})^\dagger. \tag{85}$$

From this the effect measure of this measurement is such that

$$\begin{aligned} \hat{F}(\Delta) &= \int_\Delta d\hat{F}(\overline{Q}), \\ d\hat{F}(\overline{Q}) &= d\overline{Q}\, |\Psi(\overline{Q}1 - \hat{x})|^2 = d\overline{Q} \int dx\, |\Psi(\overline{Q} - x)|^2 |x\rangle\langle x|. \end{aligned} \tag{86}$$

From Eq.(86), it is obvious that this measurement satisfies the compatibility condition (60) and the kernel function $G(\overline{Q}, x)$ representing the noise in the result can be written as

$$G(\overline{Q}, x) = |\Psi(\overline{Q} - x)|^2. \tag{87}$$

The following computations shows that the unbiasedness condition (67) holds:

$$\begin{aligned} \int d\overline{Q}\,\overline{Q}P(\overline{Q}|\psi) &= \int d\overline{Q}\,dx\,\overline{Q}|\psi(x)|^2|\Phi(\overline{Q} - x)|^2 \\ &= \int dx\,x|\psi(x)|^2 \\ &= \langle\psi|\hat{x}|\psi\rangle. \end{aligned} \tag{88}$$

Thus this measurement satisfies the conditions for approximate position measurement. The precision $\varepsilon(\psi)$ of this measurement is given by

$$\begin{aligned} \varepsilon(\psi)^2 &= \int dx\,|\psi(x)|^2 \int d\overline{Q}\,(\overline{Q} - x)^2|\Psi(\overline{Q} - x)|^2 \\ &= \int d\overline{Q}\,a^2|\Psi(\overline{Q})|^2 \\ &= \langle\Psi|\hat{Q}^2|\Psi\rangle. \end{aligned} \tag{89}$$

From definition and Eqs. (82)–(83), the resolution $\sigma(\psi)$ of this measurement is given by

$$\begin{aligned} \sigma(\psi)^2 &= \int d\overline{Q}\,P(\overline{Q}|\psi) \int dx\,(\overline{Q} - x)^2|\psi(x|\overline{Q})|^2 \\ &= \int d\overline{Q}dx(\overline{Q} - x)^2|\Psi(\overline{Q} - x)|^2|\psi(x)|^2. \\ &= \langle\Psi|\hat{Q}^2|\Psi\rangle. \end{aligned} \tag{90}$$

Let ΔQ be the uncertainty of the probe observable just before the measurement, i.e., $(\Delta Q)^2 = \langle\Phi|\hat{Q}^2|\Phi\rangle - \langle\Phi|\hat{Q}|\Phi\rangle^2$. Then in von Neumann's model we have just obtained

$$\varepsilon = \varepsilon(\psi) = \Delta Q \quad \text{for all } \psi, \tag{91}$$

$$\sigma = \sigma(\psi) = \Delta Q \quad \text{for all } \psi. \tag{92}$$

Thus the assumptions of Theorem 3 hold from the following computation (cf. Eq.(69))

$$[\varepsilon(\hat{U}_\tau\psi_{\overline{Q}})^2] = \Delta Q = \sigma(\psi) = [\Delta x(\psi_{\overline{Q}})^2] + [(\overline{Q} - \langle\psi_{\overline{Q}}|\hat{x}|\psi_{\overline{Q}}\rangle)^2] \;\geq\; [\Delta x(\psi_{\overline{Q}})^2].$$

Thus the SQL holds for von Neumann's model of approximate position measurement.

4.3 Realization of measurement breaking SQL

We shall now turn to Yuen's proposal [4]. His observation is that if the measurement leaves the free mass in a contractive state ψ_a for every readout a then we can get

$$\Delta x(\tau)(\psi_a) \ll (\hbar\tau/2m)^{1/2} \ll \Delta x(0)(\psi_a). \tag{93}$$

Thus the SQL breaks if such a measurement has a good precision

$$\varepsilon(\hat{U}_\tau\psi_a) \ll (\hbar\tau/2m)^{1/2}. \tag{94}$$

In fact, from the combination of Eqs.(76) and (93)–(94), we get

$$\Delta(\tau, \psi)^2 = [\varepsilon(U_\tau\psi_a)^2] + [\Delta x(\tau)(\psi_a)]^2 \ll \hbar\tau/m \tag{95}$$

In this subsection, I shall show that such statistics of measurement can be realized by a measurement considered first by Gordon and Louisell [24].

In [24], Gordon and Louisell considered the following statistical description of measurement. Let $\{\Psi_a\}$ and $\{\Phi_a\}$ be a pair of families of wave functions with real parameter a. The Gordon-Louisell measurement, denoted by $\{|\Psi_a\rangle\langle\Phi_a|\}$, is the measurement with the following statistics: For any prior state ψ,

$$\text{measurement probability:} \quad P(a|\psi) = |\langle\Phi_a|\psi\rangle|^2, \tag{96}$$

$$\text{state reduction:} \quad \psi \to \psi_a = \Psi_a. \tag{97}$$

One of the characteristic properties of the Gordon-Louisell measurement is that the posterior state Ψ_a depends only on the measured value a and not at all on the prior state ψ. For the condition that Eq.(96) determines the probability density, we assume that $\{\Phi_a\}$ is so normalized as

$$\int da\, |\Phi_a\rangle\langle\Phi_a| = 1. \tag{98}$$

From Eq.(98), it is provided that Φ_a may not be a normalizable vector such as position eigenstate $|a\rangle$. However, Ψ_a is assumed to be a normalized vector. The measurement statistics Eqs.(96)–(97) yields the following operation measure $\mathcal{I}$ and effect measure F:

$$\mathcal{I}(\Delta)\hat{\rho} = \int_\Delta da\, |\Psi_a\rangle\langle\Phi_a|\hat{\rho}|\Phi_a\rangle\langle\Psi_a|, \tag{99}$$

$$F(\Delta) = \int_\Delta da\, |\Phi_a\rangle\langle\Phi_a|. \tag{100}$$

The following computation shows that the operation measure $\mathcal{I}$ in Eq.(99) is completely positive: For any finite sequences of vectors $\xi_1, \xi_2, \ldots, \xi_n$ and $\eta_1, \eta_2, \ldots, \eta_n$, we obtain

$$\begin{aligned}
\sum_{i,j=1}^{n} \langle\xi_i|d\mathcal{I}(a)(|\eta_i\rangle\langle\eta_j|)|\xi_j\rangle &= da \sum_{i,j=1}^{n} \langle\xi_i|\Psi_a\rangle\langle\Phi_a|\eta_i\rangle\langle\eta_j|\Phi_a\rangle\langle\Psi_a|\xi_j\rangle \\
&= da \left|\sum_{i=1}^{n}\langle\xi_i|\Psi_a\rangle\langle\Phi_a|\eta_i\rangle\right|^2 \\
&\geq 0.
\end{aligned}$$

Thus from Theorem 1, we get the following

Theorem 4 *Every Gordon-Louisell measurement is realizable.*

As mentioned above, Gordon-Louisell measurements controls the posterior states independently of the prior states and this properties are suitable for our purpose of realization of measurement which leaves the free mass in a contractive state. The following Theorem is an immediate consequence from Theorem 4, which asserts that the state reduction of position measurement can be arbitrarily controlled.

Theorem 5 *For any Borel family $\{\Psi_a\}$ of unit vectors, the following statistics of position measurement is realizable:*

$$\text{measurement statistics:} \quad P(a|\psi) = |\psi(a)|^2, \tag{101}$$

$$\text{state reduction:} \quad \psi \to \psi_a = \Psi_a. \tag{102}$$

This measurement corresponds to the Gordon-Louisell measurement $\{|\Psi_a\rangle\langle a|\}$ *with the following operation measure* $\mathcal{I}$ *and effect measure* $\hat{F}$*:*

$$\mathcal{I}(\Delta)\hat{\rho} = \int_{\Delta} da\, |\Psi_a\rangle\langle a|\hat{\rho}|a\rangle\langle\Psi_a|, \tag{103}$$

$$\hat{F}(\Delta) = \int_{\Delta} da\, |a\rangle\langle a|. \tag{104}$$

Let $|\mu\nu 0\omega\rangle$ be a fixed contractive state with $\langle x\rangle = \langle p\rangle = 0$ and let Ψ_a be such that $\langle x|\Psi_a\rangle = \langle x - a|\mu\nu 0\omega\rangle$. Then $\{\Psi_a\}$ is the family of contractive states $|\mu\nu a\omega\rangle$ with $\langle x\rangle = a$ and $\langle p\rangle = 0$, which satisfies the assumption in Theorem 5. Thus the following measurement statistics of Gordon-Louisell measurement $\{|\mu\nu a\omega\rangle\langle a|\}$ is realizable:

$$\text{measurement probability:} \quad P(a|\psi) = |\langle a|\psi\rangle|^2, \tag{105}$$

$$\text{state reduction:} \quad \psi \to \psi_a = |\mu\nu a\omega\rangle. \tag{106}$$

Now we shall examine the predictive uncertainty $\Delta(\tau)$ of the repeated measurements of this measurement. From Eqs.(104) and (105), this measurement satisfies the compatibility condition (60) and the unbiasedness condition (67). Further, it is an exact measurement of the position observable, i.e., $G(a,x) = \delta(a-x)$ and $\varepsilon(\psi) = 0$ for all ψ. From Eqs.(76) and (16)–(17), we have

$$\begin{aligned}
\Delta(t,\psi)^2 &= \int da\, P(a|\psi)\left(\varepsilon(\hat{U}_t\psi_a)^2 + \Delta x(t)(\psi_a)^2\right), \\
&= \int da\, P(a|\psi)\Delta x(t)(\psi_a)^2, \\
&= (1/4\xi)(\hbar\tau/m) + (2\hbar/m\omega)(|\mu+\nu|\omega/2)^2(t-\tau)^2,
\end{aligned}$$

where

$$\tau = 2\xi/(\omega|\mu+\nu|^2) = \xi\hbar m/\Delta p(0)^2.$$

Thus for $t = \tau$ and large ξ, we have

$$\Delta(\tau,\psi)^2 = (1/4\xi)(\hbar\tau/m) \ll \hbar\tau/m.$$

We have therefore shown that the Gordon-Louisell measurement $|\mu\nu a\omega\rangle\langle a|$ is realizable measurement and it breaks the SQL. In the rest of this subsection, I shall give a realization of this measurement with an interaction Hamiltonian of the object-probe coupling [19].

The model description is parallel with that of von Neumann's measurement in 4.2. The free mass (the object system) is coupled to a probe system which is a one dimensional system with coordinate Q and momentum P. The coupling is turned on from $t = -\tilde{\tau}$ to $t = 0$ $(0 < \tilde{\tau} \ll \tau)$ and it is assumed to be so strong that the free Hamiltonians of the mass and the probe can be neglected. We choose the following interaction Hamiltonian

$$H = \frac{K\pi}{3\sqrt{3}}\{2(\hat{x}\hat{P} - \hat{Q}\hat{p}) + (\hat{x}\hat{p} - \hat{Q}\hat{P})\}, \tag{107}$$

where K is the coupling constant chosen as $K\tilde{\tau} = 1$. Then if $\Psi_0(x, Q) = f(x, Q)$, the solution of the Schrödinger equation is

$$\Psi_t(x, Q) = f\left(\frac{2}{\sqrt{3}}\left\{x \sin\frac{(1-Kt)\,\pi}{3} + Q\sin\frac{Kt\pi}{3}\right\},\right.$$
$$\left.\frac{2}{\sqrt{3}}\left\{-x\sin\frac{Kt\pi}{3} + Q\sin\frac{(1+Kt)\,\pi}{3}\right\}\right). \tag{108}$$

At $t = -\tilde{\tau}$, just before the coupling is turned on, the unknown free-mass wave function is $\psi(x)$, and the probe is prepared in a contractive state $\Phi(Q) = \langle Q|\mu\nu 0\omega\rangle$, so that the total wave function is $\Psi_0(x, Q) = \psi(x)\Phi(Q)$; expectation values for this state is $\langle\hat{Q}\rangle_0 = \langle\hat{P}\rangle_0 = 0$. At $t = 0$, the end of the interaction, the total wave function becomes

$$\Psi(x, Q) = \psi(Q)\Phi(Q - x). \tag{109}$$

Compare with Eq.(81); the statistics is much different. At this time, a value $\overline{Q}$ for Q is obtained by the detector of the probe observable in the subsequent stage of the measuring apparatus, from which one infers a value for x. Thus the probability density $P(\overline{Q}|\psi)$ to obtain the value $\overline{Q}$ as the result of this measurement is given by

$$P(\overline{Q}|\psi) = \int dx|\Psi(x, \overline{Q})|^2 = |\psi(\overline{Q})|^2. \tag{110}$$

The free-mass wave function $\psi_{\overline{Q}}(x) = \psi(x|\overline{Q})$ just after this measurement ($t = 0$) is obtained (up to normalization) by

$$\begin{aligned}\psi(x|\overline{Q}) &= [1/P(\overline{Q}|\psi)^{1/2}]\Psi(x, \overline{Q})\\ &= [\psi(\overline{Q})/|\psi(\overline{Q})|]\Phi(\overline{Q} - x)\\ &= C\langle x|\mu\nu\overline{Q}\omega\rangle,\end{aligned}$$

where C ($|C| = 1$) is a constant phase factor.

Thus we have just obtained the measurement statistics of this measurement as follows:

$$\text{measurement probability:}\quad P(\overline{Q}|\psi) = |\psi(\overline{Q})|^2, \tag{111}$$
$$\text{state reduction:}\quad \psi \to \psi_{\overline{Q}} = |\mu\nu\overline{Q}\omega\rangle. \tag{112}$$

This shows that this measurement is a realization of Gordon-Louisell measurement $\{|\mu\nu\overline{Q}\omega\rangle\langle\overline{Q}|\}$.

Now I have shown everything I promised before. From our analysis, we can conclude that there are no general reasons in physics which limits the accuracy of the repeated measurement of the free-mass position such as the standard quantum limit for monitoring the free-mass position. In [27], Bondurant analyzed the performance of a interferometric gravity-wave detector which has a Kerr cell in each arm used to counter the effects of radiation pressure fluctuation and has a feedback loop used to keep the interferometer operating at the proper null. He succeeded in showing that this measurement realizes a monitoring the free-mass position which breaks the SQL. It will be an interesting problem to show that the role of the Kerr cell and the feedback loop in his analysis has some corresponding part in the interaction scheme realizing the Gordon-Louisell measurement discussed above.

References

[1] R. Weiss, in *Sources of Gravitational Radiation*, edited by L. Smarr (Cambridge, Univ. Press, Cambridge, 1979).

[2] V. B. Braginsky and Yu. I. Vorontsov, Ups. Fiz. Nauk **114**, 41 (1974) [Sov. Phys. Usp. **17**, 644 (1975)].

[3] C. M. Caves, K. S. Throne, R. W. P. Drever, V. D. Sandberg, and M. Zimmermann, Rev. Mod. Phys. **52**, 341 (1980).

[4] H. P. Yuen, Phys. Rev. Lett. **51**, 719 (1983).

[5] K. Wodkiewicz, Phys. Rev. Lett. **52**, 787 (1984); H.P. Yuen, *ibid.* **52**, 788 (1984).

[6] R. Lynch, Phys. Rev. Lett. **52**, 1729 (1984); H.P. Yuen, *ibid.* **52**, 1730 (1984).

[7] R. Lynch, Phys. Rev. Lett. **54**, 1599 (1985).

[8] C. M. Caves, Phys. Rev. Lett. **54**, 2465 (1985).

[9] E. B. Davies, *Quantum Theory of Open Systems* (Academic, London, 1976).

[10] A. S. Holevo, *Probabilistic and Statistical Aspects of Quantum Theory* (North-Holland, Amsterdom, 1982).

[11] A. Barchielli, L. Lanz and G. M. Prosperi, Nuovo Cimento B **72**, 79 (1982).

[12] A. Barchielli, L. Lanz and G. M. Prosperi, Found. Phys. **13**, 779 (1983).

[13] G. Ludwig, *Foundations of Quantum Mechanics I and II* (Springer, Berlin, 1983).

[14] K. Kraus, *States, Effects, and Operations: Fundamental Notions of Quantum Theory* (Springer, Berlin, 1983).

[15] M. Ozawa, J. Math. Phys. **25**, 79 (1984).

[16] M. Ozawa, Pub. R.I.M.S., Kyoto Univ. **21**, 279 (1985).

[17] M. Ozawa, J. Math. Phys. **26**, 1948 (1985).

[18] M. Ozawa, J. Math. Phys. **27**, 759 (1986).

[19] M. Ozawa, Phys. Rev. Lett. **60**, 385 (1988).

[20] J. von Neumann, *Mathematical Foundations of Quantum Mechanics*, (Princeton University Press, Princeton, N.J., 1955).

[21] W. -T. Ni, Phys. Rev. A **33**, 2225 (1986).

[22] E. Arthurs and J.L. Kelly, Jr, Bell Syst. Tech. J. **44**(4), 725 (1965).

[23] H. P. Yuen, Northwestern University, Evanston, IL, (unpublished).

[24] J. P. Gordon and W. H. Louisell in *Physics of Quantum Electronics*, edited by P. L. Kelly, B. Lax and P. E. Tannenwald (McGraw-Hill, New York, 1966).

[25] C. M. Caves, Phys. Rev. D **33** 1643 (1986).

[26] B. L. Schumaker, Phys. Rep. **135**, 317 (1986).

[27] R. S. Bondurant, Phys. Rev. A **34**, 3927 (1986).

THE CORRELATED SPONTANEOUS EMISSION LASER: THEORY AND RECENT DEVELOPMENTS*

M. Orszag[a], J. Bergou[b], W. Schleich and M. O. Scully

Center for Advanced Studies and
Department of Physics and Astronomy
University of New Mexico
Albuquerque, New Mexico 87131
and
Max-Planck Institut für Quantenoptik,
D-8046 Garching bei München, W. Germany

I. CORRELATED EMISSION IN A LASER

As originally conceived a correlated spontaneous emission laser showed quenching of spontaneous emission quantum fluctuations in the *relative phase angle* of a *two mode laser*. It has been shown by several approaches (e.g. quantum noise operator, Fokker-Planck equation, etc.) that such devices can, in principle, have vanishing noise in this relative phase angle. A geometric pictorial analysis along these lines has been given and provides a simple intuitive explanation for this quantum noise quenching which has also been supported by recent experimental investigations.

In recent work, we have found that the spontaneous emission fluctuations in the absolute phase of *single mode laser* can likewise be quenched in the two-photon CEL. It has also been shown that such a device can demonstrate squeezing in the phase quadrature.

It is noted that the CEL provides, in principle, a device which allows one to supersede the conventional quantum limits of sensitivity in, for example, a laser gravity wave detector or a laser gyroscope.

In many areas of modern physics ultrasmall displacements are detected optically. The small displacement is converted into a change of the optical path length in an interferometer. In the so-called passive detection scheme a laser light, which is generated outside, is sent through the cavity and the change in the path length results

* Work supported by the ONR.

[a] Permanent address: Facultad de Fisica Pontificia Universidad Catolica de Chile, Casilla 6177, Santiago, Chile

[b] Permanent address: Central Research Institute for Physics, H-1525 Budapest 114, P. O. Box 49, Hungary

in a phase shift. The shift is then detected by beating or homodyning the output beam with the reference beam. The phase shift to be detected this way is generally small since the light spends only a finite time in the cavity, given by the cavity lifetime. The limiting noise source in this type of measurement is the photon counting error, or shot noise, due to the fluctuations in photon number at the detector. In the so-called active detection scheme the light is generated inside by placing an active medium into the cavity. The operating frequency of the laser changes due to the change in the path length. This results in a phase shift which is proportional to the measurement time, leading to a much bigger signal than that of the passive detection scheme. This shift is then detected by heterodyning the output light with that from a reference laser. The limiting quantum noise source in this type of measurement is spontaneous emission fluctuations in the relative phase between the two lasers or, in other words, laser-phase diffusion noise. This noise is much larger than the shot noise and the sensitivity of the active and passive detection schemes is ultimately the same.

In the present work we focus our attention on the active schemes of detection. Examples include, e.g. the laser gyroscope, laser gravity-wave detectors, etc., which will be discussed in some detail in Section 3. The central question, we are addressing here, is: "under what conditions can one quench spontaneous emission quantum noise from the relative phase of two lasers?"

Our studies in the area of correlated emission lasers[1] (CEL), which are the subject of this review, are directed toward this question. In this section we introduce the idea of correlated spontaneous emission by using simple physical and geometrical arguments. In the next section we review the concrete laser systems, proposed so far, which potentially exhibit CEL behavior. In the last section we discuss the application of CEL systems in the active schemes of detection, on the example of CEL gravity-wave detectors and CEL gyroscopes.

The quantum noise quenching in a correlated emission laser (CEL) is due to both the correlation of the quantum noise in the two modes and phase locking. However, phase locking alone is not enough to totally quench the relative phase fluctuations, leading to the complete suppression of spontaneous emission noise.

A geometrical representation of the CEL effect is shown in Fig. 1, where $\delta\epsilon_1$ and $\delta\epsilon_2$ are the contributions to the fields 1 and 2 by the spontaneous emission of a photon in the two respective modes. In the CEL, the relative phase between the two fields, after this spontaneous emission process, is unchanged. That is, the fields E_1 and E_2 experience phase diffusion in such a way that the relative phase is unchanged.

Recently there have been reports of experimental evidence supporting the CEL effect. P. E. Toschek and J. Hall,[2,3] reported the first observation of the CEL effect. In a He-Ne Zeeman laser, operating at 633 nm, they had a two mode laser system. The beat signal, in the free-running case, between the σ_+ and σ_- polarizations gave a linewidth of $\frac{1}{4}$ Hz for a measurement time of a few seconds. When the r.f. magnetic field H_1 (perpendicular to the axial Zeeman field H_0) was supplied to the gain medium, the fluctuations of the beat signal were reduced by a large factor, to less than 1/10 of the Schawlow-Townes limit. (ν_3 in the experiment was 70 KHz, $\Delta \approx 100$ MHz and $\gamma = 20$ MHz.) The linewidth of the beat signal was reduced at least by one order of magnitude. A more recent second experiment[4] was performed in a semiconductor laser with an extended cavity. As a result, the heterodyned spectral width between two lasing modes in a grating extended-cavity laser was reduced to below the spontaneous emission noise level.

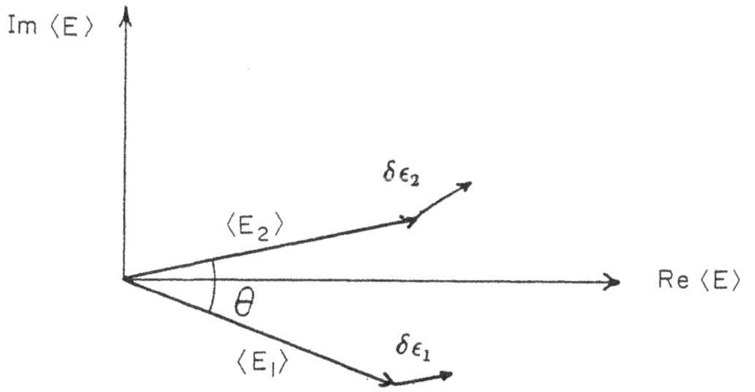

Fig. 1. Geometrical representation of the CEl effect. $\Delta\epsilon_1$ and $\delta\epsilon_2$ are the contributions to the electric field due to the emission of an extra photon in the modes 1 and 2, respectively.

Finally we note that the quantitative treatment of CEL operation is based on the following decomposition of the phase noise[5]

$$\langle(\delta\theta)^2\rangle = \frac{1}{4\bar{n}} + \langle: (\delta\theta)^2 :\rangle. \tag{1}$$

Here $1/4\bar{n}$ represents the shot noise which has its origin in the vacuum fluctuations of the reservoir, : : stands for normal ordering, and the second term represents the added noise of the active system. In this normally ordered form it can be obtained from a Fokker-Planck (FP) treatment of the system. It is uniquely related to the phase diffusion constant, as obtained from the FP equation, of the CEL system.

The key point in the theory of CEL operation is that this diffusion constant is an explicit function of the relative phase θ. From Fig. 1 it can be seen that $D(\theta) \sim \theta^2$ for small θ, which is an example for a system with multiplicative noise.

This consequence of the simple geometric picture is also supported by the full FP analysis. In all cases, to be discussed in the next section, $D(\psi) \sim (1 - \cos\psi)$, with $\Psi = \Theta - (\phi_a - \phi_b)$, which for small values of ψ corresponds to the multiplicative form of the geometric picture.

II. CORRELATED EMISSION LASER SYSTEMS

A. The Quantum Beat Laser

In a Quantum Beat Laser, an active medium, consisting of a collection of three level atoms, where the upper two levels are coherently excited, drives a doubly resonant cavity, as shown in the Fig. 2. The coherence of the upper two levels is achieved by shining a strong microwave signal E_3 onto the system, resonant with the $|a\rangle - |b\rangle$ transition.

A recent publication[6] shows that in a linear theory, one can write a FP equation for the Glauber's P distribution in polar coordinates $P(\rho_1, \rho_2, \theta, \mu)$, when $\alpha_1 = \rho_1 e^{i\theta_1}$, $\alpha_2 = \rho_2 e^{i\theta_2}$, $\theta = \theta_1 - \theta_2$, $\mu = (\theta_1 + \theta_2)/2$. The main result of this analysis shows that the injected atomic coherence generates a strong correlation between the two signals

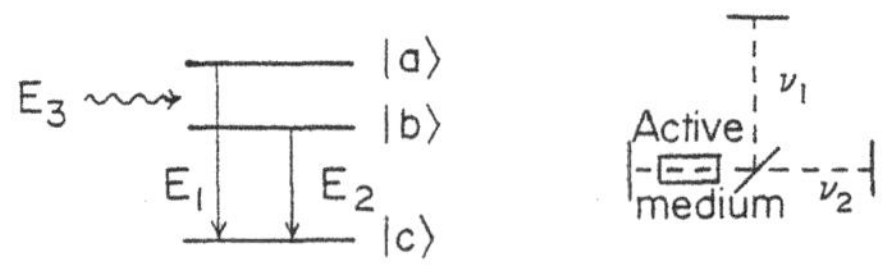

Fig. 2. The quantum beat laser. A collection of three level atoms inside a double cavity. A coherent superposition of the upper two levels is achieved by pumping resonantly with a microwave field E_3.

at frequencies ν_1 and ν_2 in such a way that the diffusion constant of the relative phase of the two signals will vanish under certain detuning conditions.

As a matter of fact, if we define $\Delta_1 \equiv \omega_{ac} - \nu_1$ and $\Delta_2 \equiv \omega_{bc} - \nu_2$, as the detunings of the two atomic transitions with respect to the cavity modes, then if $\Delta_1 = \Delta_2 = \Omega/2$, Ω being the Rabi frequency of the driving microwave field at frequency ν_3, then the diffusion coefficient of the relative phase $D(\theta)$ is given by[6]:

$$D(\theta) = \frac{g^2 r_a}{4\gamma^2 \bar{n}}(1 - \cos\psi), \tag{2}$$

where in Eq. (2) we have assumed that the two coupling constants of the atomic transitions with their respective electromagnetic fields are equal ($g_1 = g_2 = g$), γ is the inverse atomic relaxation time of the two levels (taken to be equal) to the lower level c, r_a is the atomic injection rate and $\psi \equiv -\theta + \phi$, where ϕ is a characteristic phase of the input microwave pump. From Eq. (2) it is easy to see that $D(\theta) = 0$ if $\psi = 0$. Also, from the FP equation, it is simple to show that the phase ψ locks to zero, $\psi = 0$ being a stable solution.

The physical picture that emerges from this analysis is the following:

If we represent the two fields as phasors in the complex plane, the tips of the two fields follow a random trajectory in such a way that the relative phase of the two phasors lock to a fixed angle θ_0, while the sum phase diffuses with a nonvanishing diffusion constant (Fig. 3). A full nonlinear theory of the quantum beat laser has been developed,[7] using a dressed atom-dressed mode description.

From the nonlinear master equation, one can generate a FP equation, now with nonlinear terms included (up to g^4). The relative phase diffusion constant has a nonlinear contribution which also vanishes when $\psi = 0$. Therefore, the CEL effect persists in higher orders.

B. The Hanle Laser[8]

The active medium can also be prepared in a coherent excitation of the states $|a\rangle$ and $|b\rangle$, which decay to the state $|c\rangle$ via emission of radiation of different polarization states. This is the Hanle effect and it is achieved with a polarization sensitive mirror to couple the doubly resonant cavity, as shown in Fig. 4.

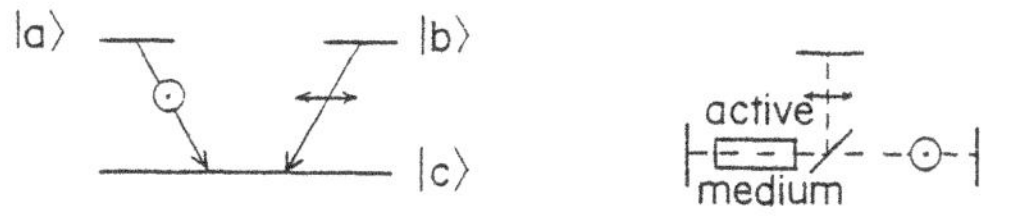

Fig. 3. In the quantum beat laser, the tips of the electric field phasors of both modes follow a random motion in such a way that the relative angle between the two fields is unchanged.

It is possible[9] to show that a linear theory gives a vanishing relative phase diffusion constant $D(\theta)$, for identical detunings. Furthermore, a nonlinear theory can be developed, using the "dressed atom-dressed mode" master equation approach, assuming, again, equal detunings. It is shown that the quenching of the relative phase noise remains valid in the nonlinear theory and that it's operation is stable above threshold

It is also a fact that if one considers the effects of the atomic motion[10] or inhomogeneously broadened atomic lines, the CEL effect is not destroyed and persists.

Under equal coupling and detuning conditions, it is possible to show that:

$$D(\theta) = \frac{rg^2}{8\rho^2} \int_{-\infty}^{\infty} dk exp(-k^2/\kappa^2) \left[\frac{1}{\gamma^2 + \Delta^+(k)^2} + \frac{1}{\gamma^2 + \Delta^-(k)^2} \right] (1 - \cos\psi), \quad (3)$$

where

$$\Delta^{\pm} = \Delta \mp \frac{\hbar k K}{m} - \frac{\hbar K^2}{2m}, \quad (4)$$

$$\kappa = \frac{\sqrt{2_m k_B T}}{\hbar},$$

and, $\hbar k$ and $\hbar K$ represents the atom's and electromagnetic mode's momentum, respectively.

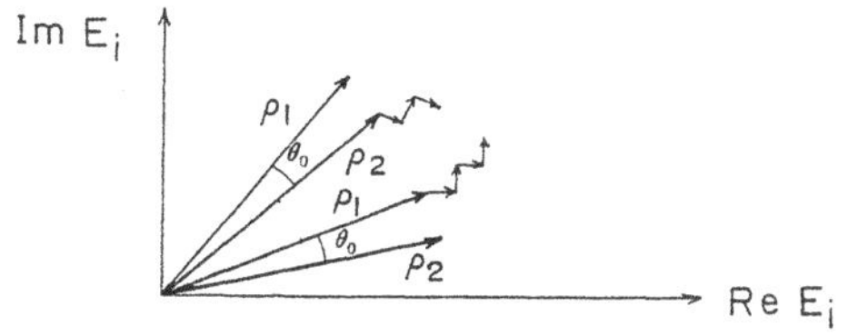

Fig. 4. In a Hanle Laser, the radiation emitted from the transitions $|b\rangle \rightarrow |c\rangle$ and $|a\rangle \rightarrow |c\rangle$ differ in polarization. In a double cavity, the two waves interact with the active medium with the help of a polarization sensitive mirror.

C. The Holographic Laser[11]

Consider a ring laser whose counterpropagating modes are coupled by a spatial modulation in the gain medium.

In the holographic laser,[11] each beam is reflected in part by the thin atomic layers of the gain medium. When the reflected light interferes constructively with the light of the counterpropagating beam, noise quenching is achieved. This is shown pictorially in Fig. 5. From a nonlinear master equation, one can write a FP equation, giving a diffusion constant for the relative phase of the two counterpropagating modes:

$$D(\theta) = \left[\frac{\alpha}{2\rho^2} - 2\beta(1+\cos\psi)\right](1-\cos\psi), \tag{5}$$

where α and β are the gain and saturation coefficients, respectively.

D. The Correlated Emission Free-Electron Laser

As we saw in the examples (B) and (C), the strong microwave signal that mixes the upper two levels of a three level atom is not necessary. One could also prepare initially the atom in a coherent superposition of the two upper states, like in the Hanle laser. The central idea of the correlated emission free electron laser is to quench the relative phase noise of two laser signals, generated in a Compton regime free electron laser with either a magnetostatic or electromagnetic wiggler, when the electron is prepared in a coherent superposition of two pure momentum states.[12]

Assuming a helical wiggler,[13] one assumes an initial momentum state as a superposition of two pure momentum states:

$$|p\rangle_{in} = \alpha_1|p_1\rangle + \alpha_2|p_2\rangle. \tag{6}$$

If one assumes that the quantum recoil of the electron is very small, an approximate expression for the evolution operator can be derived and therefore, and equation of motion for the photon density operator, in the coarse-grained time approximation can be written.

The main result of this theory is the following. If the momentum difference between the two states is such that:

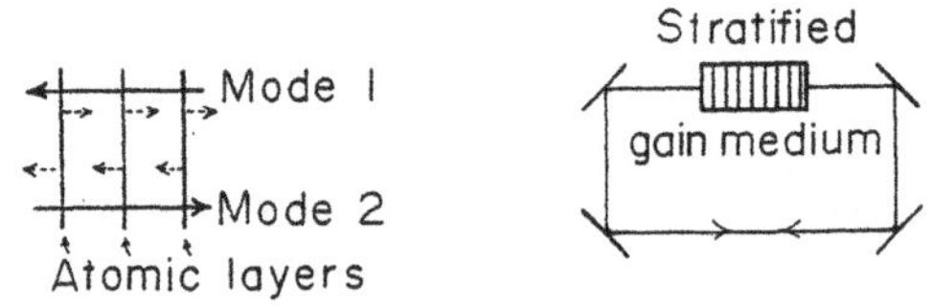

Fig. 5. The holographic laser. A stratified medium is inside a ring cavity geometry. The two traveling modes are partially reflected in the various atomic layers. The reflected light can interfere constructively with the counterpropagating mode, quenching the phase noise.

$$p_1 - p_2 = 2\hbar k_1, \tag{7}$$

k_1 ($\approx k_2$) being the wavevector of one of the modes then a vanishing diffusion constant can be achieved. However, if the momentum difference is larger, like $4\hbar k_1$, then the quenching of the diffusion constant becomes negligible.

If Eq. (7) is satisfied, one obtains:

$$D(\theta) = \frac{r}{4N_{1S}^{\frac{1}{2}}} g\sqrt{N_W} \left[\frac{\sin(\beta_1(p_2)T/2)}{\beta_1(p_2)/2}\right] \cos\left[\frac{\pi}{2} - \theta_{\Delta_1} + \theta\right], \tag{8}$$

where:

$$\beta_1(p) = \left[\hbar(k_w + k_1)^2 + 2(k_W + k_2)p\right]/2m + \omega_W - \omega_{L_1}, \tag{9}$$

$$\theta_{\Delta_1} = \beta_1(p_2)T/2,$$

N_W and N_{1S} are the average photon numbers in the wiggler and mode 1, and if $\alpha_1 = \rho e^{i\theta_1}$ and $\alpha_2 = \rho e^{i\theta_2}$, for the two laser modes, then $\theta = \theta_1 - \theta_2$.

It should be pointed out that this result may appear academic, since the two required electron momenta are so close that they are hardly discernible on a practical scale. However the results are interesting, since they indicate the possibility of correlated emission into the different modes of a free-electron laser.

E. The Two-Photon Correlated Emission Laser

The two-photon CEL is the extension of the CEL principle from the case of a two-mode, phase insensitive interaction to the two-photon, phase-sensitive interaction. In the different two-mode CEL systems (quantum beat, Hanle, holographic, FEL) coherence in the active medium is established between the upper two lasing levels which are, in some sense, the most distant levels of the system since they are only coupled by transitions via the common ground state. In the two-photon CEL the active medium, consisting of three-level atoms in the cascade configuration, drives a cavity resonant with $\nu_1 = (\omega_a - \omega_c)/2$, as shown in Fig. 6.

We are again interested in finding out the role of atomic coherence between the most distant levels a and c, in quenching the quantum noise. As it turns out, this system is not only capable of quenching the added quantum noise of an active system (second term on the r.h.s. of Eq. (1)) but also, under certain conditions, reduces the phase noise below the shot noise level, i.e. produces an output in the form of squeezed light. Thus, it is the first active system, operated as an oscillator as opposed to an

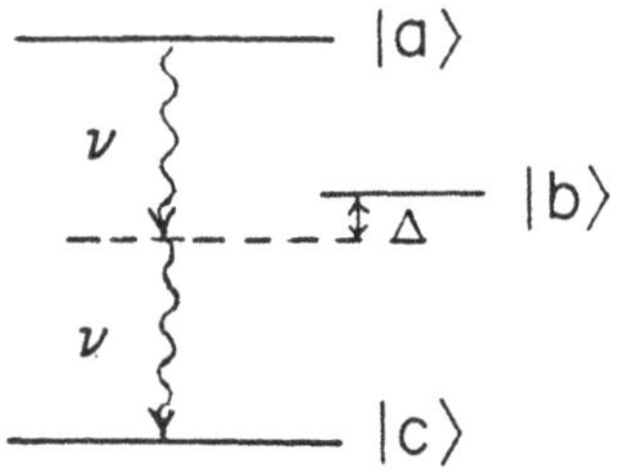

Fig. 6. The two photon correlated emission laser. The active medium are three level atoms in a cascade configuration. The cavity is tunned at ν. The intermediate level is off resonance with respect to the center of the transition.

amplifier, that has the potential of producing a cw squeezed output starting from an ordinary vacuum. In achieving this, a fine interplay between the injected atomic coherence and detuning Δ on the intermediate transition appears to be crucial.

We assume that the atoms are prepared into the coherent superposition outside the laser resonator and then injected into the laser cavity with the coherent initial condition. As we have shown in a recent publication[14], in a linear theory one can write a FP equation for the Glauber-Sudarshan P-distribution function in terms of polar coordinates. The main result of this analysis shows that the injected atomic coherence renders the phase diffusion coefficient D, a function of the phase itself

$$D(\theta) = \frac{\alpha|L|^2}{4\bar{n}} \left[\rho_{aa} + \frac{|\rho_{ac}|}{|L|} \cos\left(\phi_{ac} - 2\theta + \tan^{-1}\left(\frac{\Delta}{\Gamma}\right)\right)\right]. \tag{10}$$

Here α is the usual linear gain coefficient, $\bar{n}$ is the photon number in steady-state, $|L|$, is the Lorentzian, $|L|^2 = \Gamma^2/(\Gamma^2+\Delta^2)$, Γ is the linewidth of the atomic levels (for simplicity taken to be equal for all levels, a, b, and c, involved), Δ is the detuning on the intermediate transition, ρ_{aa} is the initial population of the upper state and $\rho_{ac} = |\rho_{ac}|e^{i\phi_{ac}}$ is the initial coherence.

We have found conditions when the argument of the cosine in (10) locks to π. Under these conditions (10) simplifies to

$$D = \frac{\alpha|L|^2}{4\bar{n}} \left[\rho_{aa} - \frac{|\rho_{ac}|}{|L|}\right]. \tag{11}$$

The important feature of this expression is that the contribution from the atomic coherence is affected by the detuning differently from that of the population. The relative weight of the initial coherence is enhanced by the detuning, since $|L|^{-1} \geq 1$. From this expression it is easy to see that one can get complete noise quenching, i.e. CEL operation, when $\rho_{aa} = |\rho_{ac}|/|L|$. Since $|L|^{-1}\rangle 1$, this is possible in an inverted system, $\rho_{aa}\rangle\rho_{cc}$. Beyond that, D can also be negative, if $\rho_{aa}\langle|\rho_{ac}|/|L|$. This is still possible in an inverted system if, simultaneously with the above condition, we also satisfy $\rho_{aa}\rangle\rho_{cc}$. In this case, as we see from (1), the added noise of the active system is negative and the total noise in the phase is below the shot-noise level. The radiation generated this way is nonclassical, squeezed in the phase variable. The extra noise is put in the amplitude component which now acquires a noise level exceeding vacuum level. With a different set of locking conditions the noise can be reduced below the vacuum level in the amplitude component on the expense of enhancing the noise in the phase component. This corresponds to the generation of amplitude squeezed (antibunched) light.

The physical picture, emerging from the analysis is the following . If we represent the fields emitted in the two subsequent transitions as phasors in the complex plane we obtain the picture shown in Fig. 7.

The average field of the two-photon CEL is denoted by ϵ. The uncertainty $\delta\epsilon_1$ in the first transition $a \rightarrow b$ contains the phase $\phi_a - \phi_b$ where ϕ_b is random. The uncertainty in the second transition $b \rightarrow c$ contains the phase $\phi_b - \phi_c$, where again ϕ_b is random. From the two consecutive spontaneous emission events the random element ϕ_b cancels. If θ locks to $\phi_a - \phi_c$ or $\phi_a - \phi_c + \pi$ then the entire noise appears as an enhanced amplitude noise, and the noise from phase θ cancels. The above situation corresponds to $\Delta = 0$. $\Delta \neq 0$ enhances the role of the atomic coherence and renders squeezing possible in an active oscillator system.

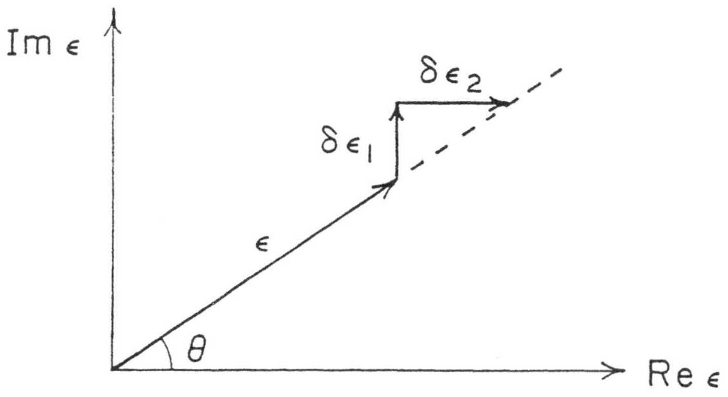

Fig. 7. In the two photon CEL, if we represent the electric field contribution of two extra photons by $\delta\epsilon_1$ and $\delta\epsilon_2$ in the complex ϵ-plane, then the phase of the final electric field remains unchanged. Only amplitude fluctuations are present.

III. APPLICATIONS

As mentioned in the Introduction the main application we have in mind for CEL systems, is in the area of ultrahigh resolution interferometry. The main question we are addressing in this Section is the following one: what are the practical advantages of CEL's in the detection of ultrasmall displacements? Before we answer this question we briefly review the signal and noise performance of the passive and active detection schemes and their resulting limitations on sensitivity.[15]

In the passive detection scheme a change $\Delta\ell$ in the optical path length of the interferometer leads to a phase shift

$$\Delta\psi^{(p)} = \frac{\nu}{\gamma}\cdot\frac{\Delta\ell}{\ell}, \tag{12}$$

where ν is the frequency of the injected field, γ is the resonator's linewidth and ℓ is the round trip path length. When we detect this signal the main noise source is photon counting error or shot noise $\Delta\psi^{(p)} = 1/\sqrt{n}$ where n is the number of detected photons. It can be related to the detected power P as $n = Pt_m/\hbar\nu$ where t_m is the measurement time. A good measure of the sensitivity (or resolution) of the passive detection scheme is the minimum detectable displacement $\Delta\ell^{(p)}_{min}$ which we obtain by equating the signal to the noise ($S/N = 1$) and solving it for $\Delta\ell$. This yields

$$\Delta\ell^{(p)}_{min} = \frac{\gamma}{\nu}\ell\sqrt{\frac{\hbar\nu}{Pt_m}}. \tag{13}$$

In the active detection scheme a change $\Delta\ell$ in the optical path length of the interferometer leads to a phase shift during the measurement time t_m

$$\Delta\psi^{(a)} = \nu t_m\frac{\Delta\ell}{\ell}. \tag{14}$$

When compared to Eq. (12) we see that the active signal corresponds to replacing γ^{-1} by t_m in the passive signal. Since usually $t_m \gg \gamma^{-1}$ the active signal can be much larger than the passive one, in certain applications by as many as 10 orders of magnitude larger. When we detect this signal, the main noise source is phase diffusion noise $\delta\psi^{(a)} = \sqrt{(\gamma/n)t_m}$ where now n is the number of photons in the lasing mode. The time rate of change of photon number outside the cavity is $\dot{n}_{out} = \gamma n$. If we detect all photons escaping the cavity then $P = \hbar\nu\dot{n} = \hbar\nu\gamma n$. Expressing n from here, the relationship between n and the detected power now reads as $n = P/\hbar\nu\gamma$. Again, a good measure of the sensitivity of the active detection scheme is the minimum detectable displacement $\Delta\ell^{(a)}_{min}$ which we again obtain by solving $S/N = 1$ for $\Delta\ell$. This yields

$$\Delta\ell^{(a)}_{min} = \frac{\gamma}{\nu}\ell\sqrt{\frac{\hbar\nu}{Pt_m}} = \Delta\ell^{(p)}_{min} = \Delta\ell^{(o)}_{min}, \tag{15}$$

i.e. the ultimate sensitivity of the two detection schemes $\Delta\ell^{(0)}_{min}$ is the same.

Now, we turn our attention to detection schemes based on the correlated emission laser. A CEL gravity wave detector[15,16] is shown in Fig. 8. The coherently excited three-level atoms of Fig. 2 drive the doubly resonant cavity at ν_1 and ν_2. The dichroic mirror deflects light at ν_1 in vertical direction. Gravity wave incoming parallel to the ν_2 axis perturbs the length of the ν_1 arm, due to the transverse nature of the g-wave. A gravity wave of amplitude h_0 causes a change in the arm length

$$\frac{\Delta\ell}{\ell} = h_0. \tag{16}$$

A CEL gyroscope[17,18] is shown in Fig. 9. The striated or modulated gain medium drives the two counterpropagating modes E_1 and E_2 of a ring cavity. Rotation at a rate Ω of the ring interferometer leads to a path length difference between the two counterpropagating modes as given by

$$\frac{\Delta\ell}{\ell} = S\frac{\Omega}{\nu}, \tag{17}$$

where S is a scale factor, $S = (4A/\lambda p)$ in which A is the enclosed area, p is the perimeter of the ring and λ is the reduced wavelength.

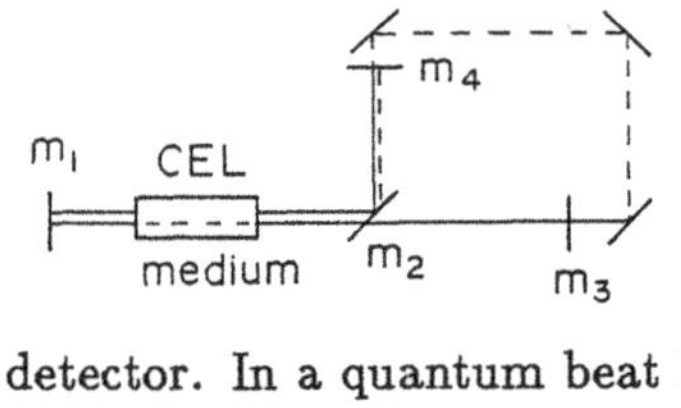

Fig. 8. CEL gravity-wave detector. In a quantum beat laser, the incoming gravity-wave perturbs the length of one of the arms.

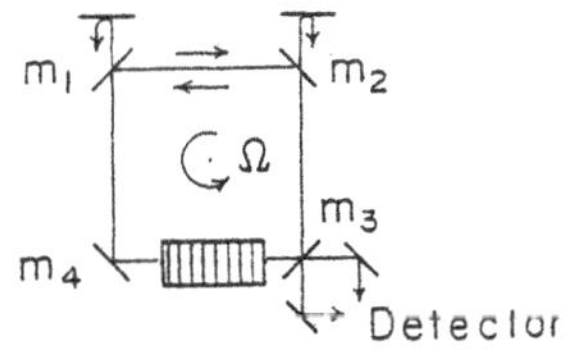

Fig. 9. CEL gyroscope in which light is extracted from mirrors m_1 and m_2 and reinjected on the top mirrors. This leads to an enhancement in the gyroscope sensitivity. The correlation is produced by a striated medium.

In the schemes of Figs. 8 and 9 controlled backscattering, i.e. extraction of light from one mode and reinjection with a controlled phase change into the other enhances the sensitivity of the detection scheme with respect to the standard limit. The extraction-reinjection system is schematically represented by the dashed lines in Figs. 8 and 9, for the case of CEL g-wave detectors and gyroscopes, respectively.

The following Table summarizes the sensitivities of the different detection schemes.

Table 1. Comparison of sensitivites for the different detection schemes. Mirror losses due to absorption are ignored. The CEL detection system has a sensitivity superseding standard quantum limit by a factor $\epsilon = \sqrt{\omega_g/2\gamma}$ for g-wave detection[15] (ω_g is the frequency of g-wave) and $\epsilon = \sqrt{\Omega/2\gamma}$ for rotation sensing[17].

LIMITS OF SENSITIVITY			
	Active Scheme	Passive Scheme	CEL
g-wave detector	$h_0^{(min)} = \frac{\gamma}{\nu}\sqrt{\frac{\hbar\nu}{P t_m}}$	$h_0^{(min)} = \frac{\gamma}{\nu}\sqrt{\frac{\hbar\nu}{P t_m}}$	$h_0^{(min)} = \sqrt{\frac{\omega_g}{2\gamma}} \cdot \frac{\gamma}{\nu}\sqrt{\frac{\hbar\nu}{P t_m}}$
gyroscope	$\Omega^{(min)} = S^{-1}\gamma\sqrt{\frac{\hbar\nu}{P t_m}}$	$\Omega^{(min)} = S^{-1}\gamma\sqrt{\frac{\hbar\nu}{P t_m}}$	$\Omega^{(min)} = \sqrt{\frac{\Omega}{2\gamma}} S^{-1}\gamma\sqrt{\frac{\hbar\nu}{P t_m}}$

We also point out, that in CEL detectors the signal appears directly as a shift in the relative phase between the two modes. Thus, these schemes are free of the usual "dead band" effect of the active scheme which is due to frequency locking. The application of the two-photon CEL[11] in the above detection schemes may yield a further improvement of the sensitivity since this device produces a squeezed output and operates below the shot noise level. This system is presently under study to establish its limits of sensitivity.

REFERENCES

1. M. O. Scully, Phys. Rev. Lett. **55**, 2802 (1985).
2. P. E. Toschek and J. Hall, Abstract in XV International Conference on Quantum Electronics, J.O.S.A. B **4**, 124 (1987).
3. For a theoretical account on the experiment of reference 2, see: J. Bergou and M. Orszag (to appear in J. O. S. A. B).
4. M. Ohtsu and K. Y. Liou (preprint).
5. M. O. Scully, and K. Wodkiewicz (to be published).
6. M. O. Scully and M. S. Zubairy, Phys. Rev. A, **35**, 752 (1987). For a geometrical argument of the noise quenching in the correlated spontaneous emission see: W. Schleich and M. O. Scully (to be published).
7. J. Bergou, M. Orszag, and M. O. Scully (to be published).
8. See W. Hanle, Z. Phys. **30**, 93 (1924) for the original experiment. Also V. Weisskopf, Ann. d. Phys. 9,23 (1931) and G. Breit, Rev. Mod. Phys. **5** 91 (1933). Also M. O. Scully in Atomic Physics I, ed. B. Benderson, V. W. Cohen and F. M. Pichanick (Plenum, N. Y. 1969) p. 81.
9. J. Bergou and M. O. Scully (to be published).
10. J. Bergou, M. Orszag and M. O. Scully (to be published)
11. J. Krause, M. O. Scully, Phys. Rev. A **36**, 1771 (1987).
12. M. Orszag, W. Becker, and M. O. Scully, Phys. Rev. A, **36**, 1310 (1987).
13. For general discussions on the compton regime free-electron laser see: For the classical theory see for example: F. A. Hopf, P. Meystre, G. T. Moore and M. O. Scully in Novel Sources of Coherent Radiation, Physics of Quantum Electronics, Vol. 5 (Addison-Wesley, Reading, Mass, 1978) p. 41; N. M. Kroll, ibid, p. 115; W. B. Colson, ibid, p. 157. For the quantum theory and many-body effects, see: J. M. J. Madey, J. App. Phys. **42**, 1906 (1976). A Bambini and A. Renieri, Lett. Nuovo Cimento **31**, 399 (1978). W. Becker and M. S. Zubairy, Phys. Rev. A **25**, 2200 (1982), W. Becker and J. K. McIver, Phys. Rev. A **28**, 1838 (1983), M. Orszag and R. Ramirez, J. Opt. Soc. Am. B **3**, 895 (1986), E. Fernandez and M. Orszag, J. Opt. Soc. Am. B **4**, 512 (1987).
14. M. O. Scully, K. Wodkiewicz, M. S. Zubairy, J. Bergou, Ning Lu and J. Meyer ter Vehn (to be published).
15. M. O. Scully and J. Gea-Banacloche, Phys. Rev. A **34**, 4043 (1986).
16. For a general discussion of laser interferometer detection of gravitational radiation, see K. Thorne, Rev. Mod. Phys. **52**, 285 (1980); for a more recent account: H. Billing, W. Winkler, R. Schilling, A. Rüdiger, K. Maischberger and L. Schnupp, in Quantum Optics, Experimental Gravitation and Measurement Theory, Vol. 94, NATO ASI Series, edited by P. Meystre and M. O. Scully (Plenum, New York 1983), p. 525.
17. M. O. Scully, Phys. Rev. A **35**, 452 (1987).

18. For a good introduction to the subject, see: F. Arnowitz in Laser Application, edited by M. Ross (Academic, New York, 1971) p. 172. Also: L. Menegozzi and W. Lamb, Phys. Rev. A8, 2103 (1973). A more recent review is found in: W. W. Chow, J. Gea-Banacloche, L. M. Pedrotti, V. E. Sanders, W. Schleich, M. O. Scully, Rev. Mod. Phys. **57**, 61 (1985).

MULTIPHOTON AND FRACTIONAL-PHOTON SQUEEZED STATES

G.M. D'Ariano *
Dipartimento di Fisica 'A. Volta ', Universitá di Pavia, 27100 Pavia, Italy

M.G. Rasetti*
Dipartimento di Fisica, Politecnico di Torino,10129 Torino, Italy

J. Katriel
Department of Chemistry, Technion - Israel Institute of Technology, Haifa 32000 Israel

A.I Solomon
Faculty of Mathematics, The Open University, Milton Keynes, U.K.

1. Introduction

There are several motivations, both physical and mathematical to pursue the construction of multi-photon squeezed states. Besides describing many photon processes which are more and more interesting from the point of view of quantum optics, they lead to non-Gaussian vawepackets which are of great relevance for other branches of physics as well (for instance the theory of phase transitions or the description of certain collective effects in nuclei). Moreover, since in several cases higher order moments can be independently squeezed, such states depend very often on a larger number of parameters and a finer tuning of the related probability distributions can be achieved.

The origin of the set of papers ([1], [2], [3], [4], [5], [6], [7], [8], [9], [10]) which motivated the present work, and which are in part briefly reviewed in it, was the puzzling paper by Fisher, Nieto and Sandberg [11]. In it the authors, trying to generalize the customary 2-photon squeezed states of Stoler and Yuen [12][13] to multiphoton states by the simplest possible ansatz, run into unexpected difficulties connected with the non-analyticity of the vacuum. Even though such difficulties could be partly overcome from the computational point of view (Ref. [14]), the problem is a very deep one.

A *non-naïve* way out of it was found on the basis and in terms of a number of observations : *i)* it is straightforward to check that the conventional squeezed states are generalized coherent states (in the sense of [15]), corresponding to the algebra $su(1,1)$. (In this framework they naturally fit into the general definition given by Glauber [16]). *ii)* The generalization proposed leads to an infinite-dimensional algebra whose coherent states are unknown – indeed they are most probably undefinable in the usual sense – and which is anyway endowed with a structure not rigid enough for such a fine effect as squeezing. *iii)* One should have therefore to preserve the two ingredients which appear to control the whole phenomenon : on the one hand the Weyl-Heisenberg (W.H.) group skeleton, responsible for the bosonic character of the many photon states, on the other the structure of a group, compact or non-compact but of finite rank, to generate squeezing. *iv)* The price one should be ready to pay is the recourse to non-linear realizations of the algebra, which necessarily introduce into

* also : Gruppo Nazionale di Struttura della Materia, CNR, Italy and Istituto Nazionale di Fisica Nucleare, Italy

play infinite power series of bosonic operators (rigid enough though not to break the delicate balance leading to squeezing).

The two main tools to realize the above program are the Brandt-Greenberg [17] multiphoton creators and annihilators, and the Holstein-Primakoff (H.P.) [18] realization of the $su(2)$ and $su(1,1)$ algebras.

The new set of multiphoton squeezed states thus constructed has several interesting features. They promise to be the best candidates as the quantum states in which the light entering the input ports of devices such as the conventional Mach-Zehnder and Fabry-Perot interferometers, or the active lossless interferometers of Yurke, McCall and Klauder should be prepared. It was in fact shown in [19] that such devices can be quite naturally characterized by an action on a Lie group space, respectively of $SU(2)$ and $SU(1,1)$, and we expect that the states described here should be able to achieve better phase sensitivity than the highest weight vector states the authors in [19] propose.

Moreover they constitute a family of quantum states whereby, with great flexibility one can achieve an arbitrary reduction of the photon number noise, a variety of different number of photon distribution laws, a finer tuning in the control of higher moments squeezing.

Finally they lead to the notion, discussed in this paper with some detail, of fractional photon states. These are mixed states, to be realized in terms of suitable density matrices, which describe the same physical output that one should have were one able to generate photon states corresponding to a fractional eigenvalue of the number operator ([8]). Indeed they describe synthetically the complex canonical transformation and projection operation (in Hilbert space of states) one should perform when describing the observable properties of a k-photon dynamical variable in the framework of a k'-photon state, when k and k' are not multiple of one another. We show here (see Ref. [10]) that in this description squeezing corresponds to fractioning, a suggestive physical image of this elusive phenomenon.

The paper is organized as follows. In Sect. 2 we briefly review the algebraic background necessary for the theory, namely the automorphism of the W.H. algebra realized by the multiphoton operators, and both the H.P. realization and the multiphoton H.P. realization of $SU(2)$ and $SU(1,1)$. In Sect. 3 we describe the whole set of new states which can thus be obtained, and discuss the related probability distribution functions, higher order moments and squeezing properties. There appear a set of very interesting scaling properties, which exhibit unexpected universality features. In Sect. 4 we introduce the notion of fractional photon states, and discuss the possibility of their physical realization. A few conclusive comments are given in Sect. 5.

2. Algebraic background

2.1 Multiphoton operators

The new type of multiphoton squeezing operators is constructed resorting to the generalized Bose operators of Brandt and Greenberg [17] $b_{(k)}$ and $b^\dagger_{(k)}$. The latter satisfy the commutation relations

$$[b_{(k)}, b^\dagger_{(k)}] = 1 \quad , \tag{2.1}$$

$$[N, b_{(k)}] = -k b_{(k)} \quad , \tag{2.2}$$

where $N = a^\dagger a$ is the usual number operator.

Equations (2.1) and (2.2) lead to interpreting $b_{(k)}$ and $b^\dagger_{(k)}$ as annihilation and creation operators of k photons simultaneously, even though it should be noted that $b_{(1)} = a$, but $b_{(k)} \neq a^k$ for $k \geq 2$.

From (2.1) and (2.2) one can derive the normal-ordered representation

$$b_{(k)} = \sum_{j=0}^{\infty} \alpha_j^{(k)} (a^\dagger)^j a^{j+k} \quad , \tag{2.3}$$

where

$$\alpha_j^{(k)} = \sum_{l=0}^{j} \frac{(-)^{j-l}}{(j-l)!} \left(\frac{1+[[\frac{l}{k}]]}{l!(l+k)!} \right)^{\frac{1}{2}} e^{i\phi_l} \quad . \tag{2.4}$$

In (2.4) $[[x]]$ denotes the maximum integer $\geq x$, whereas the phases $\phi_m, m = 0, \ldots, j$ are arbitrary real numbers.

In the Fock space $b_{(k)}$ and $b^\dagger_{(k)}$ operate as follows:

$$b_{(k)}|sk+\lambda\rangle = \sqrt{s}|sk+\lambda\rangle \quad , \tag{2.5a}$$

$$b^\dagger_{(k)}|sk+\lambda\rangle = \sqrt{s+1}|(s+1)k+\lambda\rangle \quad ; \tag{2.5b}$$

where $0 \leq \lambda \leq k-1$.

One can notice from (2.5) that the Fock space splits into k orthogonal subspaces which are invariant under the action of the k-photon operators:

$$\begin{aligned} \mathcal{F} &= \bigoplus_{\lambda=0}^{k-1} \mathcal{F}_\lambda^{(k)} \quad , \quad \mathcal{F}_\lambda^{(k)} = \bigoplus_{s=0}^{\infty} span\{|sk+\lambda\rangle\} \quad , \\ b_{(k)}\mathcal{F}_\lambda^{(k)} &\subset \mathcal{F}_\lambda^{(k)} \quad , \quad b^\dagger_{(k)}\mathcal{F}_\lambda^{(k)} \subset \mathcal{F}_\lambda^{(k)} \quad . \end{aligned} \tag{2.6}$$

The generic Fock state $|sk+\lambda\rangle$ is thus labeled by two quantum numbers s and λ, which are the eigenvalues of the complete set of commuting operators $b^\dagger_{(k)}b_{(k)}$ and $\hat{D}_{(k)} = a^\dagger a - kb^\dagger_{(k)}b_{(k)}$:

$$b^\dagger_{(k)}b_{(k)}|sk+\lambda\rangle = s|sk+\lambda\rangle \quad , \tag{2.7a}$$

$$\hat{D}_{(k)}|sk+\lambda\rangle = \lambda|sk+\lambda\rangle \quad . \tag{2.7b}$$

In Sec. 4 we shall equivalently consider a different set of commuting operators, namely $\hat{D}_{(k)}$ itself together with the canonical operator $\hat{Q}_{(k)}$

$$\hat{Q}_{(k)} = \frac{1}{\sqrt{2}}(b_{(k)} + b^\dagger_{(k)}) \quad . \tag{2.7c}$$

2.2 Holstein-Primakoff realizations of Lie algebras

The new set of squeezed states is defined by means of multiboson realizations of Lie algebras. In the papers [2-10] almost all simple Lie algebras and the usual solvable Weyl-Heisenberg algebra defined in (2.1) have been considered.

According to the Levi theorem, all these algebras are essentially the building blocks of every Lie algebra: this means that we can deal with a generic Lie algebra by decomposing it into its fundamental blocks. On the other hand in ref. [9] it is proved that the main squeezing properties for higher rank $SU(n)$ states reduce to those of $SU(2)$ and one can recover all the interesting features limiting the attention to the lowest rank Lie algebras. We thus consider only $W.H.$ (Weyl-Heisenberg), $SU(2)$ and $SU(1,1)$ groups which are the simplest examples of solvable, compact and non compact respectively, unitary Lie groups.

Therefore we now briefly summarize their defining commutation relations and their multiboson H.P. (Holstein-Primakoff [18]) realizations.

$SU(2)$

The commutation relations of $SU(2)$ are

$$\begin{aligned} &[J_+, J_-] = 2J_3 \\ &[J_3, J_\pm] = \pm J_\pm \quad . \end{aligned} \tag{2.8}$$

The UIR (unitary irreducible representation) corresponding to the eigenvalue $\sigma(\sigma + 1)$ of the Casimir operator $J_3^2 + \frac{1}{2}(J_+J_- + J_-J_+)$ can be realized on a $2\sigma + 1$ dimensional subspace of a fixed $\mathcal{F}_\lambda^{(k)}$ sector by means of the following generalized H.P. transformations:

$$\begin{aligned} &J_+^{(k)} = (2\sigma + 1 - b_{(k)}^\dagger b_{(k)})^{\frac{1}{2}} b_{(k)}^\dagger = [J_-^{(k)}]^\dagger \quad ; \\ &J_3^{(k)} = b_{(k)}^\dagger b_{(k)} - \sigma \quad . \end{aligned} \tag{2.9}$$

The special case $k = 1$ corresponding to the usual H.P. transformation has been considered too.

$SU(1,1)$

The commutation relations of $SU(1,1)$ are

$$\begin{aligned} &[K_+, K_-] = -2K_3 \\ &[K_3, K_\pm] = \pm K_\pm \quad . \end{aligned} \tag{2.10}$$

The UIR representation corresponding to the eigenvalue $\sigma(\sigma - 1)$ of the Casimir operator $K_3^2 - \frac{1}{2}(K_+K_- + K_-K_+)$ is now infinite dimensional and can be realized on a whole fixed $\mathcal{F}_\lambda^{(k)}$ sector by means of the H.P. trasformations

$$\begin{aligned} &K_+^{(k)} = (2\sigma - 1 + b_{(k)}^\dagger b_{(k)})^{\frac{1}{2}} b_{(k)}^\dagger = [K_-^{(k)}]^\dagger \quad ; \\ &K_3^{(k)} = b_{(k)}^\dagger b_{(k)} + \sigma \quad . \end{aligned} \tag{2.11}$$

As for $SU(2)$ also the $k = 1$ case has been considered. For the $SU(1,1)$ case a bilinear realization of the algebra is also possible:

$$\begin{aligned} &K_+ = \frac{1}{2} a^{\dagger 2} = [K_-]^\dagger \quad ; \\ &K_3 = \frac{1}{4}(2a^\dagger a + 1) \quad . \end{aligned} \tag{2.12}$$

There are two UIR acting on the $\mathcal{F}_0^{(2)}$ and $\mathcal{F}_1^{(2)}$ sectors of the $k = 2$ splitting of the Fock space, corresponding to $\sigma = \frac{1}{4}$ and $\sigma = \frac{3}{4}$ respectively. As we shall see in Sect.3.1, the usual Gaussian states [12],[13],[20],[21] are related to the $\sigma = \frac{1}{4}$ case.

Weyl-Heisenberg

The $W.H.$ algebra is the algebra of particle operators (or equivalently of the position and momentum operators). The commutation relations are given in (2.1) and the UIR is the usual infinite dimensional Fock representation, realized on a whole fixed $\mathcal{F}_\lambda^{(k)}$ sector.

One can notice that both the $SU(2)$ and $SU(1,1)$ H.P. realizations reduce to the $W.H.$ algebra in the limit $\sigma \to \infty$ (in Sect.3.1 we shall give an intuitive geometrical interpretation of such limit).

In conclusion we recall that one can relize both the $SU(2)$ and $SU(1,1)$ UIR using bilinear products of bose operators corresponding to more than one oscillator mode (see for example ref. [22]).

3. The new states, their distributions and moments

3.1 Definition of the states

We focus our analysis on states $|\omega\rangle$ corresponding to zero-average position and momentum, since such an average can be arbitrarily changed to any desired value by a simple traslation

$$|z\rangle_\omega = D(z)|\omega\rangle \quad , \tag{3.1}$$

where $D(z)$ is the unitary displacement operator:

$$D(z) = \exp(za^\dagger - z^* a) \quad . \tag{3.2}$$

$a^\dagger$ and a denote the usual creation and annihilation operators $[a, a^\dagger] = 1$. In fact, for, say, the position $\hat{q} = \frac{1}{\sqrt{2}}(a + a^\dagger)$, one has

$$_\omega\langle z|\hat{q}|z\rangle_\omega = \sqrt{2} Rez \quad . \tag{3.3a}$$

Thus, the generic nth moment is given by

$$_\omega\langle z|(\hat{q} - \langle\hat{q}\rangle)^n|z\rangle_\omega = \langle\omega|\hat{q}^n|\omega\rangle \equiv \chi_\omega^{(n)} \quad . \tag{3.3b}$$

Analogous result hold for the momentum operator $\hat{p} = \frac{i}{\sqrt{2}}(a - a^\dagger)$.

The property that $|\omega\rangle$ is a zero average state is guaranteed if one assumes

$$|\omega\rangle = \hat{S}_\omega|0\rangle \quad , \tag{3.4}$$

where $\hat{S}_\omega$ is a unitary squeezing operator, which is an analytic function of multiparticle operators.

Furthermore, in view of the comparison we are interested in, between squeezed states and the customary coherent states (which have a Gaussian distribution for the canonical variables), we construct even distributions using an even number of particle creators. The usual squeezing operator [12],[20], which gives rise to a gaussian distribution,

$$\hat{S}(\zeta)_{Gauss} = \exp\left[\frac{1}{2}(\zeta a^{\dagger 2} - \zeta^* a^2)\right] \quad , \tag{3.5}$$

satisfies both the above requirements. Fisher, Nieto, and Sandberg [11] have proposed generalizations of the operator (3.5) in the form

$$\hat{S}_{(k)}(\zeta) = \exp(\zeta a^{\dagger k} - \zeta^* a^k + h_k) \tag{3.6}$$

where $h_k = h_k^\dagger$ is a polynomial in a and $a^\dagger$ with powers up to $(k-1)$. The resulting squeezed states cannot be treated in general by analytic methods. Indeed, for example, the Taylor expansion of the vacuum expectation value $\langle 0|\hat{S}_k|0\rangle$ leads to a series with zero radius of convergence (see also ref. [1]) and numerical computations could be performed only resorting to Padé approximants [14]. Only very special cases of operators of the form (3.6) can be analytically handled, i.e., for example, when $\hat{S}_k$ is the evolution operator corresponding to an hamiltonian which is a power of a bilinear operator [23].

One can easily understand the appearence of the formal analytical divergencies induced by the operator (3.6) trying to compute its action on the vacuum vector. Doing that requires dealing with the B.C.H. (Backer-Campbell-Hausdorff) factorization of $\hat{S}_{(k)}(\zeta)$ in the form:

$$\hat{S}_{(k)}(\zeta) = \exp[f(\zeta)a^{\dagger k}]\exp(\mathcal{O}_0) \quad , \tag{3.7}$$

where $f(\zeta)$ is some suitable function of ζ and $\mathcal{O}_0$ is an operator which stabilizes the vacuum and gives only a normalization factor. Adopting this method one needs to compute, for example, iterated commutators of the form ($m \geq l$):

$$[a^l, a^{\dagger m}] = p(a^\dagger a)a^{\dagger l-m} \quad , \tag{3.8}$$

where $p(x)$ is a polynomial function. If $k > 2$ this procedure never ends, and infact explodes into an infinite dimensional Fock algebra which one is in general unable to handle. On the contrary if $k = 2$, as for the usual Gaussian squeezed states, the finite dimensional Lie algebra (2.12) is obtained, and the factorization (3.7) can be explicitly written. More precisely one can see that the usual Gaussian squeezed states are nothing but the generalized group theoretical coherent states of $SU(1,1)$ according to the general definition for an arbitrary Lie group given by Perelomov and Rasetti [15]. (Indeed it is well known that the usual harmonic-oscillator coherent states themselves are group theoretical coherent states for $W.H.$ group).

The last observation suggests that group theoretical coherent states are good candidates for a *non naive* generalization of the squeezed states. We recall their general definition.

The set of coherent states for a Lie group G is obtained using a UIR of the group, chosing a fixed vector $|\Omega\rangle$ in the representation space, and acting on it by the whole group. It turns out that the coherent states are labeled by means of the left cosets of the group G with respect to the soubgroup leaving $|\Omega\rangle$ invariant up to a phase factor. Resorting to the above definition we construct the generalized squeezed states for the $SU(2)$, $SU(1,1)$ and $W.H.$ groups using the H.P. realizations of Sect.2.2:

$$|\zeta; k, \sigma\rangle^{SU(2)} = \exp[\zeta J_+^{(k)} - \zeta^* J_-^{(k)}]|0\rangle \quad ; \tag{3.9a}$$

$$|\zeta; k, \sigma\rangle^{SU(1,1)} = \exp[\zeta K_+^{(k)} - \zeta^* K_-^{(k)}]|0\rangle \quad ; \tag{3.9b}$$

$$|\zeta; k\rangle^{W.H.} = \exp[\zeta b_{(k)}^\dagger - \zeta^* b_{(k)}]|0\rangle \quad ; \tag{3.9c}$$

$$|\zeta\rangle^{Gauss} = \exp[\frac{1}{2}(\zeta a^{\dagger 2} - \zeta^* a^2)]|0\rangle \quad . \tag{3.9d}$$

For the sake of completeness we have written in (3.9d) also the usual Gaussian squeezed states which correspond to the $\mathcal{D}^+(\frac{1}{4})$ discrete series UIR of SU(1,1) and not to the H.P. realization.

By the B.C.H. formula one can rewrite the states (3.9) in the more convenient form :

$$|\xi; k, \sigma\rangle^{SU(2)} = (1 + |\xi|^2)^{-\sigma} \exp[\xi J_+^{(k)}]|0\rangle \quad ; \tag{3.10a}$$

$$|\xi; k, \sigma\rangle^{SU(1,1)} = (1 - |\xi|^2)^{\sigma} \exp[\xi K_+^{(k)}]|0\rangle \quad ; \tag{3.10b}$$

$$|\xi; k\rangle^{W.H.} = e^{-\frac{1}{2}|\xi|^2} \exp[\xi b_{(k)}^\dagger]|0\rangle \quad ; \tag{3.10c}$$

$$|\xi\rangle^{Gauss} = (1 - |\xi|^2)^{\frac{1}{4}} \exp[\frac{1}{2}\xi a^{\dagger 2}]|0\rangle \quad . \tag{3.10d}$$

The relation between the parameter ξ labeling the states in (3.10) and the parameter ζ in (3.9) is reported in the first column of Table I.

In the Fock basis the squeezed states read as follows:

$$|\xi;k,\sigma\rangle^{SU(2)} = (1+|\xi|^2)^{-\sigma}\sum_{n=0}^{\infty}\binom{2\sigma}{n}^{\frac{1}{2}}\xi^n|kn\rangle \quad ; \tag{3.11a}$$

$$|\xi;k,\sigma\rangle^{SU(1,1)} = (1-|\xi|^2)^{\sigma}\sum_{n=0}^{\infty}\binom{2\sigma+n-1}{n}^{\frac{1}{2}}\xi^n|kn\rangle \quad ; \tag{3.11b}$$

$$|\xi;k\rangle^{W.H.} = e^{-\frac{1}{2}|\xi|^2}\sum_{n=0}^{\infty}\frac{\xi^n}{\sqrt{n!}}|kn\rangle \quad ; \tag{3.11c}$$

$$|\xi\rangle^{Gauss} = (1-|\xi|^2)^{\frac{1}{4}}\sum_{n=0}^{\infty}\binom{2n}{n}^{\frac{1}{2}}\left(\frac{1}{2}\xi\right)^n|2n\rangle \quad . \tag{3.11d}$$

Equations (3.11) manifestly show that the squeezed states thus constructed are indeed multiphoton states.

From eqs. (3.11) the probability distribution of the number operator is easily obtained:

$$\mathcal{N}^{SU(2)}_{(\xi,k,\sigma)}(kn) = (1+|\xi|^2)^{-2\sigma}\binom{2\sigma}{n}|\xi|^{2n} \quad ; \tag{3.12a}$$

$$\mathcal{N}^{SU(1,1)}_{(\xi,k,\sigma)}(kn) = (1-|\xi|^2)^{2\sigma}\binom{2\sigma+n-1}{n}|\xi|^{2n} \quad ; \tag{3.12b}$$

$$\mathcal{N}^{W.H.}_{(\xi,k)}(kn) = e^{-|\xi|^2}\frac{|\xi|^{2n}}{n!} \quad ; \tag{3.12c}$$

$$\mathcal{N}^{Gauss}_{\xi}(2n) = (1-|\xi|^2)^{\frac{1}{2}}\binom{2n}{n}\left(\frac{|\xi|}{2}\right)^{2n} \quad ; \tag{3.12d}$$

$$\mathcal{N}^{SU(2)}_{(\xi,k,\sigma)}(p) = \mathcal{N}^{SU(1,1)}_{(\xi,k,\sigma)}(p) = \mathcal{N}^{W.H.}_{(\xi,k)}(p) = 0 \quad ; \quad p \neq kn \tag{3.12e}$$

$$\mathcal{N}^{Gauss}_{\xi}(2n+1) = 0 \quad . \tag{3.12f}$$

Equations (3.12a)-(3.12e) represent, respectively, the binomial, negative-binomial and Poisson distributions in the many-photon variable kn.

We want now to compare the statistical properties inherent in the different states. The squeezing parameter ξ introduced in eqs.(3.10) does not lend itself to a trasparent physical interpretation and appears therefore somewhat ambiguous. On the other hand, one can see from Table I that the quantity describing the number fluctuations,

$$\delta = \frac{\Delta n}{\langle n\rangle} \quad , \tag{3.13}$$

of the number operator is inversely proportional to $|\xi|$ with a coefficient depending only on the group representation. We adopt therefore δ^{-1} as a good independent variable to compare the squeezing properties of the different states.

3.2 Squeezing properties of the new states.

In this section we give a detailed analysis of the second moment of the position variable $\hat{q}$ normalized to the value corresponding to the vacuum state

$$\chi^{(2)} = \frac{\langle\omega|\hat{q}^2|\omega\rangle - \langle\omega|\hat{q}|\omega\rangle^2}{\langle 0|\hat{q}^2|0\rangle - \langle 0|\hat{q}|0\rangle^2} \quad , \tag{3.14}$$

by varying the state $|\omega\rangle$ in the set (3.11) and in the direction of the maximal squeezing, i.e. for negative real ξ.

Figure 1 reports the second moments $\chi^{(2)}$ for various two-photon squeezed states as functions of δ^{-1} (the latter two in the $\sigma = 3$ representation). For the sake of comparison, the results for Gaussian states are also shown. One can notice that among all states the Gaussian ones exibit the best squeezing for a fixed value of δ^{-1}. However, they cannot attain a fluctuation in the observable number of photons lower than $\sqrt{2}$;in other words, the Gaussian states are *photon noisy*. Furthermore, as $\chi^{(2)}_{Gauss}$ is a monotonic decreasing function of δ^{-1} , the best squeezing corresponds to the lowest $\hat{n}$ fluctuation. On the oter hand, all the other non-Gaussian states give rise to functions $\chi^{(2)}(\delta^{-1})$ which are not monotonic but exibit a local minimum. Among them only the $SU(1,1)$ states can be completely squeezed ($\chi^{(2)} = 0$).

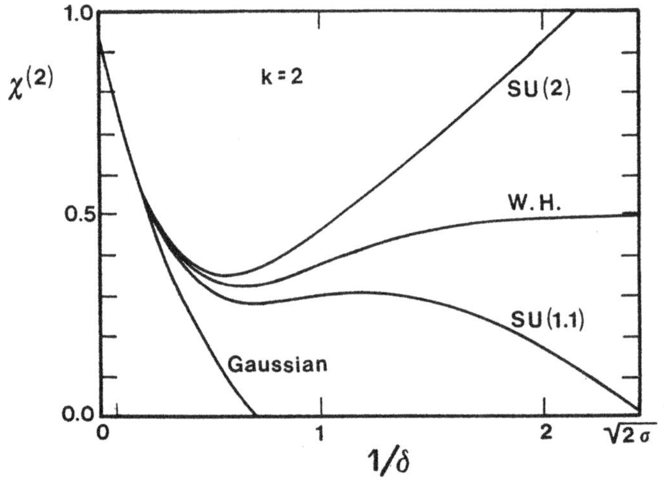

Fig.1 - Squeezing (i.e., second moment for negative squeezing parameter) vs the inverse $\hat{n}$ fluctuation parameter δ^{-1}, for various two-photon squeezed states; $SU(2)$ and $SU(1,1)$ states correspond to the $\sigma = 3$ representation (from ref.[7]).

One can notice as well that, in general, non-Gaussian states can attain a photon-number fluctuation smaller than those of the Gaussian states. In particular, the $W.H.$ states can have an arbitrarily small photon noise, but they are *squeezed limited* in that the second moment $\chi^{(2)}_{W.H.}$ exibits an *absolute* minimum $\chi^{(2)}_{W.H.,min} = 0.31744$ corresponding to $\delta^{-1} = 0.64675$. The $SU(1,1)$ states can be squeezed to zero second moment in correspondence to the optimal value $\delta^{-1} = \sqrt{2\sigma}$. Therefore one can simultaneously reduce to zero both $\hat{n}$-noise and $\hat{q}$-noise in the limit $\sigma \to \infty$. It is worth pointing out that whereas for $W.H.$ states the local minimum is also a global one, for the $SU(1,1)$ states the absolute minimum is zero (numerical values of relative minima for large σ are given in Refs.[7] and [4]). Finally, the $SU(2)$ states are no longer squeezed ($\chi^{(2)}_{SU(2)} > 1$) for small $\hat{n}$ fluctuations.

Figure 2 shows the reduced absolute fluctuations $(\Delta n/k)^2$ vs δ^{-1} for the same states considered in Fig. 1. It appears from this figure, comparing it with the previous one, that the better the squeezing the higher the photon-number fluctuations. In particular, the Gaussian states exibit the highest photon noise.

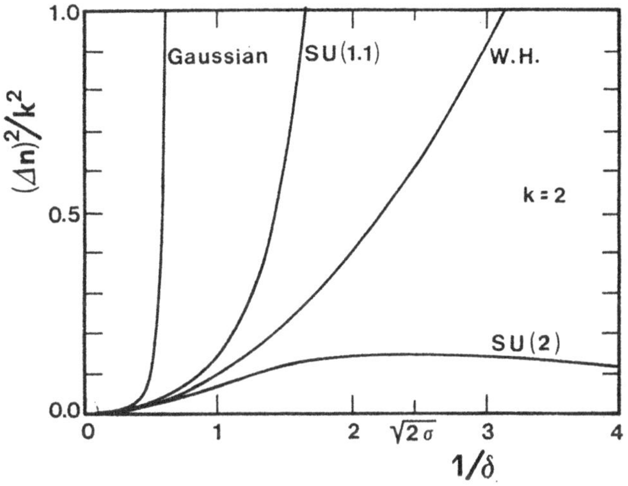

Fig.2 - Reduced $\hat{n}$ square variance vs the inverse $\hat{n}$-fluctuation parameter δ^{-1} for the same two-photon squeezed states of Fig. 1 (from ref.[7]).

In the limit of squeezing to zero second moment, $\delta \to \sqrt{2}$ for the Gaussian states or $\delta \to 1/\sqrt{2\sigma}$ for the $SU(1,1)$ states, the $\hat{n}$-variance increases asymptotically to infinity for both states. $(\Delta n/k)^2_{W.H.}$ grows parabolically with δ^{-1}, whereas $(\Delta n/k)^2_{SU(2)}$ shows a maximum, and decreases to zero as δ^{-1} tends to infinity, as $\sim (\delta^{-1})^{-2}$.

From Figs. 1 and 2, one can then conclude that the local minimum of $\chi^{(2)}$ for the non-Gaussian states can be considered as an optimum situation as it provides the best compromise between the requisite of maximum squeezing and that of minimum absolute noise in the photon number.

3.3 Probability distributions

In this section we show some numerical results (ref.[5]) concerning the position probability distribution

$$\mathcal{Q}_\omega(q) = |\langle q|\omega\rangle|^2 \quad , \tag{3.15}$$

and the number distribution

$$\mathcal{N}_\omega(n) = |\langle n|\omega\rangle|^2 \quad , \tag{3.16}$$

for some states in the set (3.11).

Position probability distribution

Figures 3(a)-3(h) represent $\mathcal{Q}_\omega(q)$ for the $W.H.$ states for various choices of k and for different values of $\zeta = \rho e^{i\phi}$ (the probability distributions for the $SU(2)$ and $SU(1,1)$ states are analogous).

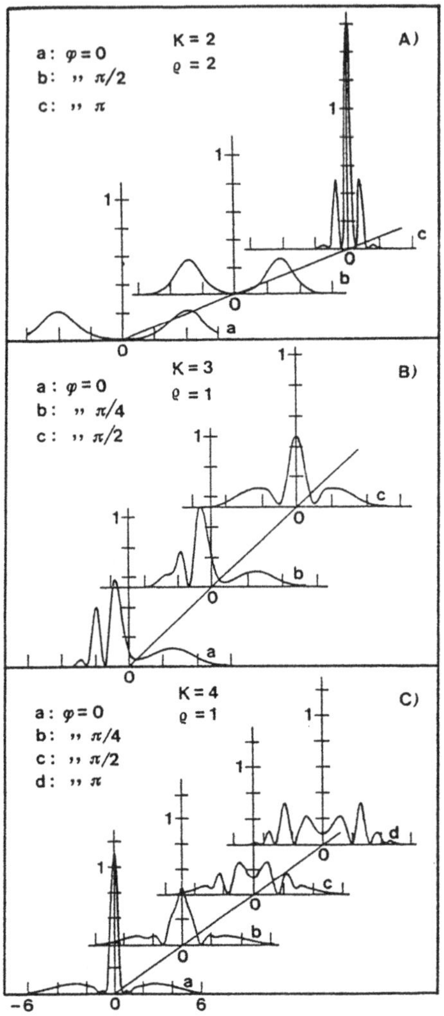

Fig.3 - The probability distribution function $\mathcal{Q}_{(k,\zeta}$ vs the (dimensionless) position q, for different values of k and different choices of $\zeta = \rho e^{i\phi}$ (from ref.[5]).

One notices how such distributions exhibit a sensible deviation from the Gaussian behavior. The functions corresponding to even k are symmetric under the exchange $q \longrightarrow -q$, whereas there is no symmetry for odd k except for $\phi = \pi/2$. A characteristic feature of the functions $\mathcal{Q}_{(k,\zeta)}(q)$ is that they show an increasing number of zeros when ρ is increased at fixed ϕ, for any k. The same effect, i.e., a richer structure corresponding to a larger number of nodes, appears when k is increased keeping ζ fixed.

As for the moments of these probability distributions, a general theorem is proven in ref. [8] concerning the probability distribution of every multiphoton state (i.e. a state which is a surposition of eigenstates $|kn\rangle$ with varying n and fixed k):

For a k-photon state $|\omega\rangle$, only the moments $\chi_\omega^{(2N)}$ corresponding to $2N \geq k$ can be squeezed for even k, $N \geq k$ for odd k .

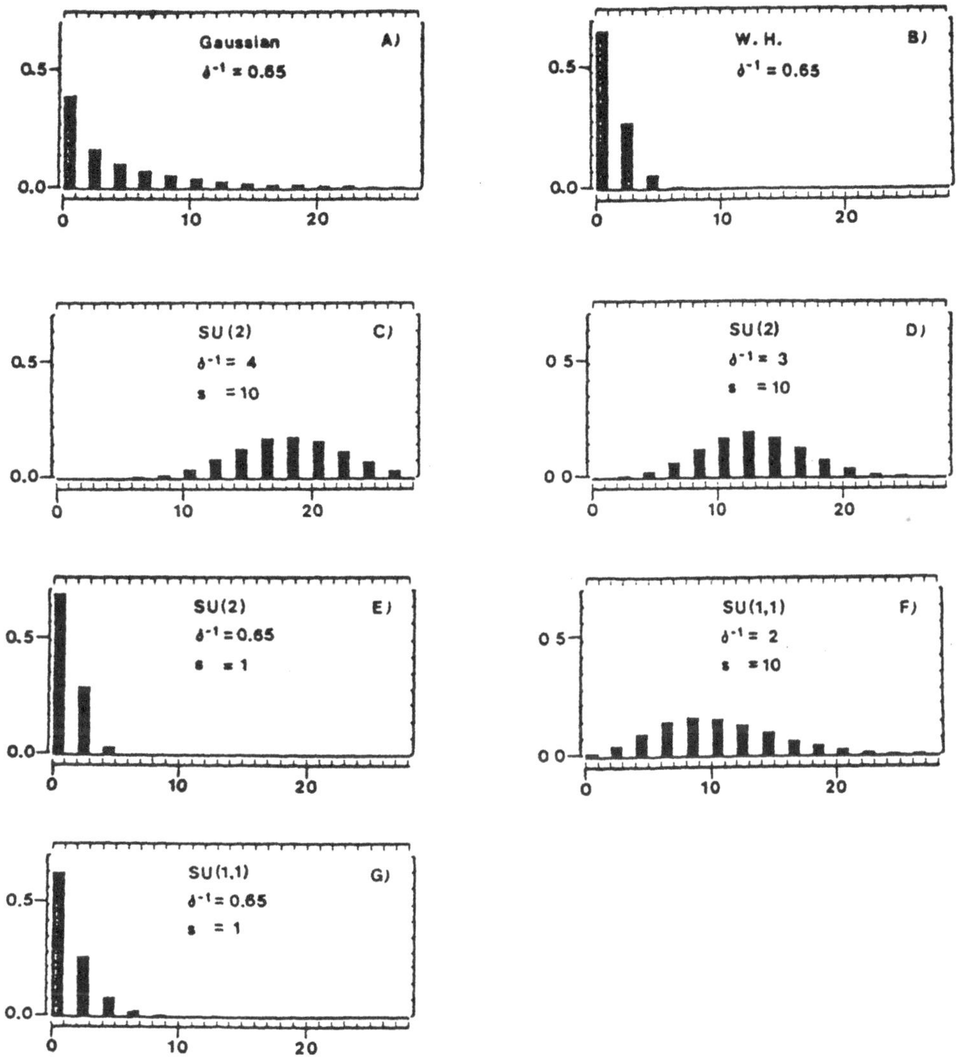

Fig.4 - The probability distribution function $\mathcal{N}(n)$ of the $\hat{n}$ operator for the $k = 2$ states for different values of σ and $|\xi|$.

Number probability distribution

Figure 4 shows some $\mathcal{N}_\omega(n)$ distributions for all the different types of states and a few values of

σ (for the $SU(2)$ and $SU(1,1)$ cases). Notice how the negative binomial distribution appears as the slowest decaying one for large n: as $|\xi|^2$ and/or σ are increased, one obtains a maximum for larger and larger n and more and more peacked functions, getting as a result smaller δ .

Restricting our attention to the Poissonian sub-Poissonian shape of the distributions, one can easily check from Table I that the only sub-Poissonian distribution is the binomial distribution, related to $SU(2)$ states: all other distributions are super-Poissonian (included that of the Gaussian squeezed vacuum).

TABLE 1 - Backer-Hausdorff parameter ξ and its range and photon-number average, variance, and fluctuation vs ξ, for the different squeezed states considered (from ref.[7]).

Squeezed state	$\xi = \xi(\varsigma)$	Range of $\lvert\xi\rvert$	$\langle n\rangle$	$\langle \Delta n\rangle^2$
Gauss	$\xi = \frac{\varsigma}{\lvert\varsigma\rvert}\tanh(\lvert\varsigma\rvert)$	$0-1$	$\frac{\lvert\xi\rvert^2}{1-\lvert\xi\rvert^2}$	$2\frac{\lvert\xi\rvert^2}{(1-\lvert\xi\rvert^2)^2}$
WH	$\xi = \varsigma$	$0-\infty$	$k\lvert\xi\rvert^2$	$k^2\lvert\xi\rvert^2$
SU(2)	$\xi = \frac{\varsigma}{\lvert\varsigma\rvert}\tan(\lvert\varsigma\rvert)$	$0-\infty$	$2k\sigma\frac{\lvert\xi\rvert^2}{1+\lvert\xi\rvert^2}$	$2k^2\sigma\frac{\lvert\xi\rvert^2}{(1+\lvert\xi\rvert^2)^2}$
SU(1,1)	$\xi = \frac{\varsigma}{\lvert\varsigma\rvert}\tanh(\lvert\varsigma\rvert)$	$0-1$	$2k\sigma\frac{\lvert\xi\rvert^2}{1-\lvert\xi\rvert^2}$	$2k^2\sigma\frac{\lvert\xi\rvert^2}{(1-\lvert\xi\rvert^2)^2}$

3.4 Scaling laws

The existence of the two vertical asymptotes for the Gaussian and $SU(1,1)$ states in Fig. 2, corresponding to the vanishing of $\chi^{(2)}$, suggests that we look at the dependence of $\chi^{(2)}$ vs $(\Delta n/k)^2$. One expects a scaling relation – that in the limit of large $(\Delta n)^2$ should give a generalized uncertainty relation – in the form

$$\chi^{(2)} \sim (\Delta n)^{-2\gamma} \tag{3.17}$$

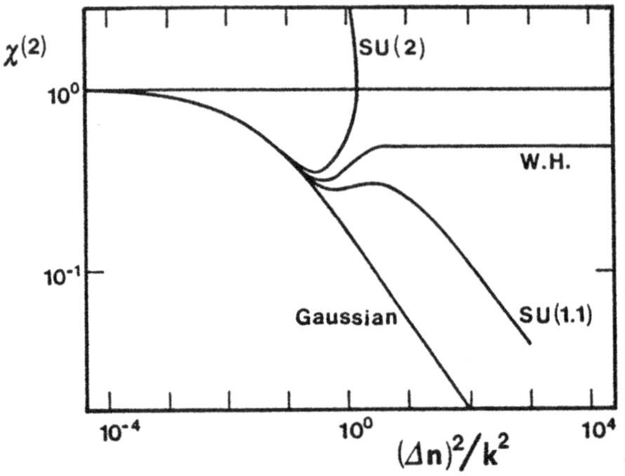

Fig.5 - Log-log plot of squeezing vs reduced absolute $\hat{n}$ fluctuation for the same squeezed states of Figs. 1 and 2 (from ref.[7]).

Figure 5 shows the log-log plot of squeezing versus reduced absolute photon-number fluctuation for all the two- photon states of Fig. 1. One can notice how $\gamma = \frac{1}{2}$ for both the Gaussian and the $SU(1,1)$ states; in the latter case γ being independent of the value of σ, provided it is finite. The proportionality constant depends on both the state [Gaussian or $SU(1,1)$] and on the representation $[\sigma]$. Thus the parameter γ can be thought of as a *universal* scale exponent. One should observe that, considering the $W.H.$ states as the $\sigma \to \infty$ limit of $SU(1,1)$, the universal behavior is broken in the same limit, and we have $\gamma = 0$.

Scaling laws analogous to (3.17) can be found for higher moments as well. Somewhat unexpectedly, scaling laws for second- and higher-order moments appear as well for all the states corresponding to the local minima of the moments themselves versus δ^{-1}.In this case, the parameter whereby the two uncertainties $\chi^{(2N)}$ and $(\Delta n)^2$ can be connected is the representation label σ, which is the only remaining free variable. Generalized scaling laws of the form

$$\chi_{(k)}^{(2N)} \sim (\Delta n)^{-2\gamma_k(N)} \tag{3.18}$$

can be obtained by eliminating σ^{-1} between $\chi^{(2N)}$ and $(\Delta n)^2$.

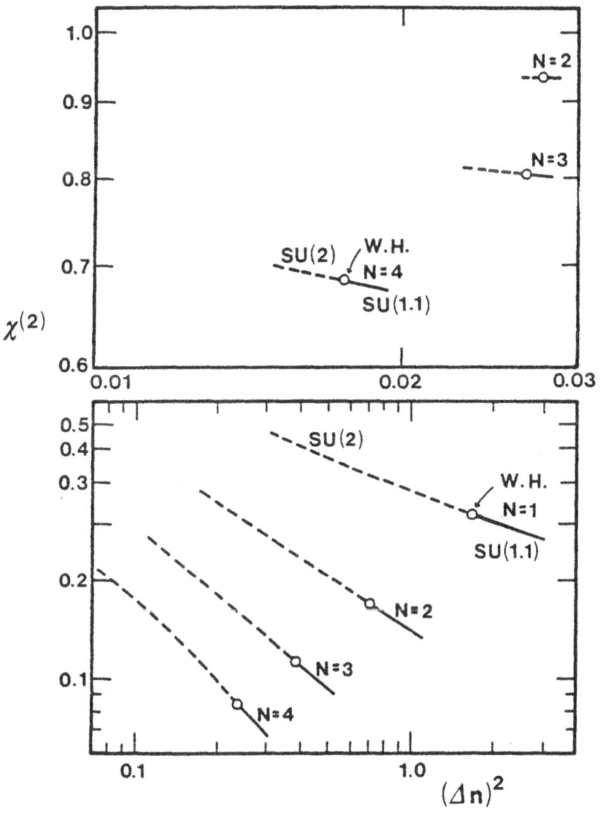

Fig.6 - Generalized squeezing for the $2N$th moments vs absolute $\hat{n}$ fluctuation at the local minimum (log-log plot): (a) $k = 4$, $N = 2 \div 4$; (b) $k = 2$, $N = 1 \div 4$ (from ref.[7]).

Figure 6 shows the log-log plots of the optimal moments $\chi_{min}^{(2N)}$ versus the corresponding $\hat{n}$-variance, which manifestly exhibit a power-law behavior of the form (3.18). It is interesting to point out how, in this representation, all states [$W.H.$, $SU(2)$, and $SU(1,1)$] lie on the same straight lines. The exponents $\gamma_k(N)$ are positive numbers less than 1, whose dependence on N and k is shown in Fig. 7. Notice that $\gamma_k(N)$ is monotonically increasing with N and decreasing with k, (on the contrary one obtains that the proportionality constant is decreasing with N and increasing with k).

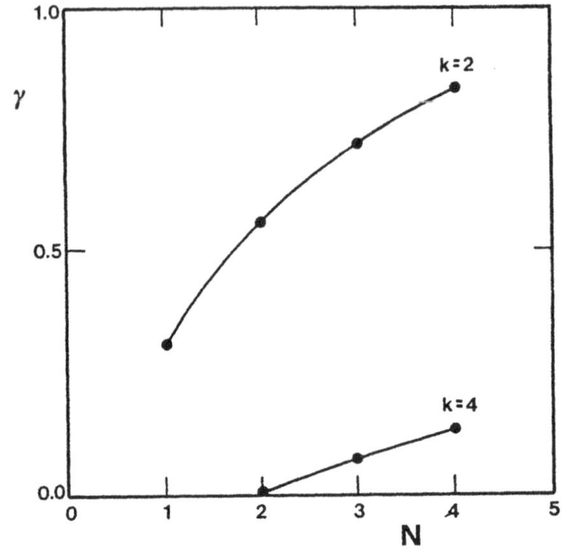

Fig.7 - The exponent $\gamma_k(N)$ of the scaling laws (3.18), corresponding to plots of Fig. 6, vs N for $k = 2, 4$ (from ref.[7]).

4. Fractional photons

4.1 The algebraic definition

In Sect 2.1 we have defined the transformation $F_{(k)}$ of (2.3),(2.4):

$$F_{(k)} : a^\dagger \longrightarrow F_{(k)}(a^\dagger) = b^\dagger_{(k)} \quad , \quad F_{(k)}(a) \equiv [F_{(k)}(a^\dagger)]^\dagger \quad , \tag{4.1}$$

for positive integer k only. We note that $F_{(1)}(a^\dagger) = a^\dagger$, and, with a little algebra,

$$F_{(k)} \circ F_{(k')}(a^\dagger) = F_{(kk')}(a^\dagger) \quad .$$

We may at least formally (Ref. [8]), extend the semigroup of nonlinear trasformations (4.1) to the Abelian group $\{F_{(k)} \text{rational} k > 0\}$ (one should really think of this group as a group of canonical transformations acting on pairs of conjugate operators), by defining the inverse trasformation $F_{(k)}^{-1}$ by

$$F_{(k)}{}^{-1} \circ F_{(k)}(a^\dagger) = a^\dagger = F_{(1)}(a^\dagger) \quad .$$

We may therefore equate $F_{(k)}^{-1} = F_{(1/k)}$, whence

$$F_{(k)}{}^{-1} \circ F_{(k')} = F_{(k'/k)} = F_{(r)}$$

where $r \equiv k'/k$ is a positive rational number. It is this extension which allows us to define the notion of *fractional photons*.

The above formal structure is equivalent to considering the action of k'-boson operators on a $\mathcal{F}$ sector with $k \neq k'$. Focusing our attention on the particular sector $\mathcal{F}_0^{(k)}$, the k'-boson action is given by the following matrix elements:

$$\langle km|(b^\dagger_{(k')})^u(b_{(k')})^v|km'\rangle = \left(\frac{[[km/k']]![[km'/k']]!}{[[km/k'-u]]![[km'/k'-v]]!}\right)^{\frac{1}{2}} \delta_{m,m'+s} \quad ,$$

where we have defined s by $s = (k'/k)(u-v)$. When $u = v$, then $s = 0$ and the expectation (2.3) always has nonzero values (for $m = m'$). When $u \neq v$, the expression (2.3) vanishes unless s is an integer; that is, $(k'/k)(u-v)$ is an integer.

Note that expression (2.3) depends only on k' and k through their ratio: $r = k'/k$. Here r is the positive rational fraction of the fractional trasformation $F_{(r)}$. We may equate expression (2.1) *formally* to an expectation involving fractional photons:

$$\langle km|(b^\dagger_{(k')})^u(b_{(k')})^v|km'\rangle = \langle m|(b^\dagger_{(r)})^u(b_{(r)})^v|m'\rangle$$

Thus the claim is not that such fractional photon modes really exist, but that physical experiments involving integral numbers of photons can be interpreted as behaving in such fractional mode.

4.2 Physical states, their distributions and squeezing properties.

In ref.[10] physical quantum states are constructed which have the same probability distributions of fractional photon states.

The definition of the probability distribution for fractional photon states is based on the construction of a complete set of eigenvectors for the two mutually commuting operators $\hat{Q}_{(k)}$ and $\hat{D}_{(k)}$ defined in Sect.2.1. The diagonalization procedure is standard, and gives the following result:

$$\begin{aligned} |Q,\lambda\rangle_{(k)} &= \sum_{l=0}^{\infty} C_l(Q)|lk+\lambda\rangle \quad , \\ C_l(Q) &= \frac{e^{\frac{Q^2}{2}} H_l(Q)}{\sqrt{2^l l! \sqrt{\pi}}} \end{aligned} \tag{4.2}$$

where $H_l(Q)$ are the usual Hermite polynomials of degree l. One can easily check that:

$$\begin{aligned} \hat{Q}_{(k)}|Q,\lambda\rangle_{(k)} &= Q|Q,\lambda\rangle_{(k)} \quad ; \\ D_{(k)}|Q,\lambda\rangle_{(k)} &= \lambda|Q,\lambda\rangle_{(k)} \end{aligned} \tag{4.3}$$

If one considers next a k-photon state in the $\mathcal{F}_0^{(k)}$ sector:

$$|\omega\rangle_{(k)} = \sum_{m=0}^{\infty} \omega_m |km\rangle \quad , \tag{4.4}$$

one can construct the probability distribution of the canonical variable $\hat{Q}_{(k')}$ for the k-photon state $|\omega\rangle_{(k)}$ $k \neq k'$ as

$$\mathcal{P}_\omega^{(t)} = \sum_{\lambda=0}^{k'-1} |_{(k')}\langle Q, \lambda||\omega\rangle_{(k)}|^2 = \sum_{l,m=0}^{\infty} \omega_l^* \omega_m C_{[[tl]]}(Q) C_{[[tm]]}(Q) \delta_{\langle\langle tl \rangle\rangle \langle\langle tm \rangle\rangle} \quad , \tag{4.5}$$

where $\langle\langle x \rangle\rangle = x - [[x]]$ denotes the fractional part of x. Eq. (4.5) shows clearly that the probability distribution depends only on $t = 1/r = k/k'$ and can thus be referred to as fractional photon probability distribution.

As an example we select, as k-photon vector $|\omega\rangle_{(k)}$, one among the generalized k-photon states of eq.(3.11), namely the $W.H.$ coherent state:

$$|\omega\rangle_{(k)} \equiv |\xi; k\rangle^{W.H.} \quad , \qquad \omega_n = \frac{e^{-\frac{1}{2}|\xi|^2} \omega^n}{\sqrt{n!}} \quad , \tag{4.6}$$

and restrict our attention to the special case $t = 1/n$. In this case the probability distributions are almost gaussian (see for example Fig. 8). More precisely they approach the vacuum gaussian shape for $t \rightarrow 0$ (and obviously for $|\omega|^2 \rightarrow 0$) in agreement with an intuitive physical meaning of vacuum as zero-fraction photon state. On the other hand if one increases $|\omega|^2$ at fixed t, the gaussian shape changes to a richer structure, correspond ing to a larger number of local maxima and minima. For very large $|\omega|^2$ the maxima raise up more and more sharply around the average value, and in the limit $|\omega|^2 \rightarrow \infty$ the distribution merges into a generalized function, as for the usual integer-k multi-photon distributions.

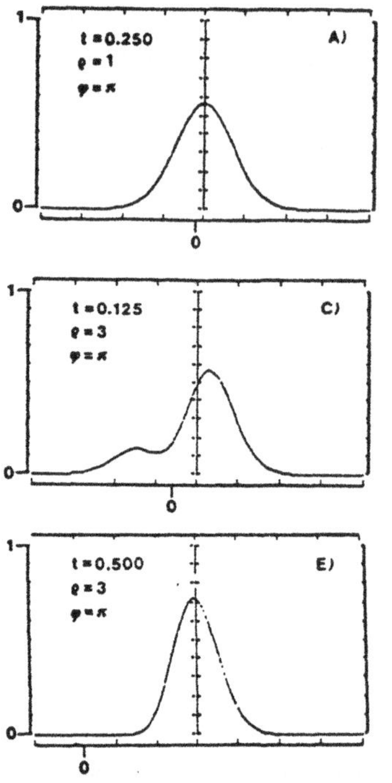

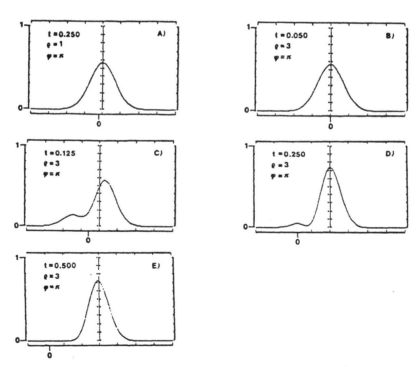

Fig.8 - The probability distribution $\mathcal{P}_\omega^{(t)}$ vs the canonical variable Q for the $W.H.$ fractional states, for some values of $\rho = |\omega|$ and $t = 1/n$.

Squeezing is obtained increasing $|\omega|$ along the negative real direction as usual. In Fig. 9 the second moment for the canonical variable $\hat{Q}_{(k)}$ is plotted versus $\rho = |\omega|$ for various values of $t = 1/n$. One can see that better squeezing is obtained for larger ρ and smaller t. One can check [10] that for increasing values of ρ the squeezing asymptotically approaches the constant value $\chi^{(2)} \sim t$. One gets thus the nice notion that *fractioning photons is equivalent to squeezing photon distributions.*

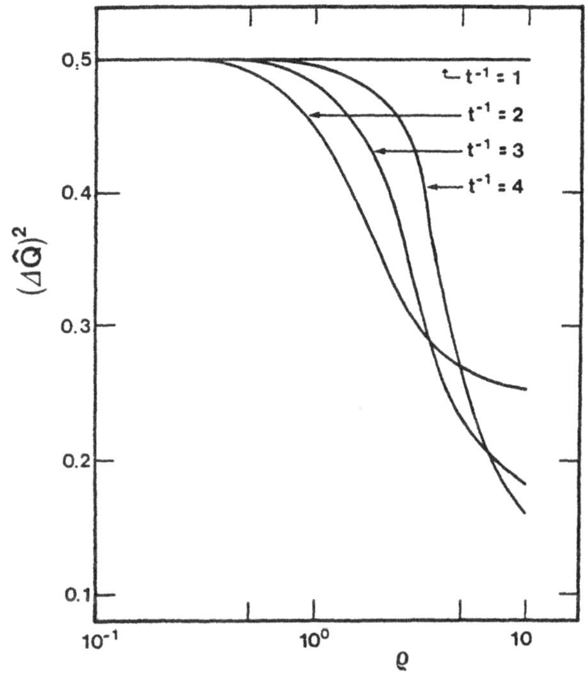

Fig.9 - Dependence of the second moment $\chi^{(2)}$ on the squeezing parameter $\rho = |\omega|$ for the *W.H.* fractional states for various values of $n = 1/t$ (from ref.[8]).

The probability distribution considered thus far refers to the canonical variable $\hat{Q}_{(k)}$ which does not have a simple physical meaning. One can inquire whether it is possible to construct a physical Fock state which has exactly both the probability distribution $\mathcal{P}_{\omega}^{(t)}$ and the number distribution $\mathcal{N}_{\omega}^{(t)}(N)$

$$\mathcal{N}_{\omega}^{(t)}(N) = \sum_{\lambda=0}^{h-1} |_{(h)}\langle N, \lambda|\omega\rangle_{(k)}|^2 = e^{-|\omega|^2} \sum_{\lambda=0}^{p-1} \frac{|\omega|^{2(pn+\lambda)}}{(pl+\lambda)!} \quad , \tag{4.7}$$

but referred to the usual position $\hat{q}$ and number $\hat{n}$ variables. The answer is positive: the physical fractional photon state is a mixed state defined by a density matrix (ref.[10]); i.e. the fractional photon is essentially a statistical object.

One can understand the physical features of the fractional states by looking at their number probability distribution (4.7). In Fig. 10 a few probability distributions are plotted for different t and various values of $\rho = |\omega|$ (for the sake of comparison, the usual coherent state correspondig to $t = 1$ is included as well). One can see how the fractional state has decreasing mean number of photons for decreasing t (it is straightforward to compute that, for large ρ, $\langle n\rangle \sim t$. Furthermore as ρ is increased a sub-Poissonian distribution is obtained (indeed one can analytically check that $(\Delta n)^2/\langle n\rangle \sim t$ for large ρ whereas $(\Delta n)^2/\langle n\rangle \sim \frac{1}{\rho^2}$.

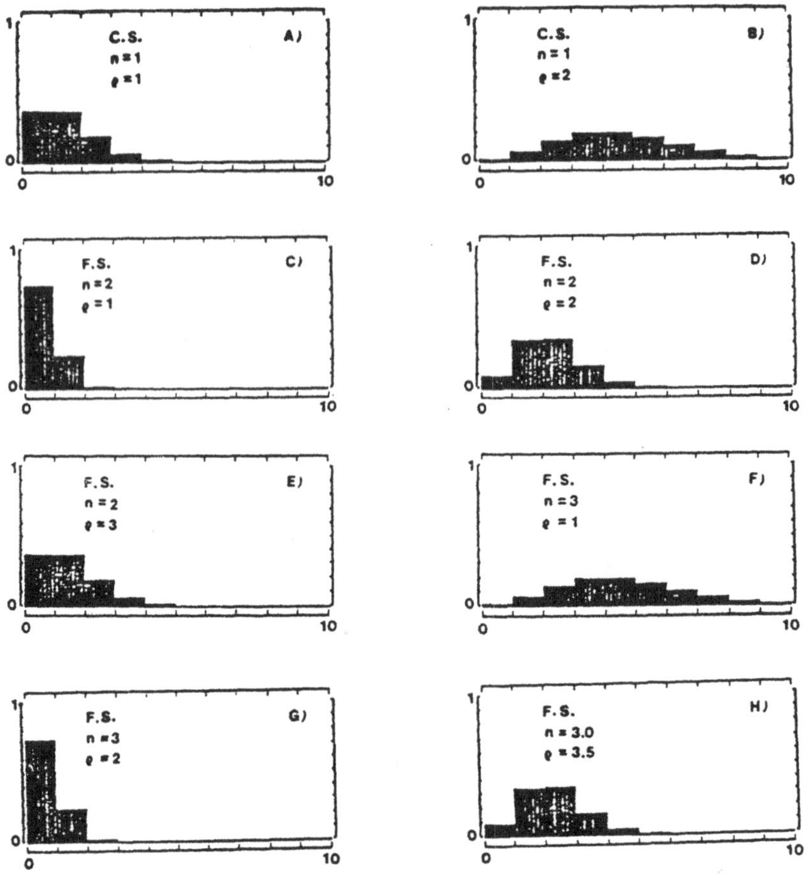

Fig.10 - Number probability distributions for the $W.H.$ fractional states for some values of $n = 1/t$ and $\rho = |\omega|$.

In conclusion, the most intriguing physical feature of the fractional photon state is that *one can simultaneously obtain complete squeezing, sub-Poissonian number distribution and very small number fluctuations* taking the limit $\rho \to \infty$ followed by the limit $t \to 0$.

5. Conclusions

We conclude with two comments related mainly with the problem of physically realizing the (formal) states described in this paper.

The information most relevant for the realization of a specified quantum state is of course the interaction whereby the state itself is generated as a pure states, stemming out of the appropriate vacuum. In all the cases described in this paper the resulting interaction – one can think of an Hamiltonian, roughly proportional to the logarithm of the squeezing operator – is very complicated in the regular phase space (it is in general described by an infinite power series of the single photon creation and annihilation operators), and typically, when expressed in terms of p and q variables does not show manifestly the structure of kinetic plus anharmonic potential energy one should expect. However, preliminary numerical analysis has shown that at least locally this in in fact the case (the potential exhibiting a characteristic double well shape). On the other hand the Hamiltonian should have the algebra corresponding to the state considered as dynamical (spectrum generating) algebra; namely there should exist non-manifest symmetries of the dynamical system resulting in the property that the Hamiltonian is generated by commutation relation in a finite rank algebraic structure. This hints to the existence of a set of action-angle variables in which the form of the interaction should be much simpler.

This is also supported by the feature – here shown explicitly for the fractional photon states – that very few single photon states are sufficient to realize the described squeezed states.

Further work along these lines is in progress.

References

[1] G. D'Ariano, M. Rasetti, and M. Vadacchino, J. Phys. **18** A, 1295 (1985)
[2] G. D'Ariano, M. Rasetti, and M. Vadacchino, Phys. Rev. D **32**, 1034 (1985)
[3] G. D'Ariano, M. Rasetti, and M. Vadacchino, in Noise in physical systems and 1/f noise 1985 A. D'Amico and P. Mazzetti eds., Elsevier Science Publ. B. V. pag.29 (1986)
[4] J. Katriel, A. Solomon, G. D'Ariano, and M.Rasetti, Phys. Rev. D **34**, 332 (1986)
[5] G. D'Ariano and M. Rasetti, Phys. Rev. D **35**, 1239 (1987)
[6] J. Katriel, A. Solomon, G. D'Ariano, and M. Rasetti, J. Opt. Soc. Am. B **4**, 1728 (1987)
[7] G. D'Ariano, S. Morosi, M. Rasetti, J. Katriel, and A. Solomon, Phys. Rev. D **36**, 2399 (1987)
[8] J. Katriel, M. Rasetti, and A. Solomon, Phys. rev. D **35**, 1248 (1987)
[9] J. Katriel, M. Rasetti, and A. Solomon, Phys. rev. D **35**, 2601 (1987)
[10] G. D'Ariano and N. Sterpi, (in preparation)
[11] R. A. Fisher, M. M. Nieto, and V. D. Sandberg, Phys. Rev. D 29, 110 (1984)
[12] D. Stoler, Phys. Rev. D **1**, 3217 (1970)
[13] H. P. Yuen, Phys. Rev. A **13**, 2226 (1976)
[14] S. L. Braunstein and R. I. Mc Lachlan, Phys. Rev. A **35**, 1659 (1987)
[15] A. M Perelomov, Commun. Math. Phys. **26**, 222 (1972); M. Rasetti, Int. J. Theor. Phys. **13**, 425 (1973); **14**, 1 (1975)
[16] R. J. Glauber (in this volume)
[17] R. A. Brandt and O. W. Greenberg, J. Math. Phys. **10**, 1168 (1969)
[18] T. Holstein and H. Primakoff, Phys. Rev. **58**, 1048 (1940)
[19] B. Yurke, S. L. McCall and J. R. Klauder, *$SU(2)$ and $SU(1,1)$ Interferometers*, AT & T preprint, 1986
[20] M. M. Nieto and L. M. Simmons Jr., Phys. Rev. D **20**,1321 (1979)
[21] J. N. Hollenhorst, Phys. Rev. D **19**, 1669 (1979)
[22] G. S. Agarwal (in this volume)
[23] P. Tombesi (in this volume)

SQUEEZING IN FOURTH-ORDER PARAMETRIC DOWNCONVERSION

Antonio Mecozzi* and Paolo Tombesi**

**Fondazione Ugo Bordoni, Via B.Castiglione 59, 00142 Roma, Italy*
***Dipartimento di Fisica, Università di Roma "La Sapienza"*
P.le A.Moro 2, 00185 Roma, Italy

1. INTRODUCTION

The systems which generate squeezing of the electromagnetic radiation[1,2], considered in the successful experiments performed in the last couple of years, are of the two-photon variety. They concern an optical medium with nonlinear polarization, with second-order[3] or third-order[4] susceptibility, interacting with an external coherent pump. Due to the low intensity of the pumps used high finesse cavity or optical fibers were employed.

It has also been shown[5] that higher-order nonlinearities, added to the Hamiltonian extensively studied by Yuen[6], can reduce the interaction time, or the length of a nonabsorbing medium, to generate squeezed light. Moreover, Tanas[7] has shown that, modeling the medium as an anharmonic oscillator, a high degree of squeezing may be achieved; however, these systems are of two-photon variety as well.

In this paper we will show that nonlinear optical effects of fourth order in the field, might be considered to obtain a great amount of squeezing in a nonabsorbing medium. We are able to exactly solve the model and show that the four-photon squeezing is different from the usual two-photon squeezing, since no squeezing at all is obtainable for an initial vacuum signal. Furthermore, the quantum probability distribution is far from being Gaussian. We note also that higher-oder squeezing, as proposed by Hong and Mandel[8], may be generated. Moreover, the squeezing strongly depends on the average number of photons in the initial state of the signal.

Squeezing from k-photon generation has also been discussed some years ago within a short time approximation[9] and Gerry[10], recently, has shown that a k-photon anharmonic oscillator gives enhanced squeezing, at least for some values of the average photon number in the incident beam. On the other hand, Fisher *et al.*[11] showed that a generalization to k-photon coherent states does not exist, whereas Braunstein and McLachlan[12] concluded that multi-photon squeezing is possible for an initial non empty signal. Finally, D'Ariano *et al.*[13] used a k-boson operators formalism to show that squeezing can be obtained, even though the squeezed states are non-Gaussian.

2. FOURTH SUBHARMONIC GENERATION

A synchronous process of fourth subharmonic generation, in crystals without inversion center[14], is connected with the nonlinear polarization which can be expressed, accurate to fourth-order terms in the field E, in the form

$$P_{nl} = \chi^{(2)} E^2 + \chi^{(3)} E^3 + \chi^{(4)} E^4 \tag{2.1}$$

where $\chi^{(k)}$ represents the k-th susceptibility tensor. Assume that the signal and pump beams are collinear and perfectly matched; then, the effective Hamiltonian describing our nonabsorbing medium, using the undepleted pump approximation, in the degenerate case is (setting $\hbar=1$)

$$H = \omega a^+ a + \kappa a^{+2} a^2 + \gamma B(t) a^{+4} + \text{h.c.} \tag{2.2}$$

where a, a^+ are the boson annihilation and creation operators for the signal at frequency ω, κ and γ are real coupling constants proportional to the 3-th and 4-th order susceptibility tensor respectively, and the external pump

$$B(t) = B_0 e^{-4i\omega t+i\phi} \tag{2.3}$$

is assumed very intense to be considered classical.

This effective Hamiltonian describes a fourth-order parametric downconversion, it gives rise to a two-photon process (the annihilation of two photons followed by the creation of two photons) and it gives also rise to a four-photon process. It represents one particular realization of the Hamiltonians considered by Fisher *et al.*[11].

By setting the phase of the external pump $\phi=\pi$, for simplicity, in the interaction picture the effective Hamiltonian Eq.(2.1) becomes

$$H = \kappa a^{+2} a^2 - \gamma B_0 a^{+4} + \text{h.c.} \tag{2.4}$$

The time evolution of the operators a and a^+ can be derived in the standard way and a short time approximation[9] may be used. However, since the intensity B_0 of the external pump can be controlled by the experimentalist, there exists a particular value of B_0 which permits an exact solution of the model. By choosing

$$\kappa = \gamma B_0 \tag{2.5}$$

we write Eq.(2.4) as

$$H = 4\kappa H_0^2 \tag{2.6}$$

where

$$H_0 = \frac{i}{2} (a^2 - a^{+2}) \tag{2.7}$$

is a particular form of the quadratic Hamiltonian fully discussed by Yuen[6].

In terms of the dimensionless time $\tau=4\kappa$, the equation of motion for the annihilation operator a becomes

$$\frac{da}{d\tau} = - ia - 2a^{+} H_o . \tag{2.8}$$

Let us introduce the two slowly varying, Hermitian, in-phase and out-of-phase, quadrature components x_1 and x_2 of the signal

$$x_1 (\tau) = a(\tau) + a^{+}(\tau)$$
$$x_2 (\tau) = i\,(a^{+}(\tau) - a(\tau)). \tag{2.9}$$

By making use of Eq.(2.8) and its Hermitian conjugate we obtain

$$\frac{d\, x_1(\tau)}{d\tau} = -x_1\,(i+2\, H_o)$$
$$\frac{d\, x_2(\tau)}{d\tau} = -x_2\,(i-2\, H_o) \tag{2.10}$$

Since H_o is a constant of motion the solutions are

$$x_1(\tau) = x_1(o)\, e^{-(i+2\, H_o)\tau}$$
$$x_2(\tau) = x_2(o)\, e^{-(i-2\, H_o)\tau} \tag{2.11}$$

3. SQUEEZING

By definition, the electromagnetic field is squeezed to the second-order if

$$\langle\, [(\Delta\, x_i\,(\tau)]^2\, \rangle \equiv \langle\, [x_i\,(\tau) - \langle\, x_i\,(\tau)\rangle\,]^2 \rangle \;<1 \qquad (i=1 \text{ or } 2). \tag{3.1}$$

Hong and Mandel[8] have generalized this definition to N-th order squeezing. It has been shown that the field is squeezed to N-th order if

$$\langle\, [(\Delta\, x_i\,(\tau)]^N\, \rangle \;<(N-1)!! \tag{3.2}$$

and higher-order squeezing is meaningful for even N only. The expectation in Eq.(3.2) is with respect to any state in which the system was initially prepared. However, since any state may be represented in terms of coherent states, we will consider only expectations with respect to these states.

First we consider normal squeezing, or second-order squeezing, thus we need

$$\langle \alpha\, |(\Delta\, x_1(\tau))^2|\, \alpha\rangle = \langle \alpha\, |\, x_1(\tau)^2|\, \alpha\rangle - (\langle \alpha\, |\, x_1(\tau)|\, \alpha\rangle)^2 \tag{3.3}$$

which represents the variance of the in-phase quadrature x_1, calculated with respect to any initial coherent state $|\alpha\rangle$. By making use of Eq.(2.11) and the resolution of identity we get

$$\langle\alpha\,|\,x_1(\tau)|\,\alpha\rangle = e^{-i\tau}\,\alpha^*\langle\alpha\,|\,e^{-2\,H_0\tau}\,|\,\alpha\rangle + e^{-i\tau}\int\frac{d^2\beta}{\pi}\,\langle\alpha\,|a(0)|\beta\rangle\langle\beta|e^{-2\,H_0\tau}\,|\,\alpha\rangle \tag{3.4}$$

with $d^2\beta = d(\mathrm{Re}\beta)\, d(\mathrm{Im}\beta)$. It is well known[15] that

$$\langle\beta\,|\,e^{-iHt}|\,\alpha\rangle = e^{-1/2(|\alpha|^2+|\beta|^2)}\; f(\beta^*,\alpha;t) \tag{3.5}$$

where $f(\beta^*, \alpha; t)$ is an entire analytical function of α and β^* and an analytical continuation to imaginary times is possible. Then, we can write

$$\int\frac{d^2\beta}{\pi}\langle\alpha\,|a(0)|\beta\rangle\langle\beta|e^{-2\,H_0\tau}\,|\,\alpha\rangle = \left(\frac{\partial}{\partial\alpha^*}+\alpha\right)\int\frac{d^2\beta}{\pi}\,f(\alpha,\beta^*,2i\tau)e^{\alpha^*\beta-|\alpha|^2-|\beta|^2} \tag{3.6}$$

where the definition of coherent states scalar product[15] has been used. Finally, Eq.(3.4) becomes

$$\langle\alpha\,|\,x_1(\tau)|\,\alpha\rangle = e^{-i\tau}\left(\alpha^*+\alpha+\frac{\partial}{\partial\alpha^*}\right)\,\langle\alpha\,|e^{-2\,H_0\tau}\,|\,\alpha\rangle\,. \tag{3.7}$$

In the same way we get

$$\langle\alpha\,|\,(x_1(\tau))^2|\,\alpha\rangle = e^{-4i\tau}\left(\alpha^*+\alpha+\frac{\partial}{\partial\alpha^*}\right)^2\langle\alpha\,|e^{-4\,H_0\tau}\,|\,\alpha\rangle \tag{3.8}$$

and we used the identity

$$x_1(0)\;e^{-(i+2\,H_0)\tau} \equiv e^{-(2\,H_0-i)\tau}\;x_1(0)\,. \tag{3.9}$$

Let us now consider the higher-order squeezing defined in Eq.(3.2) and, for sake of simplicity, we will explicitely show the N=4 case only.

$$\langle\alpha\,|(\Delta\,x_1(\tau))^4|\,\alpha\rangle = \langle\alpha\,|(x_1(\tau))^4|\,\alpha\rangle - 4\,\langle\alpha\,|(x_1(\tau))^3|\,\alpha\rangle\,\langle\alpha\,|x_1(\tau)|\,\alpha\rangle$$
$$+\,6\,\langle\alpha\,|(x_1(\tau))^2|\,\alpha\rangle\,\langle\alpha\,|x_1(\tau)|\,\alpha\rangle^2 - 3\,\langle\alpha\,|x_1(\tau)|\,\alpha\rangle^4. \tag{3.10}$$

Following the same procedure as for Eq.(3.8) it can be easily shown that, for any even N, we have,

$$\langle\alpha\,|(x_1(\tau))^N|\,\alpha\rangle = e^{-i\tau(N/2)^2}\left(\alpha^*+\alpha+\frac{\partial}{\partial\alpha^*}\right)^{N/2}\langle\alpha\,|e^{-N\,H_0\tau}\,|\,\alpha\rangle\,. \tag{3.11}$$

Thus, to obtain analytical results, we only need the expectation

$$\langle \alpha | e^{-z H_o} | \alpha \rangle = G(\alpha, \alpha^*; z) \tag{3.12}$$

which will be considered in the next Section.

4. MATRIX ELEMENT

Instead of considering the expectation value (Eq.(3.12)) we will show how is possible to obtain the analytical expression for the more general matrix element

$$G_{\alpha\beta}(t) = \langle \beta | e^{-i H_o t} | \alpha \rangle \tag{4.1}$$

when H_o is given as in Eq.(2.7). First of all we consider the Schroedinger equation of motion

$$i \frac{\partial}{\partial t} | \phi(t) \rangle = H_o | \phi(t) \rangle \tag{4.2}$$

where

$$| \phi(t) \rangle = e^{-i H_o t} | \alpha \rangle \tag{4.3}$$

represents the state at time t once the initial state was $|\alpha\rangle$. By making use of Eqs.(3.5) and (4.2) can be easily shown that the following equation

$$\frac{\partial}{\partial t} f(\beta^*, \alpha; t) = \left(-\frac{1}{2} \beta^{*2} + \frac{1}{2} \frac{\partial^2}{\partial \beta^{*2}}\right) f(\beta^*, \alpha; t) \tag{4.4}$$

holds, with initial condition

$$f(\beta^*, \alpha; 0) = e^{\beta^* \alpha} . \tag{4.5}$$

The equation (4.4) is solved by using the well known Feynman-Kac formula[16,17,18] and we obtain

$$f(\beta^*, \alpha; t) = \langle e^{\alpha z(t)} \exp[-\frac{1}{2} \int ds\, z^2(s)] \rangle \tag{4.6}$$

where the expectation is with respect to the measure associated with the stochastic process z(t) which satisfies the following stochastic differential equation

$$\begin{aligned} dz(t) &= dW(t) \\ z(0) &= \beta^* \end{aligned} \tag{4.7}$$

with dW(t) the Ito's differential of the Wiener process W(t). The process z(t) is then given by

$$z(t) = (\beta^* + W(t)) \,. \tag{4.8}$$

Finally, Eq.(4.6) becomes

$$f(\beta^*,\alpha;t) = e^{\beta^*\alpha - 1/2\, \beta^{*2} t^2} \langle \exp\{-t^2 \int_0^1 ds[y^2(s)+g(s)y(s)]\}\rangle \tag{4.9}$$

where y(s) is a Wiener process and

$$g(s) = \sqrt{2t}\ (2\alpha\delta(s-1) - \beta^* t) \,. \tag{4.10}$$

The expectations with respect to a Wiener measure of functions of the type in Eq.(4.9) were considered by Cameron and Martin[19]. We just need to apply their prescription to get

$$f(\beta^*,\alpha;t) = \frac{\exp[-1/2(\beta^{*2}-\alpha^2)\mathrm{tght}+\alpha\beta^*\mathrm{secht}]}{\sqrt{\mathrm{cosht}}} \tag{4.11}$$

and, finally,

$$G_{\alpha\beta}(t) = \frac{\exp[-1/2(|\alpha|^2+|\beta|^2)-1/2(\beta^{*2}-\alpha^2)\mathrm{tght}+\alpha\beta^*\ \mathrm{secht}]}{\sqrt{\mathrm{cosht}}} \tag{4.12}$$

This result was first obtained by Yuen[16] and then by Hillery and Zubairy[20] with completely different techniques.

5. RESULTS

By performing an analytical continuation to imaginary times in Eq.(4.12) we can write the expectation value of interest

$$\langle \alpha | e^{-N H_o \tau} |\alpha\rangle = \frac{\exp[-|\alpha|^2(1-\sec N\tau)+i/2(\alpha^{*2}-\alpha^2)\mathrm{tg}\ N\tau]}{\sqrt{\cos N\tau}} \tag{5.1}$$

with $0 \le \tau < \pi/2N$. This limitation is introduced in order to avoid any problem connected with the analytical continuation to imaginary times in the hyperbolic cosine function[20].

The normal squeezing Eq.(3.3), by making use of Eqs.(3.7), (3.8) and (5.1), is then

$$V_1(t) \equiv \langle\alpha|(\Delta x_1(t))^2|\alpha\rangle = [\sec 4\tau + \frac{4|\alpha|^2 \cos^2(2\tau-\theta)}{\cos^2 4\tau}]$$

$$\times \frac{e^{-|\alpha|^2 f(\theta,4\tau)}}{\sqrt{\cos 4\tau}} - \frac{4|\alpha|^2 \cos^2(\tau-\theta)}{\cos^3 2\tau} e^{-2|\alpha|^2 f(\theta,2\tau)} \qquad (5.2)$$

with

$$f(\theta,k\tau) = \frac{\cos k\tau - \sin 2\theta \sin k\tau - 1}{\cos k\tau} \qquad (5.3)$$

and

$$\alpha = |\alpha| e^{i\theta} \qquad (5.4)$$

The phase angle θ is controlled by the experimentalist and it may be choosen by beating the signal with a local oscillator. When $\theta = -\pi/4$ we obtain the maximum value of squeezing. Within the present model, once the scaling condition Eq.(2.5) has been fulfilled, no squeezing at all is obtained for an initial vacuum state and it strongly depends on the average number of photons in the initial state of the signal. This result is shown in Fig.1 where the variance (Eq.(5.2)) is plotted versus the dimensionless time τ for various values of $|\alpha|^2$.

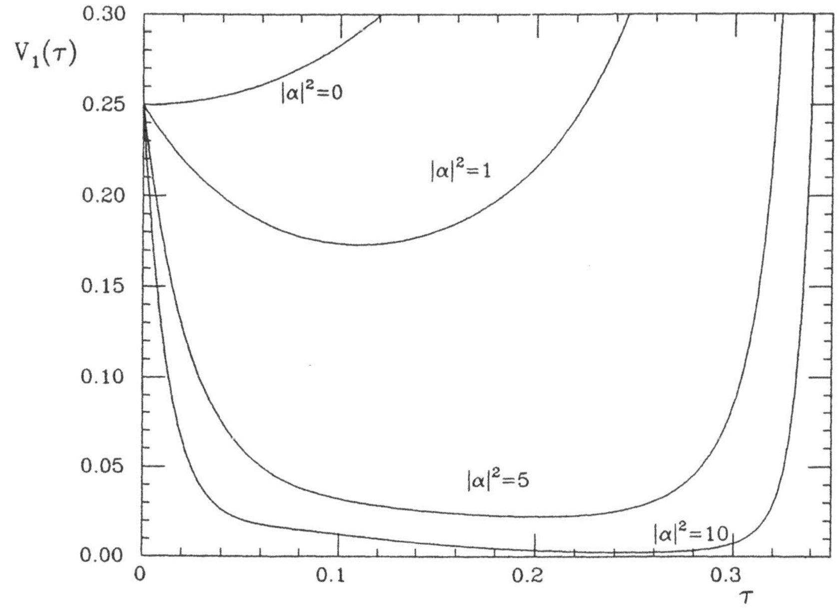

To make a comparison with the usual second-order degenerate downconversion we note that, in such a case, although the squeezing of the signal field can be obtained for an initial vacuum state, the amount of the obtainable squeezing is always the same and, for a given intensity of the external pump, depends on the interaction time only. Our model, should it be realized, would show a great enhancement of squeezing, which depends on the initial number of photons, and would therefore reduce the interaction time or the interaction length of the nonlinear medium.

Let us now consider the 4-th order squeezing Eq.(3.10). By making use of Eqs.(3.11), (5.1) and (5.4) we could write down the analytical expression; however, it appears practically unreadable and we prefer to draw the resulting expression in Fig.2. We see again that the fourth-order squeezing depends on $|\alpha|^2$. An analytical expression can be easily traced out for $|\alpha|^2 >> 1$ and short times $\tau << (4N)^{-1}$ we obtain thus

$$V_{N/2}(\tau) \approx V_{N/2}(0)\, e^{-|\alpha|^2 N\tau} \tag{5.5}$$

where

$$V_{N/2}(0) = (N-1)!! \tag{5.6}$$

represents the N-th order variance for the coherent states.

As final comment we want to stress that the probability distribution is definitely non-Gaussian. This conclusion can be inferred from the divergence of any moment as time goes on. Moreover, at a given time, there are always some higher order moments which are unbonded. Nevertheless, we cannot naively conclude that the probability distribution does not exist, because other well known distributions are well defined even though their moments are divergent, such as the Levy-flight and Cauchy distributions[21].

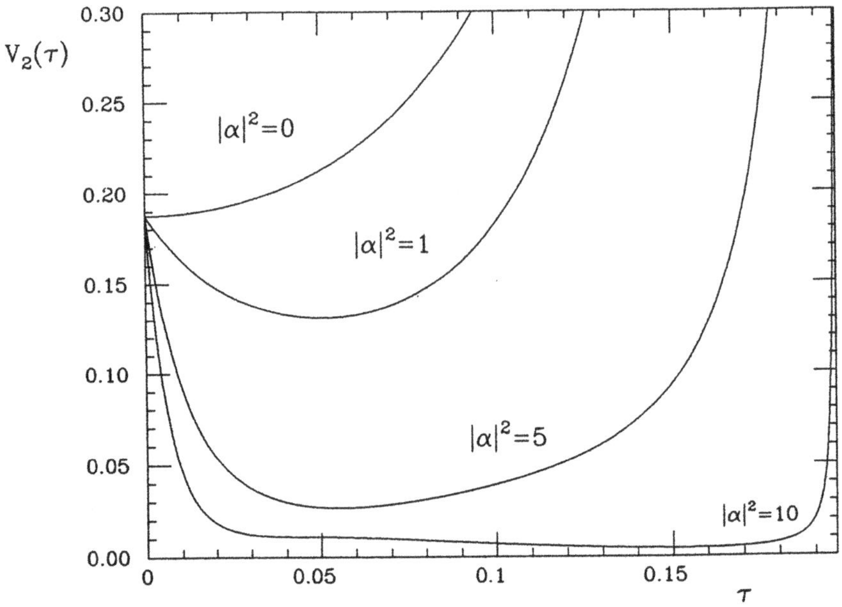

REFERENCES

1. D.F.Walls, Nature (London) **306**, 141 (1983).
2. R.Loudon and P.L.Knight, J.Mod.Opt. **34**, 709 (1987).
3. L.Wu, H.J.Kimble, J.L.Hall and H.Wu, Phys.Rev.Lett. **57**, 2520 (1986).
4. R.E.Slusher, L.W.Hollberg, B.Yurke, J.C.Mertz and J.F.Valley, Phys.Rev.Lett. **55**, 2409 (1985);
   R.M.Shelby, M.D.Levenson S.H.Perlmutter, R.G.DeVoe and D.F.Walls, Phys. Rev.Lett. **57**, 691 (1986);
   M.W.Maeda, P.Kumar and J.H.Shapiro, Opt.Lett. **12**, 161 (1987).
5. P.Tombesi and H.P.Yuen in *Coherence and Quantum Optics* Vol.5, eds. L.Mandel and E.Wolf (Plenum Press, New York 1984).
6. H.P.Yuen, Phys.Rev. A**13**, 2226 (1986).
7. R.Tanas in *Coherence and Quantum Optics* Vol.5, eds. L.Mandel and E.Wolf (Plenum Press, New York 1984).
8. C.K.Hong and L.Mandel, Phys.Rev.Lett. **54**, 323 (1985).
9. M.Kozierowski and R.Tanas, Opt.Com. **21**, 229 (1971).
10. C.Gerry, Phys.Lett. **124**A, 237 (1987).
11. R.A.Fisher, M.M.Nieto and V.D.Sandberg, Phys.Rev. D**29**, 1107 (1984).
12. S.L.Braunstein and R.L.McLachlan, Phys.Rev. A**35**, 1659 (1987).
13. G.D'Ariano, M.Ràsetti and M.Vadacchino, Phys.Rev. D**32**, 1034 (1985); see also the present volume.
14. S.A.Akmanov, A.N.Dubovik, S.M.Saltiel, I.V.Tomov and V.G.Tunkin, JETP Lett. **20**, 117 (1975).
15. J.R.Klauder and B.S.Skagerstam, *Coherent States: Application in Physics and Mathematical Physics*, (World Scientific, Singapore 1985).
16. R.P.Feynman, Rev.Mod.Phys. **20**, 367 (1948).
17. M.Kac, Proc. Second Berkely Symp. Math. Stat. and Prob. 16.
18. C.De Witt-Morette and K.D.Elworthy, Phys.Rep. **77**, 121 (1981).
19. R.H.Cameron and W.T.Martin, Bull. Amer. Math. Soc. **51**, 73 (1945).
20. M.Hillery and M.S.Zubairy, Phys.Rev. A**26**, 451 (1982).
21. E.W.Montrol and M.F.Shlesinger, in *Studies in Statistical Mechanics*, eds. J.L.Lebowitz and E.W.Montrol (North Holland, Amsterdam 1984) Vol.11.

NATO ADVANCED RESEARCH WORKSHOP ON SQUEEZED AND NON-CLASSICAL LIGHT
January 25-29, 1988, in Cortina d'Ampezzo, Italy

INDEX

CPSIA information can be obtained at www.ICGtesting.com
Printed in the USA
LVOW02s1959211113

362276LV00024B/1190/P